COMPUTER PROCESS CONTROL
with Advanced Control Applications
2nd Edition
Revised and Enlarged

Computer Process Control
With Advanced Control Applications

PRADEEP B. DESHPANDE
Professor of Chemical Engineering
University of Louisville

RAYMOND H. ASH
Manager, Control Systems Technology
Procter and Gamble Company

 Instrument Society of America

COMPUTER PROCESS CONTROL
with Advanced Control Applications Second Edition
Formerly Elements of Computer Process Control

Copyright © by Instrument Society of America 1988.
All rights reserved.
Printed in the United States of America.
No part of this publication may be reproduced, stored in a retrieval system,
or transmitted, in any form or by any means,
electronic, mechanical, photo copying, recording or otherwise,
without the prior written permission of the publisher.

INSTRUMENT SOCIETY of AMERICA
67 Alexander Drive
P.O. Box 12277
Research Triangle Park
North Carolina 27709
ISBN: 1-55617-005-X

Library of Congress Cataloging-in-Publication Data
Deshpande, Pradeep B.
  Computer process control, with advanced control
applications.

  Rev. ed. of: Elements of computer process, with
advanced control applications. c1981.
  Bibliography: p.
  1. Process control—Data processing. I. Ash,
Raymond H.   II. Deshpande, Pradeep B. Elements of
computer process control, with advanced control
applications.   III. Title.
TS156.8.D47   1988   640.72'7   88-21576
ISBN 1-55617-005-X

This book is dedicated to my parents,
who gave so much and asked for so little (PBD),
and to our wives, Meena and Joanne

PREFACE TO THE SECOND EDITION

Recent developments have combined to necessitate a revision of the first edition of this text, one of the most important being impulse response models and their use in control systems design. Early in the text we show how impulse response models are derived quite simply from easily available step response data. The reader will appreciate how the use of IR models greatly simplifies the computation of closed-loop responses by computer simulation in comparison with the traditional Z-transform method. In later chapters we will show how IR models are used effectively in predictive control formulations that accommodate operating constraints.

The next major revision is in the chapter on digital controller design. The old chapters on controller design and dead time compensation have been combined into a single new chapter in view of the fact that some of the recent algorithms contain inherent dead time compensation features. In the new chapter on controller design we begin with the notion of perfect (set point) control and show how, in the presence of nonminimum phase elements (dead viii *Preface*

time and inverse response), one must back away from the ideal of perfect control. Numerous algorithms are presented, and the exercises in the Appendix are designed to show the advantages and disadvantages of each. Robustness, which tells how well the control system will respond in the presence of modeling errors that are invariably present in real-life situations, is a new concept included in the new chapter on design.

As in the first edition, we continue to put emphasis on relatively simple methods for process identification. In this edition we have included two new and improved methods that can be applied to find the parameters of process models and digital controllers. A new chapter on adaptive control and self-tuning has also been included. The treatment is necessarily brief since a wealth of open literature on the subject is available.

The final major revision is in the chapter on multivariable control. The four phases of multivariable control systems design, (1) interaction analysis for proper pairing of variables and for finding the extent of interaction present, (2) multiloop controller design procedures for those cases where interaction is deemed to be modest, (3) explicit decoupling for noninteracting control, and (4) multivariable control strategies that inherently compensate for interaction and permit constraint handling, are covered in this chapter. Here too, the robustness issues and the importance of constraint-handling facilities for specific applications are emphasized.

Finally, we wish to mention that CAI (computer-aided instruction) software has been developed to aid in the understanding of the concepts covered in this edition, and the interested reader may contact the first author for details.

The authors wish to express their appreciation to their respective organizations for their support of the revision project. The first author expresses his special appreciation to Dean E. R. Gerhard of the University of Louisville's Speed School for his support of these and related activities.

Louisville, KY P. B. Deshpande
September 1987 R. H. Ash

Contents

PART I

Elements of
Computer Process Control

CHAPTER 1

Review of Conventional Process Control

I n this chapter we will review and highlight some of the important aspects of linear control theory and its application to conventional process control. Linear control theory forms the basis for the development of computer control concepts which is the subject of this book. A thorough review of these concepts will greatly help the reader in understanding the subsequent material on computer control.

1.1. Introduction to Process Control

Within the context of chemical engineering applications, we may define the term *process* as a collection of interconnected hardware (e.g., tanks, pipes, fittings, motors, shafts, couplings, and gauges), each doing its part toward an overall objective of producing some product or a small group of related products. This definition also fits the viewpoint of the *process engineer* whose job is to design this hardware. From the standpoint of the control engineer, the focus is on *physical variables* (e.g., temperature, pressures, levels, flows, voltages, speeds, positions, and compositions) rather than on hardware. The control engineer is interested in knowing how these variables affect one another and how they change with time. The fundamental objective in process control is to maintain certain key process variables as near their desired values, called *set points*, as much of the time as possible. From among all the process variables, certain variables are chosen as the key process variables because maintaining them at specified values means the production objectives will be satisfied. The production objectives are

—To achieve desired production, i.e. throughput.

—To produce at acceptable cost.

—To produce material of acceptable quality.

—To do all of the above in a safe manner with minimum harm to the environment.

In dealing with process control concepts, it is useful to categorize the physical variables according to the following classification:

—*Outputs*—these are the key process variables to be maintained at desired values, that is, controlled.

—*Inputs*—these are the variables that, when changed, cause one or more outputs to change. These inputs may be further subclassified as control inputs and disturbance inputs.

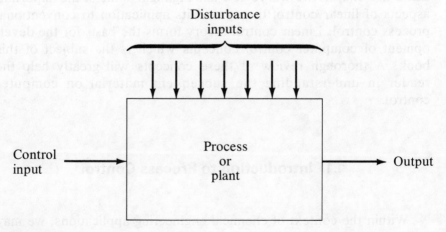

Figure 1.1
Typical Block Diagram

—a) *Control Inputs*. These are also called *manipulated variables*. These variables are changed by the controller to bring and maintain the outputs at set points. For example, the flow of process streams is often changed by actuators, such as control valves or variable speed pumps, so as to control certain outputs.

—b) *Disturbance Inputs*. All other process variables that affect the outputs in any way are called *disturbance inputs* or *process loads*. These disturbance inputs cause unwanted changes in process outputs.

1.2. Process Dynamics and Mathematical Models

The objective of control strategy development is to determine how to change a control input so as to correct a deviation between the desired value and the actual value of the process output. This development is complicated by the inertia or lag inherent in most processes. The presence of lag means that if a sudden change is made in a control input, the output will not follow immediately. There will be a time lag before the output reaches its new value.

To do a good job of controlling a process we need to know how control inputs affect outputs, quantitatively. If a control input is changed by a known amount, we need to know

—How much the output will ultimately change and in which direction.

—How long it will take for the output to change.

—What trajectory the output will follow; that is, what the pattern of output variation with time is.

An answer to these questions is provided by a dynamic mathematical model of the process. This model gives a functional relationship between an input and an output of a process and may be defined as any mathematical expression or formula that when values of the independent variables (i.e., inputs) are given, enables prediction of the value of the dependent variable (i.e., output) by plugging into the formula and calculating the result. Thus a mathematical model is an expression from which physical behavior can be predicted.

A schematic representation of the cause-and-effect relationship between input and output is referred to as a *block diagram*, shown in Figure 1.1. Generally, there is one block for each piece of the entire process. A single block usually has one output, one control input, and several identifiable and/or measurable disturbance inputs. There may also exist several unidentifiable or unknown disturbance inputs that affect the output. The block diagram also represents the quantitative cause-effect relationship (i.e., mathematical model) between an input and an output.

Any given physical variable in a process can be any one of the types shown in Figure 1.1; that is, an output or control input of one process block can be a disturbance input to another process block.

Mathematical models are derived from the laws of physics and chemistry. Some examples are

—Conservation of mass and energy (i.e., unsteady-state material and energy balances).

—Newton's laws of motion.

—Kirchoff's laws of electrical circuits.

—Equations of state for gases.

These models usually take the form of differential equations involving time rates of change of variables as well as the variables themselves. To predict the output behavior resulting from a known change in input, these differential equations must be solved, but they are more difficult to solve than algebraic equations.

Some processes are linear, that is, they are described by linear differential or algebraic equations, and the principle of superposition applies. Many others are nonlinear, and these are described by complex nonlinear differential equations. However, the behavior of the latter around some normal operating level can be approximated by linear differential equations. For processes that are under control the interest is in their behavior around some steady-state operating level, and this approximation is usually applicable.

Linear differential equations can be converted, by Laplace transforms, into algebraic equations from which transfer functions can be

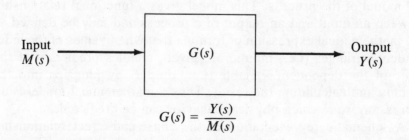

$$G(s) = \frac{Y(s)}{M(s)}$$

Figure 1.2
Transfer Function Shown on a Block Diagram

derived. Since Laplace transforms convert a differential equation into an algebraic equation, the solution is considerably simplified.

A *transfer function* $G(s)$ is a mathematical expression that represents the ratio of the Laplace transforms of a process output $Y(s)$ to that of an input $M(s)$ as shown in Figure 1.2. Recall that the output $Y(t)$ and the input $M(t)$ are expressed as deviation variables, that is,

$$Y(t) = y(t) - y_{\text{steady state}} \qquad (1.1)$$

and

$$M(t) = m(t) - m_{\text{steady state}}$$

Therefore, at the steady-state operating level Y and M will be zero. From the transfer function, the response of $Y(t)$ to a specified $M(t)$ can be obtained by inverting the equation

$$Y(s) = G(s) M(s) \qquad (1.2)$$

1.3. Types of Dynamic Processes

Processes may be categorized as being one of the following types:
—Instantaneous or steady state.
—First-order lag.
—Second-order lag.
—Dead-time or transport lag.
Of course, some processes may be of very high order, but as we shall see shortly, their behavior can often be approximated as a first- or second-order lag plus dead time.

Instantaneous Process

The dynamics of this class of processes is negligible. That is, the output follows changes in input so quickly that the process always remains at steady state, for all practical purposes. The representation of the instantaneous process is shown in Figure 1.3a. As an example, consider the operation of a control valve with a linear flow characteristic shown in Figure 1.3b. In this case

    input is valve stem position X_v 0 to 100%
    output is flow Q 0 to Q_{max}

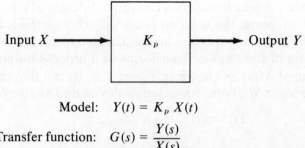

Model: $Y(t) = K_p X(t)$

Transfer function: $G(s) = \dfrac{Y(s)}{X(s)}$

where s = Laplace transform variable

Process gain $K_p = \dfrac{\text{change in output}}{\text{change in input}}$

Figure 1.3a
Instantaneous Process

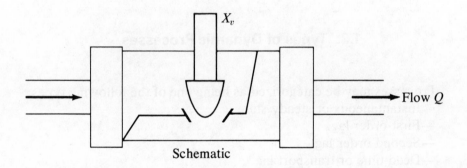

Schematic

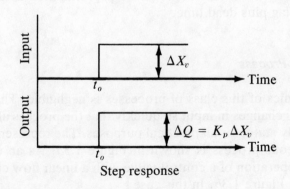

Step response

Figure 1.3b
Example of Instantaneous Process

$$K_p = \frac{\Delta Q}{\Delta X_c} = \frac{\text{gpm}}{\%} \qquad (1.3)$$

First-order Lag

The output of a first-order process follows the step change in input with classical exponential rise. The model, parameters, and transfer function for this class of processes are

$$\text{Process model:} \quad \tau \frac{dy}{dt} + y = K_p x \qquad (1.4)$$

where

$$x = \text{process input}$$
$$y = \text{process output}$$
$$K_p = \text{process steady-state gain}$$
$$\tau = \text{time constant}$$

$$\text{Transfer function:} \quad G_p(s) = \frac{Y(s)}{X(s)} = \frac{K_p}{\tau s + 1} \qquad (1.5)$$

Block diagram: $\quad X(s) \longrightarrow \boxed{\dfrac{K_p}{\tau s + 1}} \longrightarrow Y(s)$

Dynamic parameters:

$$\text{Steady-state gain } K_p = \frac{\text{final steady-state change in output}}{\text{change in input}}$$

Time constant, τ: time to reach 63.2% of final value in response to a fixed (i.e., step) change in input

The step response of a first-order lag is shown in Figure 1.4.

Examples of first-order lags include many sensor/transmitters. final control elements, and numerous processes. One simple example of a first-order process is a stirred tank where flow in equals flow out, as shown in Figure 1.5.

Input: temperature of incoming liquid T_i

Output: temperature of the effluent stream (assumed to be equal to that of the liquid in the tank because of good mixing) T_o

$$\text{Transfer function:} \quad \frac{T_o(s)}{T_i(s)} = \frac{1}{\tau s + 1} \qquad (1.6)$$

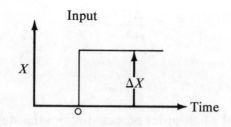

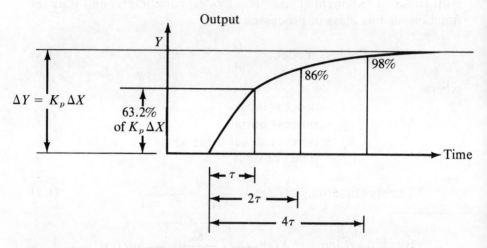

Figure 1.4
Step Response of a First-order Process

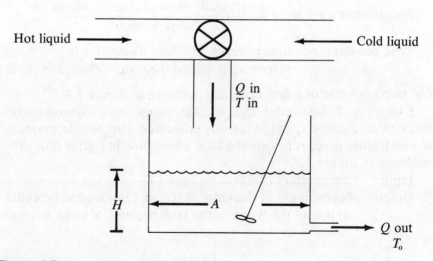

Figure 1.5
Example of a First-Order Process

where

τ = residence time of liquid in the tank

$$= \frac{\text{tank volume}}{\text{volumetric flow}}$$

$$= \frac{H\,A}{Q_{out}}$$

where

A = cross-sectional area of the tank
H = constant height of liquid in the tank

Second-order processes

Some processes are inherently second order, including a spring/mass system, an inductor/capacitor system, a U tube manometer, and so on. Such processes have the following characteristics

$$\text{Process model:} \quad \tau^2 \frac{d^2y}{dt^2} + 2\,\zeta\tau\,\frac{dy}{dt} + y = K_p\,X \qquad (1.7)$$

where

τ = time constant of the second-order process

$\omega_n = \dfrac{1}{\tau}$ = natural frequency of the process

ζ = damping factor
K_p = steady-state gain

$$\text{Transfer function:} \quad G_p(s) = \frac{Y(s)}{X(s)} = \frac{K_p}{\tau^2 s^2 + 2\zeta\tau s + 1} \qquad (1.8)$$

Block diagram: $X(s) \longrightarrow \boxed{\dfrac{K_p}{\tau^2 s^2 + 2\zeta\tau s + 1}} \longrightarrow Y(s)$

The step response of second-order processes is shown in Figure 1.6.

Two first-order lags in series, in which the output of first process is input to second, can be described by a second-order process model which may be represented as

Block diagram: $X(s) \longrightarrow \boxed{\dfrac{K_1}{\tau_1 s + 1}} \longrightarrow \boxed{\dfrac{K_2}{\tau_2 s + 1}} \longrightarrow Y(s)$

Process model: $\tau_1 \tau_2 \dfrac{d^2 y}{dt^2} + (\tau_1 + \tau_2) \dfrac{dy}{dt} + y = K_1 K_2 x$ (1.9)

A comparison of this equation with that of the inherent second-order model gives

$$\tau = \text{time constant} = \sqrt{\tau_1 \tau_2}$$

$$\zeta = \frac{1}{2} \frac{\tau_1 + \tau_2}{\sqrt{\tau_1 \tau_2}} \geq 1.0 \text{ for two first-order processes in series}$$

$$K_p = \text{process gain} = K_1 K_2$$

Dead Time or Transport Delay

For a pure dead-time process whatever happens at the input is repeated at the output Θ_d time units later, where Θ_d is the dead time. For example, consider the flow of a liquid through an insulated pipe having a cross-sectional area A and length L at a volumetric flow rate q. At steady state the temperature of the liquid at the entrance x will be the same as that of the liquid at the exit of the pipe. If the temperature at the entrance changes, the change will not be detected at the

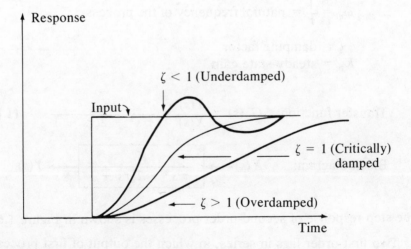

Figure 1.6
Step Response of Second-order Processes

exit until Θ_d seconds later. This dead time is simply the time required for the liquid to flow from the entrance to the exit and is given by

$$\Theta_d = \frac{\text{volume of pipe, m}^3}{\text{volumetric flow rate, m}^3/\text{s}} = \frac{L\,A}{q} \tag{1.10}$$

The process model for pure dead time is given by

$$y(t) = x(t - \Theta_d) \tag{1.11}$$

where y is the exit temperature. At steady state

$$y_s = x_s \tag{1.12}$$

The last two equations give us a model in terms of deviation variables

$$Y(t) = y(t) - y_s = x(t - \Theta_d) - x_s = X(t - \Theta_d) \tag{1.13}$$

Taking Laplace transform gives the transfer function of the pure dead-time element

$$G_p(s) = \frac{Y(s)}{X(s)} = e^{-\Theta_d s} \tag{1.14}$$

The response Y for an arbitrary input X is shown in Figure 1.7a. An

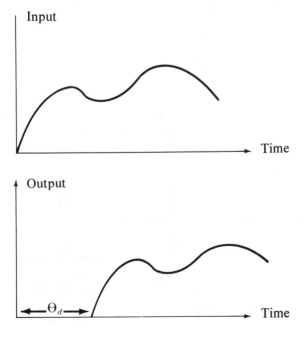

Figure 1.7a
Response of Dead-time Element

example of pure dead time in paper manufacture is shown in Figure 1.7*b*.

Higher-order Processes and Approximate Models

Most processes have many dynamic elements, each having a different time constant. For example, each mass or energy storage element in a process can provide a first-order dynamic element in the model. A 50-tray distillation column thus has 50 mass storage elements and 50 energy storage elements. The exact mathematical model relating distillate composition to feed composition would be greater than 100th order, when the dynamics of condenser and reboiler are included. Fortunately, it is possible to approximate the behavior of such high-order processes by a system having one or two time constants and a dead time. We shall demonstrate the justification for this approximation in this section.

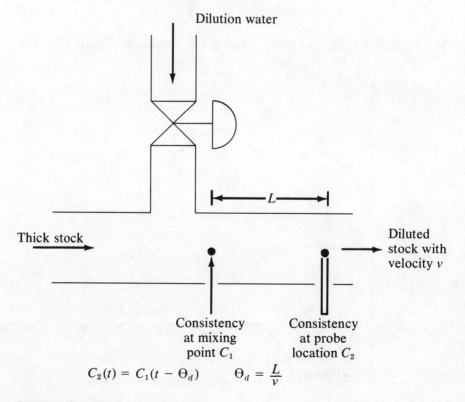

$$C_2(t) = C_1(t - \Theta_d) \qquad \Theta_d = \frac{L}{v}$$

Figure 1.7*b*
Process Having Pure Dead Time

Consider a process having N first-order elements in series as shown in Figure 1.8a. In this figure each element has a time constant τ/N. The sum of these time constants for the N elements is τ.

When one or two time constants dominate (i.e., are much larger than the rest), as is common in many processes, all the smaller time constants work together to produce a lag that very much resembles pure dead time. This is clearly seen from Figure 1.8b which shows

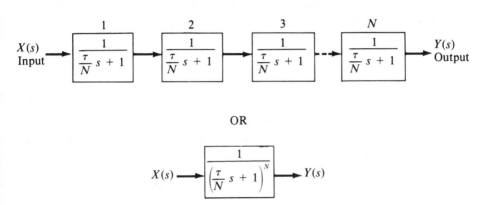

OR

Figure 1.8a
N First-order Elements in Series

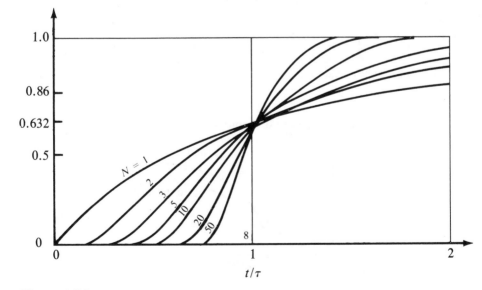

Figure 1.8b
Response of Cascaded First-order Lags

the response of the process to a unit step change in X as N is varied from 1 to ∞. This figure shows that as N gets larger, the response shifts from exact first order to pure dead time (equal to τ). It is therefore possible to approximate the actual input-output mathematical model of a very-high-order, complex, dynamic process with a simplified model consisting of a first- or second-order process combined with a dead-time element, as shown in Figure 1.8c.

The proper selection of model parameters Θ_d, τ_1, and τ_2 can make the response of the second-order model very close to the response of

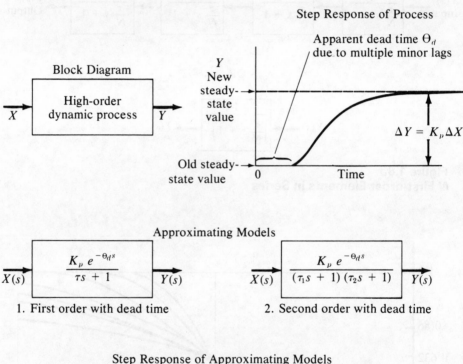

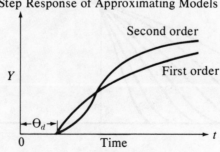

Figure 1.8c
High-order Processes and Approximating Models

many processes. The second-order model will reduce to the first-order model if one of the two time constants of the former model is much smaller than the other. Indeed, the response of some processes may be adequately described by a first-order model with dead time.

Complete Process Model

The complete process model includes the model of the process, the measuring element, and the final control element. A block diagram of the complete process model is shown in Figure 1.9.

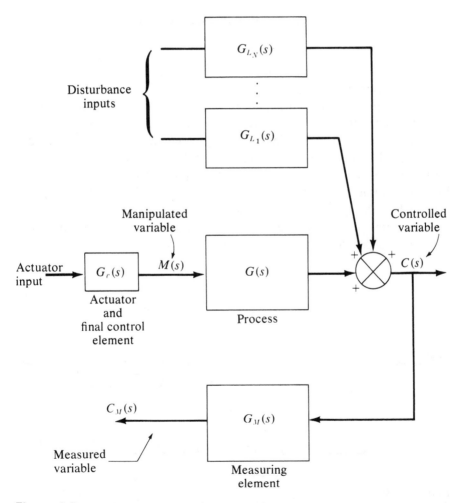

Figure 1.9
Block Diagram of the Complete Process Model

If load dynamics and process dynamics are the same, the block diagram of the process model can be as shown in Figure 1.10. Bear in mind that the notion of a transfer function applies only to linear systems. If the process is nonlinear, as most industrial processes are, the model must be linearized in the vicinity of the desired operating level of the plant. For a different operating level, new parameters will have to be established for the process model. Note also that the arrangement of the block diagram, Figure 1.10, implies that the disturbances affect output, linearly (i.e., superposition principle applies).

1.4. Basic Feedback Control

If there are no disturbances, no control is needed once steady-state operating conditions are achieved. Load upsets cause outputs to deviate from desired values; therefore, one or more control inputs must be changed so as to maintain the outputs at set point. A *control strategy* is required to achieve this objective. In this context, we define control strategy as a set of rules by which control action is determined when the output deviates from set point, that is, which control input(s) should be changed, in which direction, by how much, and when? In other words, control strategy is an algorithm or equation that determines controller output (to the actuator) as a function of the present and past measured errors. The block diagram of the control-

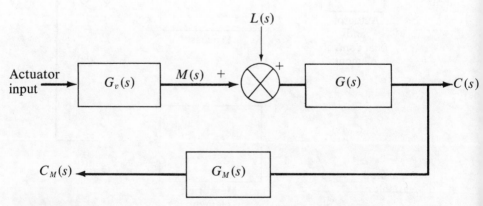

Figure 1.10
Block Diagram of the Process Having the Same Load Dynamics and Process Dynamics

ler portion of an automatic feedback control loop is shown in Figure 1.11.

A block diagram of the basic feedback control loop, which combines the process and the controller, is shown in Figure 1.12. In some applications the actuator and sensor dynamics are negligible compared with the dynamics of the process, in which case the block dia-

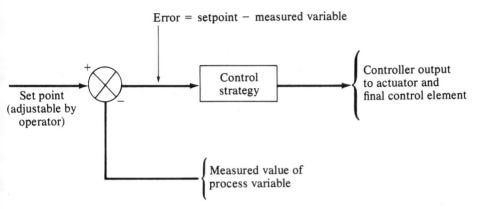

Figure 1.11
Controller Block Diagram

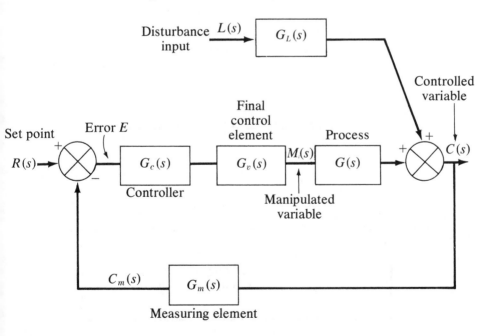

Figure 1.12
Block Diagram of a Feedback Control Loop

gram can be somewhat simplified. In some others the process transfer
function includes the actuator and sensor dynamics, as shown in the
block diagram of Figure 1.13.

Basic Control Strategy

The basic control strategy in conventional feedback control is to
compare the measured variable with the desired value of that variable
and if a difference exists, to adjust the controller output to drive the
error toward zero. The hardware that performs this function is the
automatic controller. The operation of the ideal three-mode controller
is described by the following equation:

$$m(t) = K_c \left[e(t) + \frac{1}{\tau_I} \int_0^t e(t) \, dt + \tau_d \frac{de(t)}{dt} \right] + m_s \qquad (1.15)$$

where

$$
\begin{aligned}
K_c &= \text{proportional gain} \\
e(t) &= \text{error} \\
\tau_I &= \text{integral time, seconds or minutes} \\
\tau_d &= \text{derivative time, seconds or minutes} \\
m_s &= \text{steady-state controller output that drives error to zero}
\end{aligned}
$$

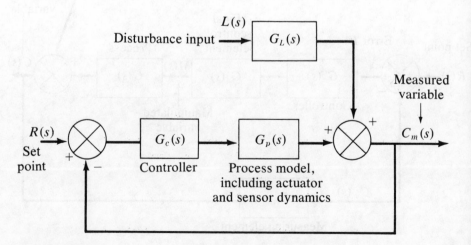

Figure 1.13
Block Diagram Showing a Composite Process Model

The transfer function of the ideal controller equation is

$$\frac{M(s)}{E(s)} = G_c(s) = K_c \left[1 + \frac{1}{\tau_I s} + \tau_d s \right] \qquad (1.16)$$

The ideal controller equation has a pure differentiator in it and therefore is not physically realizable. The transfer function of most commercial controllers is

$$G_c(s) = K_c \left[1 + \frac{1}{\tau_I s} \right] \left[\frac{\tau_d s + 1}{\beta \tau_d s + 1} \right] \qquad (1.17)$$

where β is a constant $<<1$, typically in the range of 0.01 to 0.1.

Many industrial controllers use the terms *proportional band* and *reset*. They are defined as

$$\text{Proportional band in per cent, } PB = \frac{100}{K_c} \qquad (1.18)$$

$$\begin{array}{c} \text{Reset } R_I \text{ in repeats per second} \\ \text{or repeats per minute} \end{array} = \frac{1}{\tau_I}$$

To illustrate the effect of the proportional (P), proportional integral (PI), and proportional + integral + derivative (PID) modes on control, the response of a first-order system with dead time operating under the three types of controllers to a step change of magnitude ΔL in load is sketched in Figure 1.14.

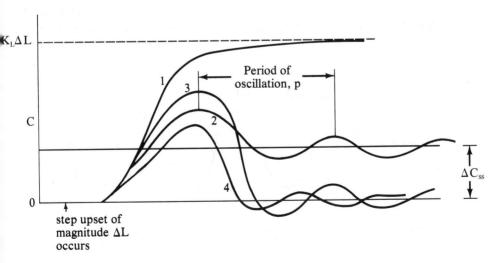

Figure 1.14
Step Response of a First-order + Dead Time System to Load Change

Figure 1.14 shows the following characteristics of the three controller modes.

Curve (1): no control: ultimate deviation in process output, $\Delta C = K_L \Delta L$

Curve (2): proportional only control: gives rise to steady-state offset,

$$\Delta C_{ss} = K_L \Delta L \left[\frac{1}{1 + K_p K_c} \right] \tag{1.19}$$

Curve (3): proportional + integral control: eliminates steady-state offset; however, has slightly higher peak offset and slightly longer period of oscillation as compared to proportional only control.

Curve (4): proportional + integral + derivative control: reduces the peak offset as well as the period of oscillation over PI control.

The PID controller is seen to give the best response of the three controller types. However, the derivative mode is sensitive to process or measurement noise, is less forgiving of process parameter changes, and is generally more difficult to tune. For these reasons PI controllers are used in a vast majority of industrial applications.

Controller Tuning

The automatic controllers can be tuned according to one of two methods. The first, called the *ultimate-cycle method*, was originally proposed by Ziegler and Nichols.[1] To use this method, the reset and

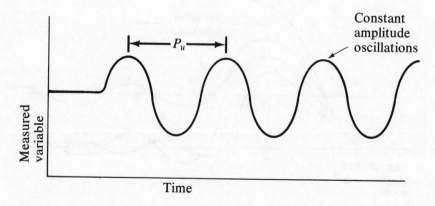

Figure 1.15
Sustained Cycling of Measured Variable under Proportional Control

derivative modes are set at their lowest values, (i.e., lowest values of R_I and τ_d). With the controller in automatic, the gain, K_c, is slowly increased until the measured variable begins to oscillate, as shown in Figure 1.15. The gain setting that results in sustained cycling is called the *ultimate gain*, K_u, and the period of the oscillation is the *ultimate period*, P_u. The Ziegler–Nichols controller settings are determined by the formulas shown in Table 1.1. The modified ultimate cycle method which incorporates the quarter amplitude criteria (see Figure 1.16) yield somewhat different controller settings, which are also shown in Table 1.1.

Recall that the ultimate gain, K_u, and the ultimate period, P_u, can also be determined from the open-loop frequency response diagram of the process, as shown in Figure 1.17.

The second method is called the Cohen–Coon method.[2] It assumes that the process can be represented as a first-order lag plus dead time. To find the tuning constants by this method, the controller is placed in manual, and a step change in its output of magnitude Δm is made. From the step response plot shown in Figure 1.18 the process gain and time constant and the dead time are estimated by the following equations:

$$K_p = \frac{\Delta C_{ss}}{\Delta m}$$

$$\tau_p = 1.5\,(t_{0.63} - t_{0.28}) \tag{1.20}$$

$$\Theta_d = 1.5\,\left(t_{0.28} - \frac{1}{3}\,t_{0.63}\right)$$

$$\alpha = \frac{\Theta_d}{\tau}$$

The controller tuning constants are then determined by the formulas shown in Table 1.1.

Measures of Control Quality

We have mentioned earlier that many processes can be described by a first- or second-order model with dead time. In this section we show how dead time affects the control quality of even properly tuned loops.

Let us define a term, *controllability ratio*, according to the equation

Table 1.1
Controller Tuning Constant Formulas

Type of Controller	Ziegler–Nichols[1] Original Method	Modified Method	Cohen–Coon[2] Method
Proportional	$K_c = 0.5 K_u$	Adjust the gain to obtain quarter amplitude decay response to a step change in set point	$K_c = \dfrac{1}{K_p}\left(\dfrac{1}{\alpha} + 0.333\right)$
Proportional + integral	$K_c = 0.45 K_u$ $\tau_I = \dfrac{P_u}{1.2}$ (min)	Adjust the gain to obtain quarter amplitude decay response to a step change in set point $\tau_I = P_u$ (min)	$K_c = \dfrac{1}{K_p}\left(\dfrac{0.9}{\alpha} + 0.082\right)$ $\tau_I = \tau_p\left[\dfrac{3.33\alpha + 0.333\alpha^2}{1 + 2.2\alpha}\right]$
Proportional + integral + derivative	$K_c = 0.6 K_u$ $\tau_I = \dfrac{P_u}{2}$ (min) $\tau_d = \dfrac{P_u}{8}$ (min)	Adjust the gain to obtain quarter amplitude decay response to a step change in set point $\tau_I = \dfrac{P_u}{1.5}$ (min) $\tau_d = \dfrac{P_u}{6}$ (min)	$K_c = \dfrac{1}{K_p}\left[\dfrac{1.35}{\alpha} + 0.270\right]$ $\tau_I = \tau_p\left[\dfrac{2.5\alpha + 0.5\alpha^2}{1 + 0.6\alpha}\right]$ $\tau_d = \tau_p\left[\dfrac{0.37\alpha}{1 + 0.2\alpha}\right]$

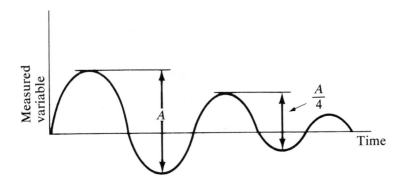

Figure 1.16
Quarter Amplitude-Decay Response

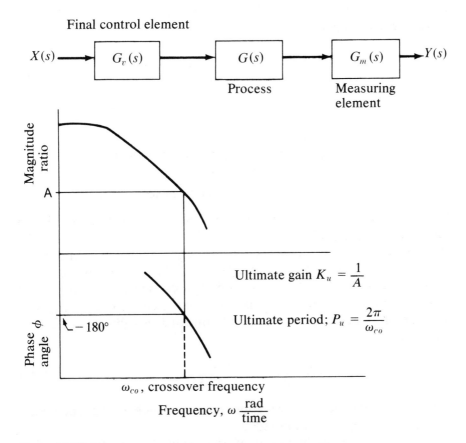

Figure 1.17
Frequency Response Diagram of $[G_v(s)] [G(s)] [G_m(s)]$

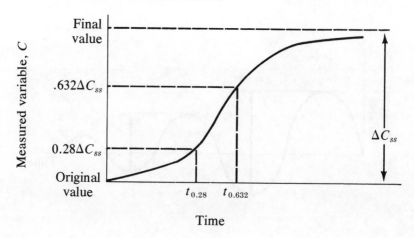

$t_{0.28}$ ($t_{0.632}$) = time at which response reaches 28% (63.2%) of the final value

Figure 1.18
Process Reaction Curve

$$\alpha = \frac{\Theta_d}{\tau_p} \tag{1.21}$$

where

Θ_d = dead time
τ_p = time constant of first-order model

or

= ($\tau_1 + \tau_2$) for the second-order model

For properly tuned loops the loop gain $K_c\,K_p$ as a function of α is shown in Figure 1.19. As seen from this figure, the loop gain decreases steadily as α increases and for large values of α approaches 0.45 for loops tuned by the Ziegler–Nichols method and 0.08 for loops tuned by Cohen–Coon method. The consequences do not end here. Figure 1.20 shows the peak offset to a step change in load as a function of α. For properly tuned PI control loops this peak offset is given by

$$\frac{\Delta C_{\text{peak}}}{K_L\,\Delta L} = \frac{1.65}{1 + K_c\,K_p} \tag{1.22}$$

For $\alpha \gg 1$ the peak offset approaches 1 which is the open loop value, an unhappy situation indeed.

Figure 1.21 shows the period of oscillation, P_u, as a function of α. For large α

$$\frac{P_u}{\tau_p} = 1 + 3\,\alpha \tag{1.23}$$

or

$$P_u = \tau_p + 3\alpha\,\tau_p \Rightarrow 3\Theta_d$$

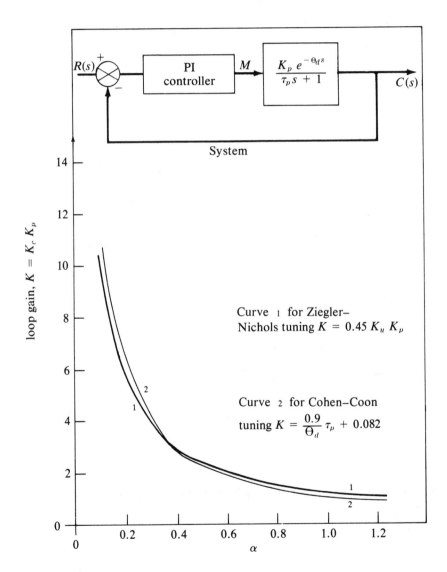

Figure 1.19
Loop Gain versus α for Properly Tuned PI Control Loops

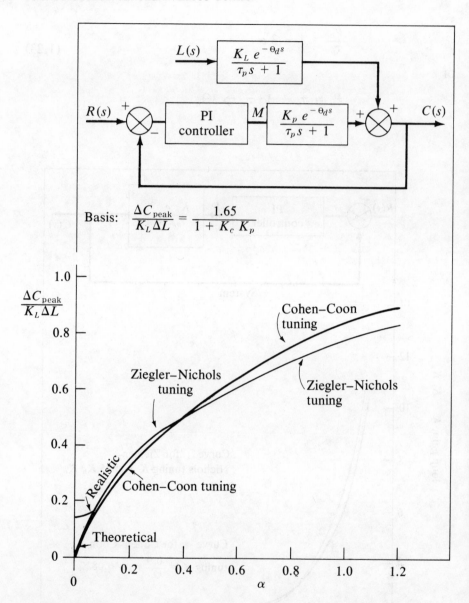

Figure 1.20
Peak offset versus α for Properly Tuned PI Control Loops

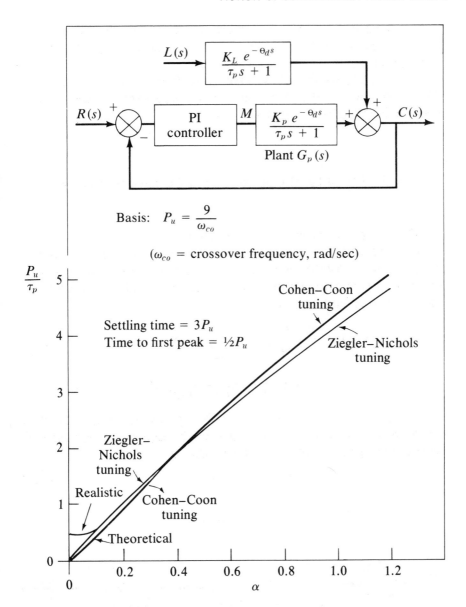

Figure 1.21
Period of Oscillation versus α for Properly Tuned PI Control Loops

Assuming the controlled variable will return to the set point in three periods of oscillation, we can say that for large α, the settling time approaches $3P_u$ or $(3)(3\Theta_d) = 9\,\Theta_d$.

These equations show that large dead times result in a large peak offset, a longer period of oscillation, and a longer settling time. These factors are all indicative of poor control.

1.5. Stability of Conventional Control Systems

Whether a conventional control system is stable or not depends on the nature of the roots of the characteristic equation. To answer the question of stability we start with the closed loop transfer function of the system shown in Figure 1.22. The transfer function relating $C(s)$ to $R(s)$ and $L(s)$ is

$$C(s) = \frac{G_c(s)\,G_1(s)}{1 + G_c(s)\,G_1(s)\,H(s)}\,R(s) \tag{1.24}$$
$$+ \frac{G_1(s)}{1 + G_c(s)\,G_1(s)\,H(s)}\,L(s)$$

In this equation $G_c(s)\,G_1(s)\,H(s)$ is the open loop transfer function which we denote as $G(s)$. The characteristic equation is the denominator of the right side of the equation, that is,

$$1 + G(s) = 0 \tag{1.25}$$

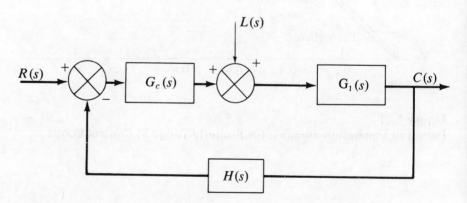

Figure 1.22
Typical Feedback Control System

The roots of the characteristic equation determine stability. For a stable system all the roots must lie to the left of the imaginary axis on the s plane, as shown in Figure 1.23.

One need not physically obtain the roots of the characteristic equation to determine stability. Routh's procedure[3] provides a simple test to determine if the control system is stable. To apply the test we write the characteristic equation $1 + G(s)$ in the form

$$a_0 s^n + a_1 s^{n-1} + a_2 s^{n-2} + a_3 s^{n-3} + \cdots + a_n = 0 \qquad (1.26)$$

If a_0 is negative, multiply both sides of this equation by -1. Then, if any of the coefficients $a_0, \ldots, a_n$ is negative, the system is unstable. If all the coefficients are positive, the system may or may not be stable. Routh's test should then be applied to determine stability. To apply the test, construct the Routh array as shown below:[4]

Row				
1	a_0	a_2	a_4	a_6
2	a_1	a_3	a_5	a_7
3	b_1	b_2	b_3	
4	c_1	c_2	c_3	
5	d_1	d_2		
6	e_1	e_2		
7	f_1			
$n + 1$	g_1			

The elements are filled in for a seventh-order characteristic equation ($n = 7$). For any other n the elements are determined using the same procedure described here.

In general, there will be $n + 1$ rows in all. If n is even, there will be one more element in the first row than the second. The remaining elements in the array are computed according to the following equations:

$$b_1 = \frac{a_1 a_2 - a_0 a_3}{a_1} ; \qquad b_2 = \frac{a_1 a_4 - a_0 a_5}{a_1} \cdots$$

$$c_1 = \frac{b_1 a_3 - a_1 b_2}{b_1} ; \qquad c_2 = \frac{b_1 a_5 - a_1 b_3}{b_1} \cdots \qquad (1.27)$$

The elements of the other rows are found from the equations that correspond to those above. According to Routh theorems the system is stable if all the elements of the first column are positive and non-zero. The Routh test does not give the roots of the characteristic equation nor the degree of stability (i.e. how far the roots are from the

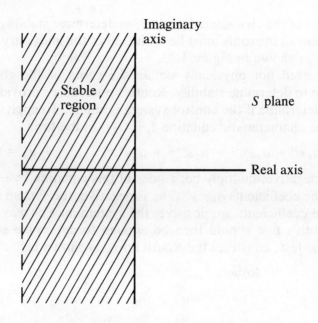

Figure 1.23
Stable Region in the *S* Plane

imaginary axis) of the control system. If this information is needed, we must determine the roots of the characteristic equation. Also, the Routh test cannot be applied to systems containing dead time. For these cases frequency response analysis can answer the questions of stability.

1.6. Problem Control Situations

PI or PID control works "adequately" well for perhaps 90% of industrial control requirements, if properly tuned (unfortunately, all loops are *not* properly tuned). In some situations PID control does not work very well, even if properly tuned.

Poor control is recognizable by a wide band painted on the process variable recorder; that is, the process variable spends a lot of time far away from its setpoint. This is usually caused by one or more of the following problem control situations:

—*Nonlinear Processes*—Gain (K_p) and/or dynamic parameters (e.g., τ_p, Θ_d) change with operating point. This causes the process to exhibit highly variable behavior—sluggish to respond at some times, wholly oscillasory (even unstable) at others; "deviation band" wide and wild at some times, narrow and well-behaved at others. Some form of adaptive control can improve this situation.

—*"Coupled" (Interacting) Processes.* These are processes in which the output is affected by the control input of one or more of the other loops as well as its own. Thus setpoint changes or load upsets in one loop affect other loops. When loops are *cross-coupled*, they "fight" each other—two loops, each stable by itself, may become unstable if strongly enough cross-coupled. Multivariable control decoupling can help this situation.

—*Problem Dynamics*—When process "apparent dead time" (Θ_d) equals or exceeds the "dominant time constant" (τ_p), (i.e., process controllability ratio $\alpha = \Theta_d/\tau_p > 1$), even with "best" PID tuning, peak offsets can approach those of the uncontrolled situation, and the settling time approaches $9\Theta_d$ as we saw earlier. Dead-time compensation can help these situations, often significantly.

—*Problem Disturbances*—These lead to a wide "deviation band," relative to desired range and accuracy of control. Disturbances that have a significant frequency content in the range around the loop's resonant frequency of oscillation are especially troublesome. Disturbances that act quickly on the outputs (i.e., load dynamics faster than process dynamics) cause problems resulting in larger deviations in the controlled variable than normal. Large inherent process noise (e.g., flow or pressure loops) or measurement noise also causes problems.

In many, if not most, of problem disturbance cases, cascade and/or feedforward control can contribute to substantial improvement in control performance.

Advanced Control Strategies

The original development of control theory was in late 1930s and early 1940s. Much theoretical investigation led to several promising approaches to better control. Some (like cascade and, to a lesser extent, feedforward) proved useful right away and have seen extensive application; others (like dead-time compensation and multivariable control decoupling) were essentially shelved because of a lack of practical hardware to implement them.

Now that digital computers are in widespread use for control and have appeared in "packaged" systems, hardware limitations have been removed, and advanced control techniques are seeing wider application in industry.

References

1. Ziegler, J. G., and Nichols, N. B., Optimum Settings for Automatic Controllers, *Trans. ASME*, **64**, 11, (November 1942) 759.
2. Cohen, G. H., and Coon, G. A., Theoretical Considerations of Retarded Control, Taylor Instrument Company's Bulletin, TDS-10A102, 1953.
3. Routh, E. J., *Dynamics of a System of Rigid Bodies, Part II, Advanced*, Macmillan, London 1905.
4. Coughanowr, D. R., Koppel, L. B., *Process Systems Analysis and Control*, McGraw-Hill, New York, 1965.

CHAPTER 2

Computer-Control Hardware and Software

In Chapter 1 we reviewed some of the important fundamentals of linear control theory and its application to the solution of conventional process control problems. In the following section we point out the basic differences between conventional control systems and computer-control systems and introduce the reader to the subject of sampled-data control. In the subsequent sections we describe the hardware and software needed in computer-control applications.

2.1 Conventional Control versus Computer Control

In a conventional control loop all the signals are continuous functions of time, as shown in Figure 2.1a. The pneumatic control loops employ air pressure for signal transmission, whereas the electronic control loops use voltage or current signals.

In a conventional control system the measuring element senses the value of the controlled variable and transmits it to the controller. This value is compared with the desired value or set point so as to generate a deviation signal, which we call *error*. The controller acts on this error to produce a control signal. The control signal is then fed to the final control element, which in many cases is an automatic positioning valve, in order to reduce the error.

In contrast to conventional control systems, the sampled-data control system involves discrete signals wherein the signal is a train of very narrow pulses of suitable height, as shown in Figure 2.1*b*. A computer-based control system is an example of a sampled-data system.

The schematic of a basic sampled-data control system is shown in Figure 2.2. The controlled variable is measured as before, and the continuous electrical signal, which represents the controlled variable, is fed to a device called an *analog-to-digital converter*, where it is sampled at a predetermined frequency. The sampling period, which is the duration between successive samples, is usually constant in process control applications. The value of the discrete signal thus produced is then compared with the discrete form of the set point in the digital computer to produce an error. An appropriate computer program representing the controller, called a *control algorithm*, is executed which yields a discrete controller output. This discrete signal is

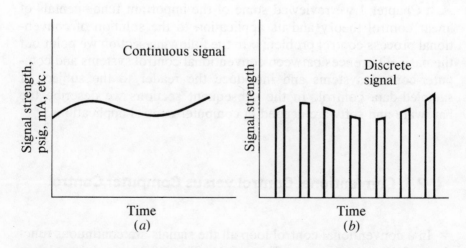

Figure 2.1
Continuous and Discrete Signals. The Discrete Signal Strength, as We Show Later, Is Proportional to the Area of the Corresponding Pulse.

then converted into a continuous electrical signal by means of a device called a *digital-to-analog converter*, and the signal is then fed to the final control element. This control strategy is repeated at some predetermined frequency so as to achieve the closed-loop computer control of the process. As mentioned earlier, this type of sampled-data control technique is referred to as direct-digital computer control, and it is the primary type of computer-control technique with which we concern ourselves in this text.

In many industrial applications the output of the measurement device and the input to the final control element are pneumatic signals. Because the computer works with electrical signals, signal transducers will be required for the conversion of electrical signals into pneumatic signals and vice versa for such applications.

From the foregoing discussion of conventional and sampled-data control systems, we note that the analog controller in the conventional control system is replaced by a digital computer, and the control action that is produced by the controller in the conventional loop is initiated by a computer program in the sampled-data control system. Thus one factor in justifying computer control for a given application is the number of conventional controllers that will be replaced by the digital computer. (It should be understood that this is only one factor in the computation of total cost of a computer control project; such things as the cost of transducers, software develop-

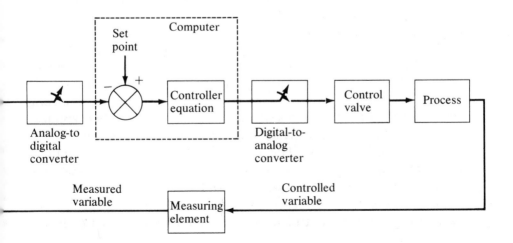

Figure 2.2
Typical Computer Control System

ment, and backup control systems for critical loops must be considered in arriving at the total cost of the project.)

Since digital computers were first used in control applications in the 1950s, their numbers have steadily increased in the last several years. With the development of microprocessors, based on LSI (large-scale integration) technology, the size and cost of digital computers have decreased. Also, with the introduction of microcomputers, the concept of distributed control, wherein small microcomputers control different parts of a large plant, has become attractive. The advantage of this approach is that in the event one of the units fails, only that portion of the plant is affected. As a result of these developments, the number of computer-control applications is expected to increase in coming years.

A possible justification for computer control may come from better performance. Owing to the computational power of digital computers, we will be able to implement control strategies that are otherwise either impractical or impossible with conventional hardware. Examples of such techniques include dead-time compensation, multivariable control decoupling, and optimal control.

The subject matter in this book can be grouped into logical subtopics. Later on in this chapter we describe the hardware and software required for computer-control applications. Chapter 3 presents a detailed procedure for converting a single conventional control loop into a computer control loop using the digital equivalent of the conventional PID (proportional + integral + derivative) controller. With an understanding of the material in that chapter, the reader should be able to operate single or multiloop processes under computer control of P, PI, or PID control algorithms.

The design and analysis of sampled-data control systems requires a knowledge of Z-transforms. At this stage it suffices to say that Z-transforms are to sampled-data systems what Laplace transforms are to conventional-control systems. Therefore, the next several chapters are devoted to the study of Z-transforms and their applications to the design and analysis of computer control systems.

As was pointed out earlier, the availability of control computers makes the implementation of advanced control strategies a relatively straightforward task. The final chapters in this book are concerned with the design and application of advanced control strategies, including feedforward control, dead-time compensation, and multivariable control.

2.2 Basic Concepts of Computers

The remainder of this chapter is directed at introducing the basic ideas of how computer systems are structured and how they work. Although our focus is on computers used in on-line systems to monitor and control processes, the principles developed apply for the most part to all digital computer systems. The scope of this text permits only a brief introduction to the basic concepts of computers and a broad overview of computer hardware and software. References 1 and 2 at the end of the chapter provide the interested reader with a much more detailed, yet still introductory level study of the vast field of digital computers. Both are highly recommended.

From the smallest microcomputer to the largest digital computer in use today, all digital computers are, in their essence, *serial number processors*. They take in data (numbers) from the outside world, process the data (make some calculations or perform some other operations based on the data), and then deliver results (also numbers) to the outside world. By *serial* we mean that the computer performs operations on numbers sequentially, one operation at a time.

Often, relative to the user (human being, process or, other outside-world device), the data are available or required in other than numerical form, for example, characters typed on a typewriter or voltages from sensors of physical variables such as temperature or flow. However, within the computer, data are processed in numerical form only and thus must be converted from the original form to numbers for processing by the computer and then back to whatever form is required for its end use in the outside world. Devices that perform the conversion of information (data) to and from numerical form are called *peripheral* devices and form a part of the overall computer system.

In basic operation a digital computer is similar to a simple electronic calculator. In a calculator the human operator enters numbers (data) and arithmetic operations to be performed on the data (e.g. add, subtract, multiply, divide) using the calculator's keyboard and receives the results of those operations on the calculator's numerical display. A computer is similar in that it performs arithmetic operations on numbers as its primary internal function and communicates with the outside world (in and out) to get data and deliver results of its internal operations. The range of peripheral or outside-world devices from or to which a computer gets data or delivers results is much wider than the simple keyboard and display of the calculator, however. Computers communicate not only with human beings but directly with processes, using a variety of devices to be discussed later.

In addition to the basic arithmetic processing and input/output function of the ordinary calculator, all digital computers have three additional features which greatly increase their power. These features are memory, automatic execution of instructions, and decision-making capability.

1. *Memory*. The memory of a digital computer can store numbers and retrieve them at some later time. Modern computer systems have very large memories capable of storing many thousands of numbers concurrently. Memory is also used to store *instructions*, making possible the next feature.

2. *Automatic Execution of Instructions*. An *instruction* is the vehicle used to tell the computer which operations to perform on what data and what to do with the result. Instructions are numbers in a special format recognizable to the computer and are put into memory by the user. A set of instructions in computer memory is called a *program*. When the computer "runs" a program, it retrieves the instructions from its memory, one-by-one, without operator intervention, and executes them. Programs are task oriented and frequently consist of many thousands of instructions.

3. *Decision-Making Capability*. A digital computer has the capability of changing its normal sequence of instruction execution as a function of the numerical value of a specified internal variable which may represent the result of a calculation or an input variable. This provides the ability of executing different sequences of instructions under varying process or program conditions.

The tremendous power of digital computers does not arise from a vast number of different available operations. A digital computer can execute only a relatively small number of different instruction types (operations), typically fewer than 100. Rather, the power of computers arises from the three features just described. There is virtually no limit on the complexity or sophistication of tasks that can be done with long sequences of basic operations having built-in alternate instruction sequences.

Conceptual Organization of a Computer System

There are three basic parts, or subsystems, in any computer

system, no matter how large or small. They are the *central processing unit* (or CPU), *memory* and the *input/output* (I/O) *interface*. Figure 2.3 shows the conceptual organization of a computer system and is a reference diagram for the discussion which follows.

Central Processing Unit (CPU)

The CPU is the "heart" of a digital computer, the subsystem which performs the actual execution of instructions. The CPU also controls sequencing of operations and controls communication with the memory and I/O interface subsystems.

Within the CPU are a number of hardware devices called *registers* which are used for temporary storage and manipulation of numbers and instructions. Registers are made up of electronic circuits called *flip-flops* which are two-state devices; that is, they operate in one of two stable states, either off or on, and they can be changed from one state to the other at any time on command from control circuits within the CPU.

Within digital computers, numbers are represented and manipulated in binary (i.e., base 2) form, in which each digit (called a *bit*) is

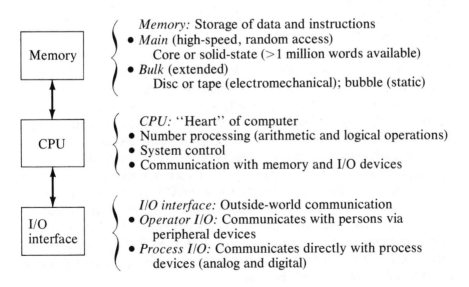

Figure 2.3
Conceptual Organization of a Computer System

either zero or one. Each flip-flop in a register holds one bit of a binary number. The number of flip-flops in a single register corresponds to the number of bits the CPU transfers to or from memory in a single operation. This number is called the *word length* of the computer. Thus a *word* is a group of binary digits used to represent a number or instruction. Sixteen bits is the most common word length for mini-computers.

Arithmetic and logical operations are performed within the CPU on numbers contained in registers according to instructions contained in other registers, and results are stored in the same or other registers.

Memory

The memory subsystem may consist of a number of different physical devices, but conceptually there are only two types of memory: high-speed, random-access memory (often called main memory) and bulk memory (often called extended memory).

High-speed random-access memory is made up of miniature magnetic cores or very-high-density solid-state flip-flop circuits. Each core or flip-flop stores one bit of information. A typical computer has many thousands of words of main memory and can access each and every word on a completely unconstrained (i.e., random) basis. The CPU can read or change the contents of any selected word in main memory in a single operation.

A high-speed random-access memory that permits both read and write operations by the CPU is called RAM (for random access memory). Another kind of memory, which the CPU can read but not change, is called ROM (for read only memory). Variations that can be reprogrammed by off-line devices are called PROM (for programmable ROM) or EPROM (for erasable programmable ROM).

Rapid advances in solid-state technology have drastically lowered the cost and reduced the size of main memory devices in recent years, and the trend will continue. Today (1980), circuit boards only a few square inches in area can contain over 64,000 16-bit words of solid-state RAM, and the entire main memory (up to 1 million or more words in many minicomputers), along with the CPU and several I/O interface boards, can be housed in a single chassis containing a dozen or so small, plug-in circuit boards.

Bulk (or extended) memory is much less expensive per word than main memory and has a much greater storage capacity. It generally resides on peripheral devices and is connected to the CPU through devices called *controllers*. The most common devices in use today for bulk storage are *disc* (rotating magnetic platters) or *magnetic tape* (cassette or reel-to-reel). Not all computer systems have bulk memory, but they all have main memory.

Bulk memory devices are much slower to access than main memory and cannot be accessed for read and write operations in the same random manner which main memory can. Bulk memory is accessed by *blocks*. A *block* is a fixed number of words which varies in size between computer manufacturers—256 words is a commonly used block size. In some systems block size can be specified by the programmer. Reading from or writing to bulk memory requires the transfer of an entire block of words between the bulk device and a predesignated area of main memory, often called a *buffer*. The CPU then reads or writes the desired word(s) in the main memory buffer.

By using removable magnetic discs or tapes, there is virtually no limit on the amount of data that can be stored in bulk memory. As this text is written (1980) most bulk memory devices in use are electromechanical magnetic devices (disc or tape), which tend to be the least reliable components in a computer system. Within the decade, however, bulk memory technology will move rapidly toward static devices built with high-density, high-capacity bubble-memory chips which have recently appeared in the marketplace. A vast improvement in performance (i.e., memory access time) and reliability should result. The equipment is almost certain to change, but the concepts of structure and operation will remain intact.

Input/Output (I/O) Interface

The I/O interface is the subsystem through which the CPU communicates with the outside world. There are two basic kinds of I/O: *operator I/O* and *process I/O*, the latter found only in real-time control and information systems.

Operator I/O communicates with people. Process operators use such devices as pushbuttons, thumbswitches, and keyboards to input data or commands to the computer, and they receive information from the computer via such devices as black-and-white or color CRT (cathode-ray tube) screens, LED (light-emitting diode) numerical displays, and pilot lights. Computer operators and program developers use *computer terminals* (a combination of CRT screen or typewriter with a keyboard) to communicate with the computer. Until recently, punched cards or punched paper tapes were widely used to store programs, and thus card and paper-tape readers and punches were commonly used computer I/O devices. Within the past 2 or 3 years, however, low-cost floppy-disc (a flexible plastic disc with a magnetic surface) and cassette-tape drives have become available and are rapidly replacing punched cards and paper tape.

Process I/O communicates directly between the CPU and all manner of process devices. Typical devices in a real-time process system include sensors, limit switches, tachometers, and shaft encoders for input; control valves, motor starters, stepping motors, and motor-speed controllers for output. Process devices, and some operator devices such as pushbuttons and pilot lights, are connected to the computer through two parts of the I/O interface called the *analog* and *digital I/O subsystems.*

To convert process input signals (usually voltages or currents from transmitters) to the digital (numerical) form needed by the CPU, the *analog I/O subsystem* uses hardware devices called *analog-to-digital* (A/D) *converters*. Because precision A/D converters are still relatively expensive, each A/D converter is generally shared by several (typically 8 to 16) individual analog input lines. A device called a *multiplexer* connects any one of these inputs to the A/D converter to be read on command from the CPU. The analog I/O subsystem also contains hardware used to "condition" the analog signals, for example, circuits to reject noise and smooth the input.

For output, the analog I/O subsystem converts numerical (digital) outputs from the CPU into analog voltages or currents by means of a *digital-to-analog* (D/A) *converter*. Since process analog output signals connected to devices such as control valves must be available to the device continuously, a separate D/A converter is used for each analog output device. A D/A converter sustains its output value until it is changed, which is done in most computer control systems on a periodic basis.

The *digital I/O subsystem* communicates with process devices which have only two possible states: on or off. Digital inputs arise most often from contact closures (e.g., limit switches). Digital outputs are "switches" (reed-relays or solid-state triacs) opened or closed by the I/O subsystem on command from the CPU. Most computer systems read or change more than one digital input or output concurrently; typically, 16 inputs (one word) are read or 16 outputs are updated in a single operation. A temporary location in main memory is used for the CPU to read or change individual digital inputs or outputs. The contents of this temporary memory location are transferred to or from the I/O subsystems by the CPU to actually change or read digital inputs and outputs.

Hardware and Software

A complete computer system, regardless of whether it is being used for process control, data processing, or any other function, consists of two basic kinds of components: Hardware and Software.

Hardware refers to all *physical* components of the system such as the CPU, main memory, bulk memory devices (e.g., disc and tape drives), the I/O interface and peripheral I/O devices such as terminals. In essence, anything which can be touched is classified as hardware. *Software* refers to all the *programs* (sets of instructions that tell the computer which operations to execute, step by step, to perform tasks) which are necessary for the operation of the system.

Tradeoffs are possible between hardware and software in computer systems. For example, the arithmetic unit in the CPU of some very small computers can only add two numbers and change the sign of a number; thus add is the only arithmetical function that can be accomplished with a single instruction. Subtraction requires two instructions (change sign of one number, then add). Multiplication and division are accomplished by repeated sequences of the basic add and change-sign operations which are controlled by programs; that is, multiplication and division are software operations in such computers. Additional hardware can be purchased that performs the multiplication and division operations in a single instruction step, thus making them hardware operations. The advantage of having hardware to perform additional arithmetic operations is simply speed. With multiply and divide hardware, these functions can be executed by the computer in a single instruction step. Without such hardware, multiplying or dividing two numbers typically requires between 10 and 30 instruction cycles, depending on the particular numbers involved.

Program Example

Many of the concepts developed in the foregoing sections can be illustrated with an example program for accomplishing a relatively simple task which is representative of the kinds of tasks performed by on-line computer systems in real-time process-control applications.

The task is as follows:

Read the value of a process sensor into memory, display this value on an LED numerical display, and set an external alarm if the value is outside the limits stored in memory.

Figure 2.4 illustrates this program in flowchart form, with a detailed description of what happens in each step depicted by the flowchart. Flowcharts are graphical descriptions of the sequential steps in a computer program. Each block represents a step in the sequence, and the arrows connecting the blocks define the order in which they are executed. The shape of the block defines the kind of operation: Rectangular blocks are CPU and memory operations, trapezoidal blocks are input/output operations involving the I/O interface, and diamond-shaped blocks are decision points in the program, at which an operation is performed and then one of two or more possible alternative paths in the program is taken, depending on the result of the operation. The oval blocks do not necessarily represent computer operations but mark the beginning and end of a program. The operations described in each block of a program flowchart are typically not single computer instructions but "subtasks" requiring many sequentially executed computer instructions to perform.

The steps required for this example program are:

1. Read the value of the process sensor voltage from the analog input subsystem into a holding register in the CPU. In typical computer systems, this requires connecting the A/D converter to the input line for the desired sensor with the multiplexer, waiting several microseconds for the A/D converter output to stabilize, and then transferring the resulting digital value into the CPU register.

2. Convert the A/D converter reading to physical units. Suppose that the sensor is a temperature-measuring instrument spanning the range 60°F to 300°F, corresponding to a 0 to 10 volt range on the analog input line. For simplicity, assume that the A/D converter output is a number between 0 and 10,000 (0 = 0 volts, 10,000 = 10 volts) and thus represents the number of millivolts present on the input line. (More typically, the A/D output span would be from 0 to 4096 to 0 to 16384, depending on the device precision). To convert the A/D output number to degrees, it is multiplied by a scale factor and then an offset is added; i.e. temperature (°F) = (scale factor) × (A/D output number) + offset. In this example the scale factor is 240/10,000 = 0.024, and the offset is 60. These conversion factors are stored in memory.

3. Store the converted value of the process temperature in a location assigned to it in memory.

4. Display the process temperature on an output LED numerical

Program to Read, Convert, Display, and Alarm Process Temperature

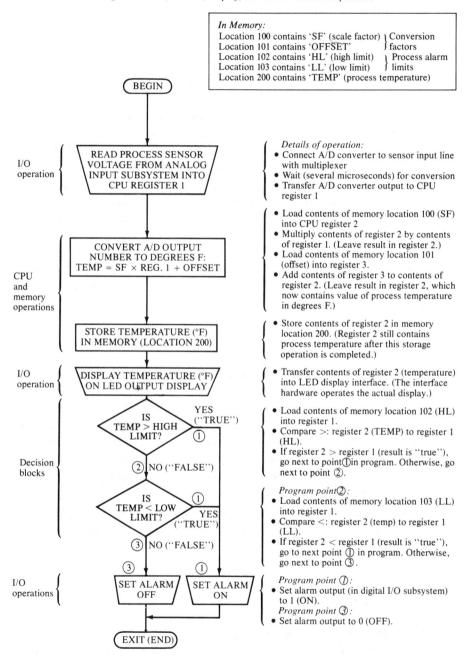

**Figure 2.4
Example Program**

display. Such displays hold their values until changed by the computer.

5. Compare the process temperature to high and low process limits stored in memory. If out of limits, set alarm (horn and/or blinking light) on. If within limits, set alarm off.

The decision or comparison steps in this program are examples of logical operations in the CPU. The result of a logical operation can have only one of two values: true or false. In a compare $> (A \to B)$ operation, true results if $A > B$, false otherwise. In a compare $< (A \to B)$ operation true results if $A < B$, false otherwise. A logical operation causes a flag (called a *condition code*) in the CPU to be set with its result (true or false). After the logical operation is complete, the next program instruction will read the internal flag and set one of two possible program routes for the ensuing calculations, depending on the flag's value.

In a typical real-time on-line computer system, a program such as this one would be executed periodically, typically once per second. Another program, called an *executive*, performs the automatic scheduling and execution of such programs.

2.3 Computer System Hardware Concepts

In this section we expand on the basic principles of computer hardware structure and operation introduced in the foregoing section. Although details of hardware structure and operation (called *architecture*) vary substantially from one computer to another, there are a number of basic principles involved which are essentially the same in any digital computer. Understanding of these basic principles will provide the student with a foundation on which to build further knowledge of specific computer systems.

Binary Numbers and Words

The hardware of a digital computer consists primarily of thousands or even millions of *two-state devices*, as discussed earlier. Each can be viewed, at any given time, as containing a single-digit number with only two possible values: 0 if the device is off, or 1 if the device is on. Use of such devices leads in a natural way to the use of binary (base 2) numbers for internal number representation and calculations.

In the binary number system there are only two digits, 0 and 1, as compared to the 10 digits, 0 through 9, in the familiar decimal (base 10) number system. Multidigit binary numbers are built on powers of two (i.e., 1, 2, 4, 8) for each higher significant digit, rather than powers of 10, as in the decimal system. Each digit in a binary number is called a *bit*. Using three-digit binary numbers, one can count from 0 to 7 as depicted in Table 2.1, which shows all possible combinations of three binary digits.

Table 2.1
Binary Numbers: 0 to 7

2^2: 4's digit	2^1: 2's digit	2^0: 1's digit	
0	0	0	$= 0 + 0 + 0 = 0$
0	0	1	$= 0 + 0 + 1 = 1$
0	1	0	$= 0 + 2 + 0 = 2$
0	1	1	$= 0 + 2 + 1 = 3$
1	0	0	$= 4 + 0 + 0 = 4$
1	0	1	$= 4 + 0 + 1 = 5$
1	1	0	$= 4 + 2 + 0 = 6$
1	1	1	$= 4 + 2 + 1 = 7$

In a similar way, four binary digits enable counting from 0 to 15. In general, n binary digits have 2^n possible combinations ranging from 0 to $2^n - 1$. The largest decimal number which can be represented with 12 bits, for example, is 4095 ($2^{12} - 1$), while with 16 bits it is 65,535 ($2^{16} - 1$).

Computers manipulate and store binary digits in groups called *words*. The number of bits in a single word (called the *word length*) varies between different computers, but 16-bit word length has become the most common for small computers. Larger systems use 32-bit words. A group of 8 bits is called a *byte*, a near-universal standard of reference. Early microcomputers used a word length of one byte, though the trend today is toward two-byte (16-bit) words.

Memory Organization and Addressing

RAM systems are organized in units of either bytes or words, with each unit (byte or word) being assigned a unique address so that it can be individually accessed for reading or writing. Addresses are assigned sequentially beginning with 0 and ending at the total number of bytes or words in main memory. Main memory sizes exceeding 1 million words are available and economically feasible in modern mini-computer systems. Memory sizes are usually expressed in units of K (for kilo) words or bytes, 1K actually being 1024 or 2^{10} units. Thus a 32K word memory segment contains 32×1024 or 32,768 words.

Bulk memory devices (disc and tape systems) are organized and addressed in terms of *blocks*, as discussed earlier. Each block, or group of words, has a unique address relative to the beginning of device. In many bulk memories, blocks can be organized into groups called *files* which are identified by unique names assigned by programmers. Such file-structured devices have directories on the device which are tables associating specific device block locations with file names.

Words or bytes in memory can represent one of two things: data or instructions. The following two sections discuss each in further detail.

Data and Number Representation

Stored data can represent either numbers in binary form or coded alphanumeric symbols. Alphanumeric symbols are letters, digits, and punctuation or mathematical symbols printable on a typewriter. Such symbols are used for computer input and output and are stored internally in coded form. The most common code used for internal representation is called ASCII, in which 7 bits are used to represent 96 different symbols and 32 control codes such as tab and carriage return. Each ASCII coded symbol is stored in 1 byte of computer memory.

Numbers are represented internally in one of two forms: fixed point or floating point.

In *fixed-point* (also called *integer*) representation, one word is used to represent a single whole number or integer. Thus with a 16-bit word length, 2^{16} or 65,536 different numbers can be represented. To handle negative as well as positive numbers, the span of the numbers is taken from −32,768 to +32,767. A negative number is represented by the "two's complement" of the corresponding positive number, which is obtained by binary subtraction of the number from 0, or (equivalently) complementing each bit in the number (i.e., all 0s are changed to 1s and vice-versa) and then adding 1 to the result. Two's complement number representation allows very efficient arithmetic operations within the CPU.

In some computers, double-precision numerical storage and

operations are possible. In the double-precision mode, two words are used to represent a single number, which greatly increases the range of integers and precision with which numerical calculations can be carried out.

In *floating-point* (also called *real-number*) representation, numbers are stored and manipulated in the equivalent of a scientific notation format, in which a number is represented by a *characteristic*, or exponent and a *mantissa*, or fraction. For example, in decimal scientific notation, the number 321.3 is represented as 0.3213×10^3, where the fraction is 0.3213 and the exponent is 3. The fraction is "normalized" to a value between 0.100 and 0.999, and the exponent can be viewed as the number of digits to the right (left for negative exponents) the decimal point must be shifted to convert the fraction to the actual number. Binary scientific notation follows the same principles, with the exponent and fraction being binary numbers. Floating-point representation in digital computers assigns a certain number of bits to the fraction and a certain number of bits to the exponent. Most 16-bit computers use two words for each floating-point number, but no standard at all applies in bit assignment for exponent and fraction between manufacturers. Typical, but by no means universal, is to use 1 bit for the sign of the number, 8 bits for the exponent, and 23 bits for the fraction. This allows a range of about $\pm 10^{\pm 38}$ with a precision equivalent to about seven significant decimal digits. Double-precision-floating point numbers, where available, add one or two additional words of precision to the fraction. Zero is a special floating-point number, because it falls outside the range of standard representation. Usually, zero is represented by all zeros (which corresponds to zero exponent and zero fraction).

Arithmetic operations on floating-point numbers are accomplished either with special, extra-cost floating-point hardware or with software (special programs that add, subtract, multiply, and divide floating-point numbers using sequences of fixed-point arithmetic and shifting operations). Floating-point software is available from most computer vendors as part of the software library provided with their hardware.

Floating-point representation is almost always used in applications requiring substantial numerical calculations, because it allows fixed precision independent of magnitude and a much wider range of numbers than fixed point. However, unless special arithmetic hardware is purchased, floating point arithmetic operations are much slower to execute than fixed-point operations and double the storage required in most minicomputers.

Instructions and Basic Computer Operations

Instructions stored in the memory of a computer are coded commands with which the programmer specifies what the computer is to do—what operation to perform, where to get the data, and what to do with the result. *Programs* are groups of instructions, usually stored in adjacent locations in memory, that are intended to be executed sequentially or in another order specified by the instructions themselves.

Each instruction defines a single operation to be performed by the CPU. The instruction is coded in a special format, shown in Figure 2.5. Although the specific operations which can be performed by the CPU and the specific instruction layout differ between computers, the structure shown is essentially universal to all computers. The operation code, or OP-code, specifies the operation and occupies from 3 to 10 bits of the instruction, depending on the number of different instruction types available in the computer's instruction set. Sixteen-bit minicomputers typically allocate 8 bits to the OP-code and have 50 to 100 instruction types in their instruction set. The total length of an instruction can be one, two, or even three words depending on the particular instruction. Bits beyond the OP-code are used for additional information defining the operation in more detail, for example, specifying CPU registers to be used or memory addresses of data to be fetched and/or stored as part of the operation.

No two computers, except possibly different models in the same family from the same manufacturer, have identical instruction sets. Nevertheless, the types of basic operations performed by the CPU are the same in all computers. It is conceptually convenient to classify instruction types into three categories: operations on data, program execution control operations, and input/output operations.

Operations on data perform arithmetic or logical operations on one or two numbers (called *operands*) located in memory or CPU registers; for example, an ADD instruction adds the contents of a specified memory location to the contents of a specified CPU register. Both the memory address and the CPU register are specified in the additional-information bits of the instruction. At the conclusion of this operation, the register would contain the sum of the contents of the memory location and the previous number contained in the register.

The following instruction types from this category are representative of basic operations performed by the CPU of a digital computer:

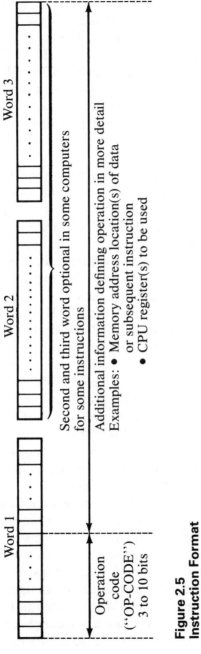

**Figure 2.5
Instruction Format**

- *Arithmetic Operations*   ADD, SUBTRACT (plus MULTIPLY and DIVIDE with added hardware) on two operands. CHANGE SIGN, IN-CREMENT (add 1), DECREMENT (subtract 1) on single operands.
- *Logical Operations*   AND, OR, EXCLUSIVE OR on two operands. NOT, SHIFT (left or right) on single operands.
- *Transfer Operations*   LOAD (memory to register), STORE (register to memory) and MOVE (register to register or memory to memory).

Program execution control operations alter the normal sequential execution of instructions and are used to provide alternate execution paths as illustrated in the program example. Instruction in this category are known broadly as BRANCH instructions.

In normal sequential operation the CPU gets the next instruction to be executed from the memory location following that of the instruction just completed. BRANCH instructions specify a different memory address for the next instruction.

There are three categories of BRANCH instructions:

—*Branch unconditional* (JUMP) instructions simply specify the memory address of the next instruction to be executed.

—*Branch on condition* instructions also specify a memory location for the next instruction, but the transfer is made only if a *condition*, specified by the instruction and based on the result of the last CPU operation, is met. For example, a BGT (branch on greater than zero) instruction will execute the transfer if the result of the last instruction executed was positive. Otherwise, normal sequential execution continues. All computers have many conditional branch instructions specifying different transfer conditions.

—*Branch and link* instructions are unconditional branch instructions that also cause the address of the next sequential instruction to be stored in a location (memory or CPU register) called a *link*. This allows one program to initiate the execution of (*call*) another group of instructions, called a *subroutine*. The last instruction in a subroutine is a RETURN instruction which executes an unconditional branch to the instruction located at the address stored in the link [i.e., the instruction following the subroutine call (branch and link) instruction], thus resuming execution of the previous program from the point at which the subroutine was initiated.

Input/Output operations transfer data between peripheral devices and the CPU or main memory. Two basic operations are provided: READ, which transfers data into the CPU or memory from a peripheral device, and WRITE, which moves the data from CPU or memory to a peripheral device.

I/O operations, and the instructions provided for them, are highly dependent on the architecture (hardware design) of the particular computer, and there is a wide variation in I/O instruction subsets between manufacturers. In many computers each peripheral device has a program (usually called a *driver*) to manage the detailed operations and timing needed for an I/O operation on that device. The driver is called (as a subroutine) whenever I/O operations are required by a user-written program.

Bulk memory devices (discs and tapes), as well as all peripheral devices, are connected to the CPU and main memory through hardware known as *device controllers*. Collectively, the device controllers comprise the I/O interface subsystem described earlier. Thus bulk memory devices are often considered to be peripherals, and transfers of data in and out of them are viewed as I/O operations. Most bulk memory devices are capable of transferring large blocks of data into or out of main memory in response to a single CPU command. This capability is known as *direct memory access*, or DMA, and allows the CPU to perform other operations while the block transfer is in progress.

The question frequently arises as to how the CPU knows whether the contents of a specific memory location is data or an instruction. The simple answer is that it doesn't, in any direct way. Once execution has begun, the sequence is inherently specified by the program itself, and the CPU assumes that the specified memory locations contain instructions. If, because of programmer error, there is data rather than an instruction at a specified memory location, the CPU will attempt to execute the data as an instruction. If the data happen to represent a valid instruction, it will be executed with, of course, unpredictable results. If it is not a valid instruction, the CPU generates an error message and stops execution of the program.

Program errors are called *bugs* and can wreak havoc in the orderly operation of a computer system, particularly if it is operating in a multitask mode in which several programs are resident in memory concurrently and execution is under control of a master executive program. An error that inadvertently causes even a single instruction in memory to be changed can lead to quick destruction of all programs in memory as more and more incorrect instructions are executed, some of which further change the instructions in memory.

Such a phenomenon is known as "bombing" the system and can be corrected only by reloading the destroyed ("bombed") programs into memory.

Writing error-free programs is a real challenge to users of computer systems. Many programming aids to the tedious and challenging job of debugging have been developed and are usually provided by vendors as part of the system software available with the hardware.

Principles of CPU Operation

Although all CPUs are different, there are certain principles of structure and operation that are essentially universal to all computers. The computer system block diagram shown in Figure 2.6 expands the conceptual diagram of Figure 2.3 and provides more detail for CPU structure.

Computers are *synchronous* devices, which means that they operate under control of an internal clock that steps the CPU circuitry through the operations necessary to execute instructions. A single instruction may require from 2 to 20 or more CPU clock steps (cycles) to execute, depending on the instruction. Most computers of 1980 vintage run at speeds exceeding one million cycles per second.

Within the CPU are located several registers that are key to its operation:

1. The *program counter* (PC), sometimes called the instruction address register, which contains the memory address of the next instruction to be executed.

2. The *instruction register*, into which instructions are loaded from memory.

3. The *accumulator*, which accumulates the results of repeated arithmetic and logical operations.

4. *General-purpose* (GP) *registers*, which are used to temporarily store numbers or memory addresses needed by the program. In many computers, all GP registers can be used as accumulators in arithmetic and logical operations, greatly increasing programming flexibility and eliminating the need for a separate accumulator.

The following steps comprise the complete execution of a single instruction:

1. Load instruction which is located at the memory address stored in the program counter into the instruction register. This sets up the control circuitry to perform the operation specified by the instruction.

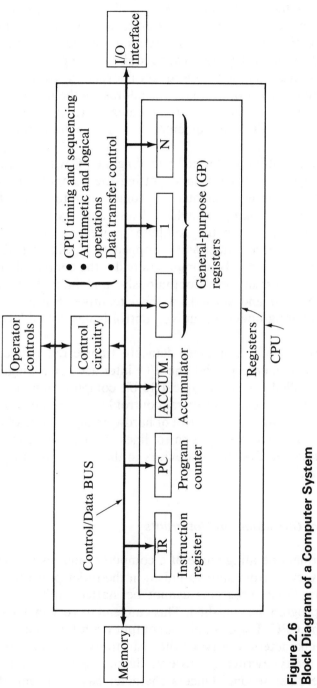

Figure 2.6
Block Diagram of a Computer System

2. Increment (i.e., add 1 to) the contents of the program counter.

3. Execute the instruction. This typically requires several sequential operations, each of which takes place within a single CPU clock cycle. Execution of multiword instructions includes fetching subsequent words of the instruction when required. The program counter is incremented again for each subsequent instruction word fetched.

Branching instructions replace the address in the program counter when executed. In the absence of branching instructions, execution is sequential, as the PC is incremented by 1 for each instruction word fetched.

Program execution is started by loading the PC with the address of the first program instruction and putting the computer into "run" mode. This can be done manually through operator controls or by automatic "boot" hardware which loads and starts an *executive* (monitor) program when the computer power is turned on. Once started, execution continues until stopped by a halt instruction, an error, or operator controls. Computers under the control of an executive program normally run continuously, and individual task programs are loaded and started by the executive. Further detail on systems operating under executive control is developed later in this chapter.

In many modern computers data are transferred between CPU registers, main memory, and the I/O interface on a single set of connectors called a *bus*. A typical 16-bit computer bus has 16 data lines, 16 address lines, and several control lines. The bus operates in a time-shared mode under control of hardware in the control circuitry of the CPU. The address and control lines are used to "connect" the proper source and destination devices to the bus in effecting a single data transfer.

Sequence of Operations and Interrupts

The normal operating mode of a computer requires that an entire program execute, from start to finish, in the order prescribed by the instructions. Other programs cannot be started until the currently executing program has finished. This can lead to very inefficient utilization of the CPU. For example, an output operation that sends a sequence of characters to a typewriter can transfer characters only as quickly as the typewriter can receive them, which can be as slow as 10 characters per second. Once a character has been sent, the CPU must wait, effectively doing nothing, until the output device is ready to receive another character.

To allow more-efficient CPU utilization and provide more flexibility in operation, computers are provided with *interrupt* capability. When a CPU interrupt line is enabled by a peripheral or other external device, normal execution stops (on completion of the currently executing instruction), and the CPU branches to a prespecified memory location for its next instruction, storing the current instruction location in the link. An *interrupt service program* is located at the specified transfer address, which executes and returns to the interrupted program on completion. An interrupt is, in effect, a hardware-initiated subroutine call (i.e., branch and link operation).

In order to successfully interrupt and later return to an executing program, the contents of all CPU registers (called the *state* of the CPU) must be stored in memory (saved) prior to beginning execution of the interrupt service program and reloaded into the same registers (restored) when control is returned to the interrupted program. This is done as part of the interrupt service program.

Every peripheral or external device capable of generating a CPU interrupt has an individual interrupt service routine. In many computers all such routines are grouped together into a single program, which first identifies the interrupting device and then branches to the proper service routine. More advanced designs can branch directly to a number of assignable memory addresses, often called *vectors*, one of which is assigned to each interrupting device.

The ability to interrupt and later resume execution of the currently running program is what enables a computer to monitor and control processes effectively. A program can run immediately in response to an external event (for example, a critical temperature detector) that occurs at a random time unrelated to internal computer operation. In addition, the time in which the CPU would be idle, waiting for an I/O device to be ready, for example, can be used to run other programs, returning to the waiting program on a "ready" interrupt from the peripheral device.

Most computers have an interrupt structure that allows the assignment of different *priority levels* to different devices. An interrupt service routine for a device of a given priority can itself be interrupted by a device of higher priority. An interrupt from a lower or equal priority device would, however, not be serviced until the current interrupt service is completed. Such a structure provides great flexibility in sequence of operations and permits the most efficient utilization of the CPU with satisfactory response to external events.

A *real-time clock* is an important hardware component in all computer systems that need to keep track of elapsed time or time intervals. It is a simple device that interrupts the CPU on a periodic basis.

Many clocks run off AC line supply and generate interrupts 60 times per second, or every 16.67 milliseconds. The interrupt service routine for a real-time clock simply counts interrupts, that is, increments a specified memory location each time the routine runs. By reading this memory location, other programs can easily determine time of day or can track elapsed time.

Hardware Configurations

Our introduction to computer hardware, structure, and principles of operation is now complete. The diagram of Figure 2.7 expands Figure 2.6 to include the many devices that are connected to the computer through the I/O interface. All real-time computer process-control systems would have some, but not necessarily all, of the peripheral devices shown in this diagram. The exact hardware configuration of any given computer system is highly dependent on the application.

2.4 Computer System Software Concepts

Software Classification

No computer system can operate without software, the programs necessary to execute the tasks performed by the system. Software can be classified into three categories:

1. *Application software* consists of programs for tasks directly related to the primary functions of the system. In a real-time process control system, the application software might include:

—A program to read analog inputs into memory.

—A program to convert these "raw" analog inputs to engineering units.

—A program to compute control outputs based on input values.

—A program to set analog outputs.

—A program to print an operating log.

Also included in this category are utility programs for functions such as calculating the square root of a number. Most application programs are written by the user, although vendor-supplied software generally includes common utility programs such as a square root program.

2. *System-support software* consists of programs that aid the user in the development of application programs. Programs in this category are almost always vendor-supplied. Included are:

—Language processors and linkers which convert programs written in high-level languages such as FORTRAN into machine-language programs.

—Editors, which facilitate the creation or modification of user-written programs.

—Programs which aid in debugging (finding program errors).

3. *Executive software*, often called the *operating system* of the

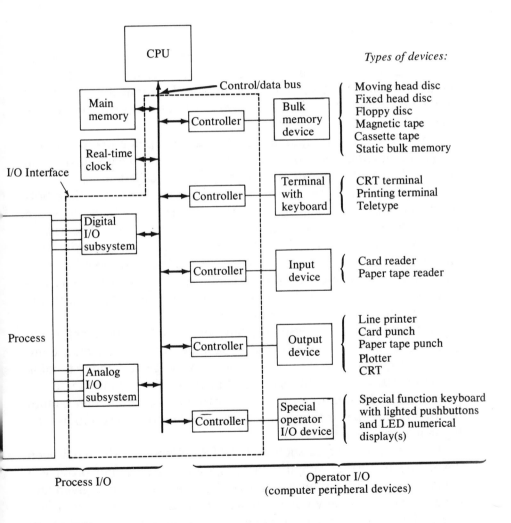

Figure 2.7
Hardware Configurations

computer, consists of programs that oversee or supervise the actual operation of the system while it is running, performing such functions as:

—Scheduling and starting the execution of system-application programs.

—Allocating main memory and loading programs into main memory from bulk memory (e.g., disc).

—Overseeing I/O operations.

—Servicing interrupts.

Programming Languages

Although it is possible to write computer programs directly in machine language (i.e., binary instructions), the extensive bookkeeping required to assign and keep track of memory locations for all data and instructions required makes this a very tedious and error-prone procedure.

To simplify the programming task, all computer manufacturers assign a mnemonic symbol of two or three letters to each instruction type and provide a system-support program called an *assembler* which translates these symbols into machine language. Symbols (names) assigned by the programmer, sometimes called *variables*, are used to designate memory locations to be used for data. The assembler assigns memory locations to such symbols. An *assembly language statement* combines an instruction mnemonic symbol with variable names to produce the symbolic equivalent of a single machine instruction. Examples of such statements are included in the task example developed in the next section.

Writing programs in assembly language enables a programmer to take maximum advantage of the capabilities of an individual computer. It requires, however, a thorough knowledge and understanding of the instruction set and assembly language of the computer being programmed. In addition, programs written for one computer cannot be used on another machine, because all instruction sets and assembly languages are different.

High-level languages such as FORTRAN and BASIC have been developed to further simplify the programming task and provide a degree of standardization that allows programs written in a given language to be run on any computer which has a compiler for that language. High-level language programs are sequences of *source statements* that specify, for example, basic arithmetic or logical operations to be performed on data, I/O operations, branching (conditional or unconditional), and subroutine calls. As in assembly language, symbols defined by the programmer are used to indicate specific variables or memory locations. High-level languages require no specific knowledge of computer architecture or instruction sets. They are easy to learn and have been routinely and widely taught for many years in university and high school technical programs. It is expected that most readers are familiar with one or more high-level languages. The process of translating a program written in a high level language into an executable machine language program is developed later in this chapter.

Example

In this subsection a very simple task is used to compare machine language, assembler language, and high-level language programming. The task is as follows:

Task: Add numbers in memory location 1121 and 1122. Store the sum in memory location 2322.

Steps:

1. Load contents of memory location 1121 into register 1.

2. Add contents of memory location 1122 to contents of register 1. (Register 1 now contains the sum).

3. Store register 1 contents in memory location 2322.

A. MACHINE LANGUAGE

Step	Memory Locations for Instruction	Instruction Word 1		Instruction Word 2
1 (Load)	121,122	1110 0101	0001 0000	0000 0100 0100 0001
2 (Add)	123,124	1110 1101	0001 0000	0000 0100 0110 0010
3 (Store)	125,126	1110 0110	0001 0000	0000 1001 0001 0010

OP-Code   First operand (Reg. 1)   Second operand (memory)   Memory address of second operand

Three two-word instructions are required for this task. In each instruction, the left 8 bits of word 1 contain the OP code, and the remaining bits specify operand addresses. In these instructions the first operand is register 1, and the second operand is the memory location at the address of the second instruction word.

B. ASSEMBLY LANGUAGE

Symbols (Variables):
I represents contents of memory location 1121
J represents contents of memory location 1122
K represents contents of memory location 2322

Assembler Program:

Step	Assembly	Language	Source Statement
1 (Load)	LDM	1	I
2 (Add)	ADM	1	J
3 (Store)	STM	1	K
	OP Code mnemonic symbol	First operand (register 1)	Second operand (symbolic memory location)

Each assembly-language statement is translated into a corresponding machine-language instruction by the assembler program. The assembler normally assigns and keeps track of memory addresses for symbols. Specific memory locations, if desired, can be assigned by the programmer with additional assembly-language statements.

C. HIGH-LEVEL LANGUAGES

Symbols: Same as assembly-language example.

	Source statement
FORTRAN:	K = I + J
BASIC:	LET K = I + J

In high-level languages, only a single statement is required for this task. The statement is translated into machine-language instructions by the language *compiler* program.

Language Compilers and Interpreters

A system-support program that converts, or translates, high-level language programs into machine-language instructions is called a *compiler*. Writing a high-level language program and converting it to an executable machine-language program actually involves several steps:

1. The *source program* is the set of high-level language statements

created by the programmer. It must be in a form which can be read as input by the computer, for example, punched cards, paper tape, floppy disc, or cassette tape. A system support program called a *source editor* is available in many systems to aid in creating or modifying source programs. Using a source editor, source program statements can be typed into a terminal, and the resulting source program is stored in a file on a bulk memory device. Such source files can also be easily modified using the source editor program.

2. The compiler program is run and reads the source program as input. The compiler produces machine instructions for the program in what is called *object* form. As noted earlier, many machine instructions contain memory addresses which locate data or other instructions as branching targets. Therefore, a machine-language program cannot be completely written until all its instructions and data are assigned specific memory addresses. A compiler generates instructions assuming that the programs will be located in memory beginning at memory address 0. The *object module* produced by the compiler contains, in addition to these instructions, data tables that can be used by another program to *relocate* the program; that is, modify all its memory references so the program can run in another assigned memory location, different from 0.

Many compilers are two-step programs which first translate the source program into assembly language. The system assembler is then run to produce the final object module. All compilers also check for and report programming errors. In addition to the object module, they can also produce printed listings of the source statements, memory "maps" for all program variables, and generated code listings, usually in assembly language.

3. The final step in conversion is to assign a specific memory location to the program and modify the object module addresses for that location. This is done by a system support program called a *linker*. The linker output is the "memory image" of executable machine code*. Actually, the linker is a powerful program that can accept many separately compiled object modules and "link" them into a single executable program, assigning memory addresses to and relocating each module. This capability allows a programmer to develop and compile individual program pieces separately, for example, a main program and the subroutines it calls. It also permits grouping

---

*The term *code* is a frequently used synonym for program instructions in various formats. Thus we speak of source code, assembler code, object code, and machine code to refer to programs, or subsets thereof, in these formats.

commonly used programs (such as square root) into a *library* in object format. The linker automatically extracts all such required programs from the library and includes them in the final run module (i.e., executable program). Linkers also produce printed memory maps of all program locations.

The oldest and most widely used high-level "scientific" language is FORTRAN IV. Virtually all computers, except the smallest microcomputer systems, have a FORTRAN compiler available. Although FORTRAN was originally created and standardized for batch processing (i.e., sequential program execution) applications, many vendors provide real-time extensions, a set of subroutines that work in conjunction with the computer's operating system to perform functions required by real-time systems, such as:

- Analog and digital process input and output.
- External interrupt handling.
- Scheduling task execution and other timekeeping functions.

Programs written in FORTRAN or other high-level languages generally require more memory and time to execute than equivalent programs written in assembly language by a skilled programmer. Usually, however, this is an acceptable price to pay for the great increase in flexibility and ease of programming provided by the high-level language.

A different approach to the use of high-level languages is a program called an *interpreter*. Unlike a compiler, an interpreter does not convert high-level source statements into machine language; rather, it reads the source statements during actual program execution, "interprets" the statements to determine actual functions required, and calls subroutines (which are part of the interpreter) to carry out the functions. Memory is assigned to variables "dynamically" (i.e., during actual execution) the first time a symbol (i.e., variable name) is encountered. The popular high-level language BASIC is usually implemented with an interpreter. BASIC is similar in structure to FORTRAN but simpler, particularly for I/O operations.

Interpreter-based programs run much more slowly than compiled programs because of the time required to interpret source statements. They can be easily modified, however, simply by changing the source code. Changes to a *compiled* program require that the program be recompiled and relinked after the source code is changed. Interpreters are advantageous, therefore, in applications such as pilot plants where programs are changed frequently, whereas compiler-based languages are more advantageous for applications that, once developed and debugged, require little or no change, or where the higher speed is required by the application.

Real-Time Operating Systems

Executive programs, usually called *operating systems*, are used in many computers to control program execution and permit more efficient utilization of the CPU. The operating system (OS) is the master program in the system. Once it is loaded into memory and started, it runs continuously. All other programs run, in effect, as subroutines that are called by the OS and return to the OS on completion. Many computers have automatic "boot" hardware that loads the OS into memory from bulk storage and starts execution when the power is turned on.

The primary function of any operating system is to schedule and execute programs, which are often called *tasks* when run under OS control. Tasks can be scheduled in two ways:

1. On *command* from an operator or system user. Commands are statements read by the OS from its primary input device, usually a console terminal into which commands are typed by the user. Although each operating system has its own command structure (called *syntax*), many vendors use a command syntax adapted from IBM's JCL (job control language) developed in the early 1960s for System/360 operating systems. For example, the following command (from DEC's RT-11 operating system) is used to compile a FORTRAN program and produce a source listing on the system printer as well as an object module. The source program is stored on bulk memory device DX1 (a floppy disc) in a file named PGM1.FOR, and the object module is to be stored on the same device in a file called PGM1.OBJ. The command is:

FORTRAN/LIST/OBJECT:DX1:PGM1.OBJ DX1:PGM1.FOR

In response to this command the operating system loads the FORTRAN compiler into memory from bulk storage, opens the specified files on device DX1, and starts execution of the compiler program. When the compiler finishes, it returns control to the OS, which then closes the specified files and reads the next command.

2. On request from a running program. Real-time operating systems include a series of subroutines that can be called to schedule other programs, either immediately, at some later time, or whenever a specified event such as an external interrupt occurs. A call to a scheduling routine makes an entry in a table of pending tasks maintained by the operating system. The task-manager portion of the OS, which runs whenever the CPU is interrupted, executes these tasks as scheduled if the CPU is free. When there is a conflict for the CPU between two or more tasks, the task manager uses task priorities, dis-

cussed in more detail later, to arbitrate the conflict. As an example, consider a program to read analog inputs that must run periodically, once per second. If the first statement of this program is a subroutine call that schedules *itself* to run after 1 second has elapsed, a single execution of the program (scheduled, say, by an initialization program) achieves the desired periodic execution.

Another important operating system function, closely related to task scheduling, is called *multitasking* or *multiprogramming*. In a multitasking system it is possible to have more than one program in memory and in some stage of execution. Although to the user the programs seem to be running concurrently, only one program is executing at any given time. To change from one executing program to another, the OS saves the state of the running program (CPU register contents) in memory and transfers to another program. Execution of the suspended program can be resumed at a later time by restoring the CPU state and restarting execution at the point where the program was suspended. This operation, called *context switching*, is identical in concept to interrupt handling described earlier.

Time-sharing systems are simple examples of multitasking operations, where each user of the system is allocated a portion of CPU time in a "round-robin" fashion. Context switches also occur when the running program is held up (bound), waiting for some event to occur, such as completion of an I/O operation. In this way the CPU is used with maximum efficiency, and is, in fact, never idle whenever there are pending or uncompleted tasks to be run.

The simplest multitasking systems can handle only two concurrent programs, generally called the *foreground* and *background* tasks. The foreground task has priority, so the background runs only when the foreground is suspended for I/O or some other reason. More elaborate systems can handle several (as many as 256) concurrent tasks. Each program is assigned a *priority level* of execution, and the program running at any given time will be the highest-priority program that is not bound. A program can, for example, schedule another program with higher priority than itself to run immediately. This would cause an immediate context switch from the calling to the called program, with the calling program resuming execution after completion of the called task. This is, in effect, identical to a subroutine call within a single task, and allows tasks to interact in a manner specified by the tasks themselves.

Two other key operating system functions are *interrupt handling* and managing *I/O operations*. In most operating systems interrupts transfer control to the OS which then calls the required interrupt service routine. Routines to service peripheral device interrupts are usually vendor-supplied and often part of the device drivers. User-written interrupt service routines, when required, are "connected" to the operating system with OS commands or program calls to vendor-supplied subroutines that perform this function.

I/O operations involving peripheral or process devices are managed by the operating system, which includes a series of I/O subroutines. Generally, a single call to the appropriate subroutine is all that is required in a user-written program to execute I/O operation. The operating system loads and calls the device driver and services device interrupts as required to complete the specified operation. I/O is a complex procedure in many computers, and a single I/O call from a user program will frequently initiate a sequence of several thousand instructions, all under control of the operating system.

Real-time operating systems (RTOS) are complex and often very large programs that require years of development by computer vendors. A complete RTOS may contain hundreds of individual programs or modules. Not all these modules will be required by most systems. The operating system for any given system is created by selecting from the total set the programs needed for desired OS functions and support of the system devices. These programs are linked into a single operating system tailored to the specific computer system in a process known as *system generation.*

Today's sophisticated real-time operating systems have many capabilities beyond the basic functions described here. Such powerful executive programs substantially ease the development of on-line computer applications.

2.5 Configurable Digital Systems and Networks

In spite of the powerful support software and operating systems available for digital computers today, it is an expensive and time-consuming task to develop application software for real-time computer control and information systems. In custom, one-of-a-kind applications it is not unusual for the software development costs to exceed the cost of the computer system hardware, sometimes by a factor of two or three. Writing such software also requires personnel with a very high level of specialized computer skills.

In an attempt to reduce software costs and permit engineers without a high level of computer skills to produce application software, several manufacturers have created software packages consisting of generalized subroutines to perform functions commonly found in computer control systems, for example, process I/O control algorithms, operator communication, and generation of reports and logs. Such systems are "programmed" by entering data and parameters into *tables* specifying which functions are to be used as well as which process or internal variables are to be used by each function. The systems run under the control of an internal, interpreter-like executive program that reads the data tables and calls the required subroutines to perform the specified functions.

The process of entering data into the tables begins in many such systems with filling in a standard form supplied by the manufacturer to specify each function; consequently, they are often called *fill-in-the-forms* packages. More generally, they are referred to as *control-oriented languages* or *table-driven software*. Generating application software using such packages does not require writing programs but rather *configuring* prewritten programs together for a specific application. Thus the combination of a digital computer system with a table-driven application software package is referred to as a *configurable digital system*. Applications are limited to those which require only functions available in the package, although some packages permit user-written programs to be added for specialized functions.

The cost of computer hardware has dropped so dramatically in recent years that several manufactuers of process-control equipment have developed configurable digital systems built around microprocessors that are cost-competitive with conventional analog control hardware. Such systems are *distributed*, which means that several individual computers are connected in a *network* and communicate with one another (i.e., transfer data and commands) on a high-speed, time-shared communications line called a *data highway*. Such systems include not only microprocessor-based controllers with built-in process I/O but also *operator stations* built around color CRTs and keyboards which replace the conventional control panel as an operator interface. They are easily configured by engineers with no specialized computer knowledge. Their total capability includes all conventional control system functions plus many capabilities not easily implemented with analog hardware, including such advanced control strategies (developed in later chapters) as feedforward, dead-time compensation, and multivariable control. Redundancy is widely used in such systems to provide reliability which often exceeds that of the conventional systems they replace.

The advent of such cost-effective and easy-to-apply systems has caused a dramatic upturn in the use of digital systems for process control as well as the application of advanced control strategies. It is likely that by 1985 or so, the great majority of new process control systems will be implemented with networks of digital computers.

References

1. Mellichamp, D.A., Ed., *CACHE Monograph Series in Real-Time Computing*, Monographs I–VI, CACHE Publication Committee, c/o Brice Carnahan, Dept. of Chemical Engineering, University of Michigan, Ann Arbor, MI 48019, 1977
2. Harrison, T. J., Ed., *Minicomputers in Process Control—An Introduction*, Instrument Society of America, Pittsburgh, 1978.

CHAPTER 3

Single-Loop Computer Control

H aving read the description of the hardware and software required for computer-control applications, let us now see how we might go about implementing computer control on a typical process loop. For simplicity, let us consider a single-loop system. The single-loop system is a flow control loop currently on conventional control. The pneumatic controller of this loop is the proportional + integral (PI) type. It is desired to place this loop under computer control, utilizing the discrete equivalent of the PI controller.

3.1. The Present System

A schematic of the analog flow control loop is shown in Figure 3.1.

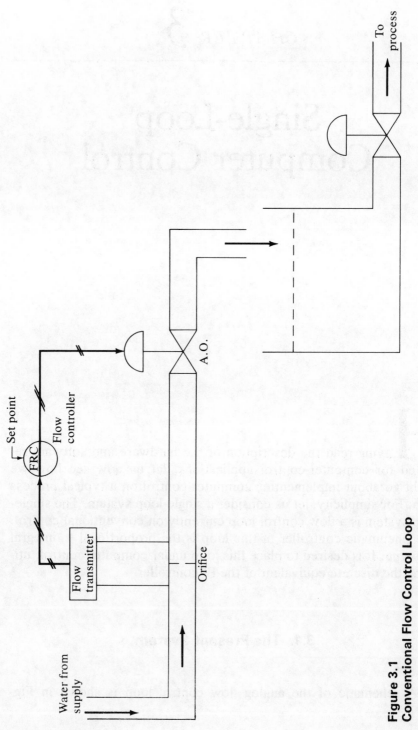

Figure 3.1
Conventional Flow Control Loop

The flow transmitter measures the pressure differential across an orifice and converts it into a 3 to 15 psig pneumatic signal. The flow controller compares this signal with the desired value, the set point. If an error exists, the controller outputs a signal (also between 3 and 15 psig) which manipulates the valve position so as to eliminate the error. Note that all instrument signals in this pneumatic loop are continuous signals. If some of the instruments were of the electronic type, their inputs and outputs would be current or voltage signals, but all would still be continuous signals.

The operation of the ideal PI controller may be described in terms of the following equation:

$$v = v_0 + \frac{100}{PB} \left\{ e + \frac{1}{\tau_I} \int_0^t e \; dt \right\} \tag{3.1}$$

where

v = output signal from the controller at time t, percentage of full-scale

v_0 = output signal from the controller at time $t = 0$ (at the time when the controller is switched from manual to automatic), percentage of full-scale

e = error signal (i.e., $R - c_m$), percentage of full-scale (c_m is the measured variable, percentage of full-scale, and R is the set point, percentage of full-scale)

PB = proportional band, percent (defined as the proportional scale change in input necessary to cause a full-scale change in controller output)

τ_I = integral time, minutes

Some literature describes Equation (3.1) in terms of the gain k_c and reset T_I. The relationship between k_c, PB, and τ_I, T_I is

$$k_c = \frac{100}{PB} \quad \text{dimensionless} \tag{3.2}$$

and

$$T_I = \frac{1}{\tau_I} \quad \text{repeats per minute} \tag{3.3}$$

Thus if PB is 50% and τ_I is 2 minutes, then

$$k_c = \frac{100}{50} = 2 \tag{3.4}$$

and

$$T_I = \frac{1}{2} = 0.5 \text{ repeats/minute}$$

Let us assume that the desired flow rate is 3.78 gallons per minute (gpm). Then the value of R and v_0 can be calculated from the instrument calibrations as follows:

1. From orifice calibration shown in Table 3.1, at 3.78 gpm, the flow transmitter output is 7.26 psig (i.e., 35.5% of full-scale range of 3 to 15 psig). Therefore, $R = 35.5$.

2. From valve characteristics shown in Table 3.1, at 3.78 gpm, the signal to the valve must be 8.9 psig (i.e., 49.17% of full-scale range of 3 to 15 psig).

3. $v_0 = 49.17$ (from Table 3.1, this corresponds to 6 on the flow recorder/controller).

Let us also assume that tests have been conducted at the desired operating level of 3.78 gpm and the following tuning constants have been found to be satisfactory:

$$PB = 50\% \tag{3.5}$$

$$\tau_I = 1 \text{ minute}$$

To start up and operate this loop, the following procedure may be used:

1. Turn on instrument air and water supply.

Table 3.1
Flow Transmitter Calibrations and Valve Characteristics

Flow (gpm)	Indicated Flow on FRC (arbitrary units)	Output Signal from Flow Transmitter, (percentage of full-scale)	Signal to Valve (percentage of full-scale)
0	0	0 (= 3 psig)	0 (= 3 psig)
0.79	1	1	14.58
1.29	2	3	21.67
1.92	3	8	29.17
2.54	4	15	34.2
3.20	5	24	40.5
3.78	6	35.5 (= 7.26 psig)	49.2 (= 8.9 psig)
4.44	7	49	59.2
4.92	8	64	65.8
5.61	9	82.5	79.6
6.31	10	100 (= 15 psig)	100 (= 15 psig)

2. Move the auto/manual knob on flow controller to manual.

3. Set the proportional band at 50 and reset dial at 1.

4. Move set-point needle to 6 on FRC (i.e., 35.5% of full-scale input).

5. Adjust the output of the controller to get 49.17% (8.9 psig) output signal. This will be achieved when the set point and measurement needles on FRC match.

6. Switch FRC to automatic.

3.2. Switchover to Computer Control

Figure 3.2 shows the schematic of the computer-controlled flow loop. The output of the flow transmitter is fed to an air-to-current (P/I) transducer. The resulting 4 to 20 mA electrical signal is connected to the terminals on the control computer that represents one of the analog-to-digital (A/D) converter channels. The discrete output of the A/D converter is available to the computer on demand. The discrete output from the computer is converted to a continuous signal, also on demand, by one of the digital-to-analog (D/A) converter channels. The D/A output, 4 to 20 mA, is available at the analog output terminals of the computer. It is connected to an I/P transducer, which produces a proportionate 3 to 15 psig pneumatic signal. This signal then operates the valve.

The control computer is instructed to sample the A/D channel every T seconds (where T is the sampling period). The computer program operates on this measurement (which represents the value of the measured variable at the sampling instant) using the discrete equivalent of the PI controller equation and computes the desired control-algorithm output. The computer is then asked to forward this output to the D/A converter and on to the valve. This procedure is repeated every T seconds so as to achieve closed-loop computer control. Now, let us develop a computer program that will accomplish this task. First, we must derive an equation that represents the discrete equivalent to the PI controller.

The output of the control equation at the nth sampling instant, using numerical integration for the integral of Equation (3.1), is

$$v_n = v_0 + \frac{100}{PB} \left\{ e_n + \frac{1}{\tau_I} \sum_{i=0}^{n} e_i T \right\} \tag{3.6}$$

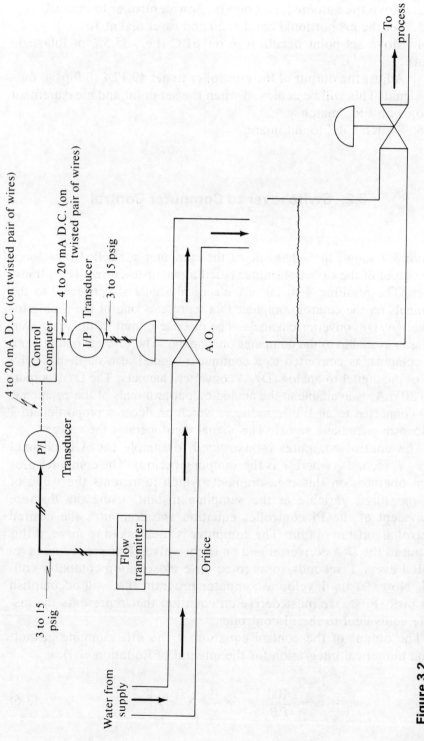

Figure 3.2
Computer-Controlled Flow Loop

Similarly, the output at the $(n - 1)$th sampling instant is

$$v_{n-1} = v_0 + \frac{100}{PB} \left\{ e_{n-1} + \frac{1}{\tau_I} \sum_{i=0}^{n-1} e_i T \right\} \tag{3.7}$$

Subtracting Equation (3.7) from Equation (3.6) gives

$$v_n = v_{n-1} + \frac{100}{PB} \left\{ (e_n - e_{n-1}) + \frac{T}{\tau_I} e_n \right\} \tag{3.8}$$

where

v_n = control-law output at the nth sampling instant
v_{n-1} = control-law output at the $(n - 1)$th sampling instant
e_n = error at the nth sampling instant
e_{n-1} = error at the $(n - 1)$th sampling instant
T = sampling period, seconds

The output v and error e are in percentages, as in the case of the conventional control system. Equation (3.8) is the digital equivalent to the conventional PI controller and is referred to as the proportional + integral (PI) control algorithm.

The output of the algorithm one sampling period after the loop is placed under computer control is computed from the equation

$$v_1 = v_0 + \frac{100}{PB} \left(e_1 + \frac{T}{\tau_I} e_1 \right) \tag{3.9}$$

where v_0 is initialized to a value which is estimated to cause the measured variable to equal its set point at steady state. The algorithm output at subsequent sampling instants is computed from Equation (3.8).

The flowchart of a computer program to implement computer control using the PI control algorithm is shown in Figure 3.3. From this flowchart the control program can be developed using a suitable programming language.

The start-up and operational procedure for the computer control loop consists of the following steps.

1. Connect the equipment as shown in Figure 3.2.
2. Turn on instrument air and water supply.
3. Turn on computer and the I/O device (e.g., teletypewriter).
4. Initialize the computer (the procedure will vary from computer to computer).

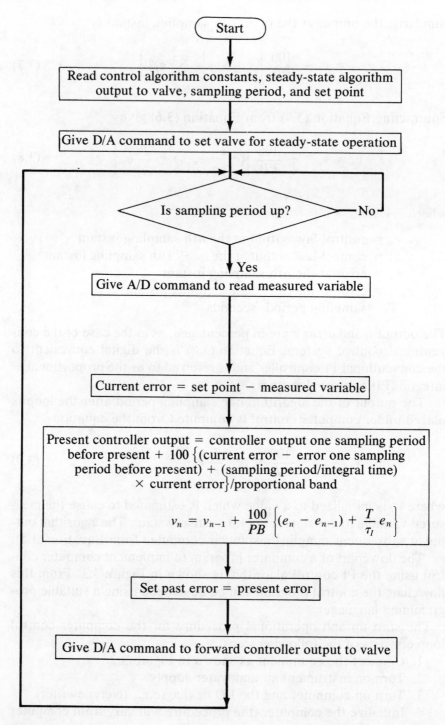

Figure 3.3
Flowchart of Computer Program to Implement Control

5. Follow the manufacturer's procedure to enter the control program into the computer.

6. Execute the program.

The procedure described in this chapter applies to single-loop computer control using the PI control algorithm. This approach can be readily extended to computer control of multiple loops using the PI (or P, or PID) control algorithm by suitable modifications using repetitive program structures. The real power of a computer in control applications lies in its ability to facilitate the implementation of improved or advanced control strategies that would be impractical with analog hardware. The development of such strategies requires that we be able mathematically to represent and analyze the computer-control loops. This type of analysis is also needed to determine the stability characteristics of computer-control loops. The next several chapters are, therefore, devoted to the mathematical analysis of the computer-control loops.

CHAPTER 4

Mathematical Representation of the Sampling Process

In a computer-control loop, the process variable is sampled every T seconds, where T is the sampling period. This function is performed by an A/D converter. For the purpose of analyzing and designing computer-control systems, it is necessary to develop a mathematical representation of the analog-to-digital converter or the *sampler*, and this is the subject of this chapter.

The basic purpose of the sampler is to obtain the values of the process variable, which is a continuous function, at regular intervals of time. Figure 4.1 is a pictorial representation of this concept.

The sampler permits the input signal to pass through only during the short interval τ but blocks it during the remaining portion of the sampling period. The signal transmission interval τ is very short in comparison with the sampling period. The values of the output function during the interval τ are equal to the corresponding values of the input function during the interval. In other words, the output from the sampler can be thought of as a train of very narrow pulses, the enve-

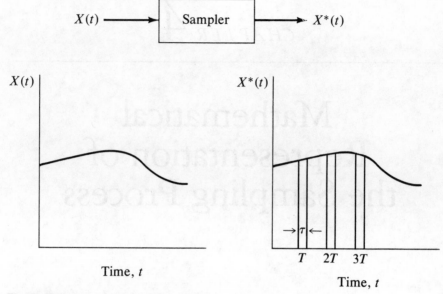

Figure 4.1
Input-Output Signals from Sampler

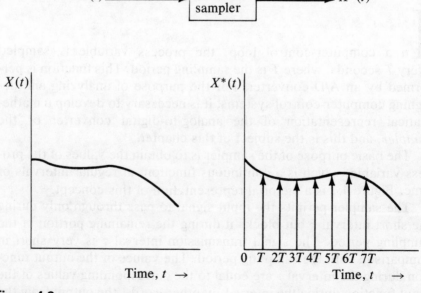

Figure 4.2
Idealized Sampling Operation

lope of which is identical to the input signal. The quantity $f_s = 1/T$ is called the *sampling frequency*.

Because the sampling duration τ (i.e., the width of the sampled pulse) is small compared with the most significant time constant of the control system as well as the sampling period T, the sampler output can be considered as a train of impulses whose strengths are equal to the values of the continuous function at the respective sampling instants, as shown in Figure 4.2. This assumption greatly simplifies the representation of the sampler. To fix the concept of representing the sampled function by impulses, let us review the mathematical representation of a rectangular pulse and the impulse.

A rectangular pulse function $f(t)$ is represented (see also Figure 4.3) as

$$f(t) = \begin{cases} 0 & t < 0 \\ \dfrac{1}{h} & 0 \leqslant t \leqslant h \\ 0 & t > h \end{cases} \qquad (4.1)$$

Now, let $h \to 0$. Then we obtain a new function that is zero everywhere except at the origin, where it is infinite. However, the area under this function always remains unity. This new function is called the *delta function* $\delta(t)$. The fact that its area remains unity means that

$$\int_{-\infty}^{\infty} \delta(t)\, dt = 1 \qquad (4.2)$$

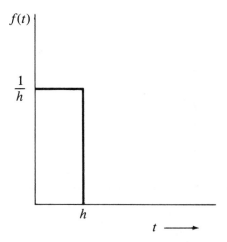

Figure 4.3
Pulse Function

The delta function is also called the *unit impulse function* (see Figure 4.4).

Mathematically, a *train of unit impulses* (whose areas or strengths are unity) can be represented (see also Figure 4.5) as

$$\delta_T(t) = \sum_{n=0}^{\infty} \delta(t - nT) \tag{4.3}$$

In this equation $\delta(t)$ is the unit impulse function at $t = 0$, and $\delta(t - nT)$ is the unit impulse function occurring at $t = nT$.

What we are interested in as the output of the sampler is a train of impulses whose areas numerically equal the values of the continuous function at the respective sampling instants. Such a relationship can be readily written with the aid of $\delta_T(t)$ as

$$X^*(t) = X(t) \; \delta(t - nT)$$
$$= X(t) \sum_{n=0}^{\infty} \delta(t - nT) \tag{4.4}$$

Note that

$$X(t) = 0 \text{ for } t < 0$$

Because the function $\delta(t - nT)$ is zero everywhere except at $t = nT$, the only values of $X(t)$ needed are those at $t = nT$. Thus Equation (4.4) can be rewritten as

$$X^*(t) = \sum_{n=0}^{\infty} X(nT) \; \delta(t - nT) \tag{4.5}$$

Note that although Equations (4.4) and (4.5) represent the relation-

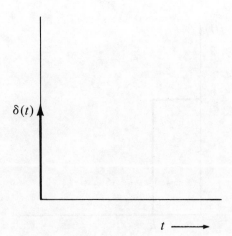

Figure 4.4
Impulse Function

ship between the continuous function and the sampled function, we cannot substitute numerical values into the right-hand side of the equation to obtain the values of the sampled function at a particular sampling instant. This is because the delta function has meaning only when it appears as an integral. The value of the Kth sample is given by

$$X(KT) = \int_0^\infty X(t)\ \delta(t - KT)\ dt \qquad (4.6)$$

The validity of Equation (4.6) should be intuitively clear. At $t = KT$ the integral $\int_0^\infty \delta(t - KT)\ dt$ has a value of unity at which instant $X(t)$ has a value of $X(KT)$.

On the basis of the discussion in this chapter, then, we are able to represent the sampled function as

$$\begin{aligned} X^*(t) &= X(t) \sum_{n=0}^\infty \delta(t - nT) \\ &= \sum_{n=0}^\infty X(nT)\ \delta(t - nT) \end{aligned} \qquad (4.7)$$

and the value of the Kth sample is given by

$$X(KT) = \int_0^\infty X(t)\ \delta(t - KT)\ dt \qquad (4.8)$$

A pictorial representation of the various functions is summarized in Figure 4.6.

It should be noted that there are some well-defined limits on how

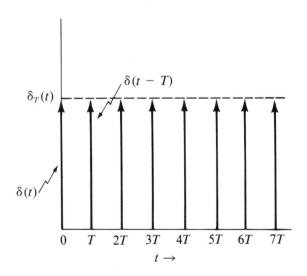

Figure 4.5
Train of Unit Impulses

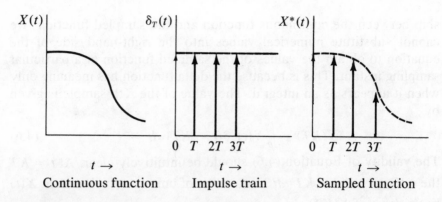

Figure 4.6
Pictorial Summary of Sampling Operation

frequently the process variable should be sampled. If we do not sample at all or sample very infrequently, the quality of control will be very poor. In fact, there is a theorem that specifies a minimum sampling rate for proper signal recovery. On the other hand, if we sample infinitely fast, the performance of the computer-control system will approach that of the conventional-control system, but this is a hypothetical concept. Somewhere, between these extremes, is the desirable sampling frequency for computer-control applications. In Chapter 7 we consider these aspects of sampling-frequency selection.

CHAPTER 5

The Z-Transformation

W̶e found Laplace transforms to be helpful in the analysis of conventional-control systems. In subsequent chapters we will see that the analysis of sampled-data control problems is conveniently handled in terms of Z-transforms. In this chapter we shall develop some of the important concepts of Z-transforms. To begin, recall that the Laplace transform of a function $f(t)$ is defined as

$$l\{f(t)\} = f(s) = \int_0^\infty f(t) \ e^{-st} dt \tag{5.1}$$

Also recall that the sampled function $f^*(t)$ and the continuous function $f(t)$ are related by

$$f^*(t) = f(t) \sum_0^\infty \delta(t - nT) = \sum_{n=0}^\infty f(nT) \ \delta(t - nT) = f(0)\delta(t) \tag{5.2}$$

$$+ f(T) \delta(t - T) + f(2T) \delta(t - 2T)$$
$$+ f(3T) \ \delta(t - 3T) + \cdots$$

where $f(T) \, \delta(t - T)$ represents an impulse at time $1T$ whose area is $f(T)$. Now take the Laplace transform of Equation (5.2)

$$
\begin{aligned}
l\{f^*(t)\} &= l\{f(0) \, \delta(t)\} + l\{f(T) \, \delta(t - T)\} + l\{f(2T) \, \delta(t - 2T)\} \\
&+ \cdots = f(0)l\left[\delta(t)\right] + f(T)l\left[\delta(t - T)\right] \\
&+ f(2T)l\left[\delta(t - 2T)\right] + \cdots = f(0) + f(T) \, e^{-sT} \, l\{\delta(t)\} \\
&+ f(2T) \, e^{-s2T} \, l\{\delta(t)\} + \cdots
\end{aligned}
\tag{5.3}
$$

because $l\{\delta(t)\} = 1$

$$
l\{f^*(t)\} = f^*(s) = \sum_{n=0}^{\infty} f(nT) \, e^{-nsT}
$$

If we introduce a new variable $Z = e^{sT}$ into Equation (5.3), we get

$$
f^*(s) \bigg|_{Z = e^{Ts}} = \sum_{n=0}^{\infty} f(nT) \, Z^{-n}
\tag{5.4}
$$

This result is defined to be the Z-transform[1] of $f(t)$ and is denoted as $F(Z)$. Thus

$$
F(Z) = Z\{f(t)\} = f^*(s) \bigg|_{Z = e^{Ts}} = \sum_{n=0}^{\infty} f(nT) \, Z^{-n}
\tag{5.5}
$$

5.1. Z-Transform of Various Functions

Let us apply the definition and evaluate the Z-transform of some common functions.

Unit Step Function

$$
f(t) = \begin{cases} 0 & t < 0 \\ u(t) & t \geq 0 \end{cases}
\tag{5.6}
$$

By definition,

$$
Z\{u(t)\} = F(Z) = \sum_{n=0}^{\infty} u(nT) \, Z^{-n}
$$

$$= \sum_{n=0}^{\infty} Z^{-n} = 1 + Z^{-1} + Z^{-2} = \frac{1}{1 - r} \qquad (5.7)$$

where

$$r < 1 \text{ is the ratio of successive terms}$$
$$= Z^{-1}$$

Therefore,

$$Z\{u(t)\} = \frac{1}{1 - Z^{-1}} \qquad |Z^{-1}| < 1 \qquad (5.8)$$

Exponential Function

$$f(t) = \begin{cases} 0 & t < 0 \\ e^{-at} & t \geq 0 \end{cases}$$

$$Z\{e^{-at}\} = \sum_{n=0}^{\infty} e^{-anT} Z^{-n} = \sum_{n=0}^{\infty} (e^{-aT} Z^{-1})^n$$

$$= \frac{1}{1 - Z^{-1} e^{-aT}} \qquad |Z^{-1}| < e^{aT} \qquad (5.9)$$

Ramp Function

$$f(t) = \begin{cases} 0 & t < 0 \\ kt & t \geq 0 \end{cases}$$

$$\begin{aligned}
F(Z) = Z\{kt\} &= \sum_{n=0}^{\infty} knT \, Z^{-n} \\
&= kT \left\{ Z^{-1} + 2Z^{-2} + 3Z^{-3} + \cdots \right\} \\
&= kTZ^{-1} \left\{ 1 + 2Z^{-1} + 3Z^{-2} + 4Z^{-3} + \cdots \right\} \\
&= \frac{kT \, Z^{-1}}{(1 - Z^{-1})^2} \\
&= \frac{kT \, Z}{(Z - 1)^2} \qquad |Z^{-1}| < 1
\end{aligned}$$

$$F(Z) = \frac{kTZ}{(Z - 1)^2}$$

Sine Function

$$f(t) = \begin{cases} 0 & t < 0 \\ \sin at & t \geq 0 \end{cases}$$

because

$$f(t) = \sin at, \quad f(nT) = \sin anT$$

and

$$\sin anT = \frac{e^{jnaT} - e^{-jnaT}}{2j}$$

We get upon substituting in the definition

$$Z\{\sin at\} = \sum_{n=0}^{\infty} \sin anT \, Z^{-n} = \sum_{n=0}^{\infty} \frac{1}{2j} \, (e^{jnaT} - e^{-jnaT}) \, Z^{-n}$$

$$= \frac{1}{2j} \left[\sum_{n=0}^{\infty} \left(e^{jaT} \, Z^{-1} \right)^n - \sum_{n=0}^{\infty} \left(e^{-jaT} \, Z^{-1} \right)^n \right]$$

Therefore,

$$Z\{\sin (at)\} = \frac{1}{2j} \left(\frac{1}{1 - e^{jaT} \, Z^{-1}} - \frac{1}{1 - e^{-jaT} \, Z^{-1}} \right)$$

Putting the right side over a common denominator, we have

$$Z\{\sin at\} = \frac{1}{2j} \left(\frac{1 - e^{-jaT} \, Z^{-1} - 1 + e^{jaT} \, Z^{-1}}{1 + Z^{-2} - Z^{-1}(e^{jaT} + e^{-jaT})} \right)$$

$$= \frac{Z^{-1}(e^{jaT} - e^{-jaT})/2j}{1 + Z^{-2} - 2Z^{-1} \dfrac{(e^{jaT} + e^{-jaT})}{2}}$$

Therefore,

$$Z\{\sin at\} = \frac{Z^{-1} \sin aT}{1 - 2Z^{-1} \cos aT + Z^{-2}}$$

$$= \frac{Z \sin aT}{Z^2 - 2Z \cos aT + 1}$$

By similar procedure, Z-transforms of other functions can be obtained. For convenience, Z-transforms are tabulated with Laplace transforms and the corresponding $f(t)$ in Appendix A.

5.2. Properties of Z-Transforms

1. Z-transforms Are Linear

$$
\begin{aligned}
Z\{f(t) + g(t)\} &= \sum_{n=0}^{\infty} \left[f(nT) + g(nT) \right] Z^{-n} \\
&= \sum_{n=0}^{\infty} f(nT) Z^{-n} + \sum_{n=0}^{\infty} g(nT) Z^{-n} \qquad (5.10) \\
&= Z\{f(t)\} + Z\{g(t)\}
\end{aligned}
$$

that is, the Z-transform of the sum of two functions is the sum of their individual transforms.

$$
\begin{aligned}
Z\{cf(t)\} &= \sum_{n=0}^{\infty} cf(nT) Z^{-n} \\
&= c \sum_{n=0}^{\infty} f(nT) Z^{-n} \qquad (5.11) \\
&= cZ\{f(t)\}
\end{aligned}
$$

Thus the Z-transform of the product of a constant and a function equals the product of the constant and the Z-transform of the function.

2. Initial Value-Theorem.
If the Z-transform of $f(t)$ is $F(Z)$, then

$$
\lim_{t \to 0} f(t) = \lim_{Z \to \infty} F(Z)
$$

Proof. By definition,

$$
F(Z) = \sum_{n=0}^{\infty} f(nT) Z^{-n} = f(0) + f(T) Z^{-1} + f(2T) Z^{-2} + \cdots \qquad (5.12)
$$

Let us take the limit as $Z \to \infty$

$$
\lim_{Z \to \infty} F(Z) = f(0) = \lim_{t \to 0} f(t) \qquad (5.13)
$$

which proves the theorem.

3. Final-value Theorem.
If the Z-transform of $f(t)$ is $F(Z)$, then

$$\lim_{Z \to 1} \left[F(Z) (1 - Z^{-1}) \right] = \lim_{t \to \infty} f(t) \tag{5.14}$$

Proof

$$Z\left[f(t + T) - f(t) \right] \equiv \sum_{n=0}^{\infty} \left\{ f\left[(n + 1)T \right] - f(nT) \right\} Z^{-n} \tag{5.15}$$

Also, by definition

$$F(Z) = \sum_{n=0}^{\infty} f(nT) Z^{-n} = f(0) + f(T) Z^{-1} + f(2T) Z^{-2} + \cdots$$

If we multiply both sides of this equation by Z, we get

$$ZF(Z) = Zf(0) + f(T) + f(2T) Z^{-1} + f(3T) Z^{-2} + \cdots \tag{5.16}$$

Now,

$$Z\{f(t + T)\} = \sum_{n=0}^{\infty} f(nT + T) Z^{-n}$$
$$= f(T) + f(2T) Z^{-1} + f(3T) Z^{-2} + \cdots \tag{5.17}$$
$$= ZF(Z) - Zf(0)$$

Substituting in Equation (5.15), we have

$$ZF(Z) - Zf(0) - F(Z) = \sum_{n=0}^{\infty} \left[f(nT + T) - f(nT) \right] Z^{-n} \tag{5.18}$$

Now, divide this equation by Z and take the limit as $Z \to 1$

$$\lim_{Z \to 1} \left[(1 - Z^{-1}) F(Z) \right] - f(0) = \lim_{k \to \infty} \sum_{n=0}^{k} \left\{ f(nT + T) - f(nT) \right\}$$
$$= \lim_{k \to \infty} \left\{ \left[f(T) - f(0) \right] + \left[f(2T) - f(T) \right] + \left[f(3T) - f(2T) \right] \right.$$
$$\left. + \cdots \left[f(kT + T) - f(kT) \right] \right\} \tag{5.19}$$
$$= \lim_{k \to \infty} f\left[(k + 1)T \right] - f(0)$$

Therefore,

$$\lim_{Z \to 1} \left[(1 - Z^{-1}) F(Z) \right] = \lim_{t \to \infty} f(t) \tag{5.20}$$

which establishes the result.

4. Translation of the Function. If

$$f^*(t) = \begin{cases} f^*(t - kT) & t \geq kT \\ 0 & t < kT \end{cases} \tag{5.21}$$

$$= u(t - kT) f^*(t - kT)$$

The function $f^*(t)$ is delayed k sampling periods. Then

$$Z\{f(t - kT) u(t - kT)\} = Z^{-k} F(Z) \qquad (5.22)$$

Proof

$$Z\{f(t - kT) u(t - kT)\} = \sum_{n=0}^{\infty} f(nT - kT) u(nT - kT) Z^{-n} \qquad (5.23)$$

Let us make a variable substitution $m = n - k$. Then, Equation (5.23) becomes

$$Z\{f(t - kT) u(t - kT)\} = \sum_{m=-k}^{\infty} f(mT) u[mT] Z^{-(m+k)}$$
$$= Z^{-k} \sum_{m=0}^{\infty} f(mT) Z^{-m} \qquad (5.24)$$
$$= Z^{-k} F(Z)$$

which proves the result.

5. Complex Translation Theorem. The complex translation theorem states that if the Z-transform of $f(t)$ is $F(Z)$, then,

$$Z\{e^{\mp at} f(t)\} = Z\{f(s \pm a)\} = F(Ze^{\pm aT}) \qquad (5.25)$$

Proof By definition

$$Z\{e^{\mp at} f(t)\} = \sum_{n=0}^{\infty} f(nT) e^{\mp anT} Z^{-n}$$
$$= \sum_{n=0}^{\infty} f(nT) (e^{\pm aT} Z)^{-n}$$
$$= F(Z e^{\pm aT}) \qquad (5.26)$$

Suppose we wish to use this theorem to obtain the Z-transform of $u(t)e^{-at}$. First, we note that the Z-transform of $u(t)$ is given by

$$Z\{u(t)\} = \frac{Z}{Z - 1} \qquad (5.27)$$

To get $Z\{u(t) e^{-aT}\}$ we substitute in Equation (5.27) the product $Z e^{aT}$ for Z. Thus

$$\frac{Z e^{aT}}{Z e^{aT} - 1} = \frac{1}{1 - Z^{-1} e^{-aT}} = Z\{u(t) e^{-at}\}$$

5.3. The Inverse Z-Transformation[2]

Three methods are outlined here for inverting the Z-transform:
1. Partial fraction expansion.
2. Long division.
3. Tables of Z-transform.
The operation of inverting the Z-transform is denoted as

$$f^*(t) = Z^{-1}\{F(Z)\}$$

Note that the inverse yields the sampled function $f^*(t)$ and not the continuous function $f(t)$. The $f^*(t)$ so obtained is unique, but it is conceivable that the sampled function $f^*(t)$ could be derived from two different continuous functions $f_1(t)$ and $f_2(t)$. It follows, therefore, that the inverse of $F(Z)$ does not necessarily yield a unique continuous function $f(t)$. Thus we should not expect any additional information about $f(t)$ from the inverse other than the values at the sampling instants. Tables of inverse transforms are available. Some simple functions $F(Z)$ for which the inverse is simply the continuous function $f(t)$ are shown in Appendix A.

Most often the inverses we encounter cannot be obtained directly from the tables, and some mathematical manipulations are necessary. One way to accomplish inversion is to expand the function in *partial fractions*. The method is best illustrated by an example.

Example 1.

Find the inverse transform of

$$F(Z) = \frac{Z^2(Z^2 + Z + 1)}{(Z - 0.5)(Z - 1)(Z^2 - Z + 0.8)}$$

The method requires the expansion of $F(Z)/Z$ into partial fractions so that the inverse Z-transform of each of the component term multiplied by Z is recognizable from the tables of Z-transforms.

$$F(Z) = \frac{Z^2(Z^2 + Z + 1)}{(Z - 0.5)(Z - 1)(Z^2 - Z + 0.8)} = ZF_1(Z)$$

Therefore

$$F_1(Z) = \frac{Z(Z^2 + Z + 1)}{(Z - 0.5)(Z - 1)(Z^2 - Z + 0.8)} = \frac{A}{Z - 0.5} + \frac{B}{Z - 1}$$

$$+ \frac{C}{Z - (0.5 + j.74)} + \frac{D}{Z - (0.5 - j.74)}$$

Multiply the entire equation by $(Z - 0.5)$ and set $Z = 0.5$. Then

$$A = \frac{(0.5)(0.25 + 0.5 + 1)}{-(0.5)(0.25 - 0.5 + 0.8)} = -3.18$$

Multiply by $Z - 1$ and set $Z = 1$. Then,

$$B = \frac{(1)(3)}{(0.5)(0.8)} = 7.5$$

Next, multiply the same equation by $Z - (0.5 + j.74)$ and set $Z = 0.5 + j.74$. Then, we will get

$$C = -1.67 + 0.52j$$

Similarly,

$$D = -1.67 - 0.52j$$

Thus $F_1(Z)$ becomes

$$F_1(Z) = \frac{7.5}{Z - 1} - \frac{3.18}{Z - 0.5} + \frac{-1.67 + 0.52j}{Z - (0.5 + j.74)}$$

$$+ \frac{-1.67 - 0.52j}{Z - (0.5 - j.74)}$$

Now, recognize that the presence of a real number other than one in the denominator of the Z-transform expression will give rise to an exponential function in its inverse. Therefore, we must express the number 0.5 in exponential form. Similarly, complex numbers give rise to exponential and sine/cosine terms. Thus, by converting from rectangular to polar coordinates,

$$e^{-0.693} = 0.5$$

and

$$e^{-0.11 \pm 0.98j} = 0.5 \pm 0.74j$$

Substitution and multiplication by Z yields

$$F(Z) = Z F_1(Z)$$

$$= 7.5 \frac{Z}{Z - 1} - 3.18 \frac{Z}{Z - e^{-0.693}}$$

$$+ (-1.67 - 0.52j) \, \frac{Z}{Z - e^{-.11+0.98j}}$$

$$+ (-1.67 - 0.52j) \, \frac{Z}{Z - e^{-.11-0.98j}}$$

Now we may look up the inverse of each term in tables. Thus

$$f^*(t) = 7.5 - 3.18e^{-0.693t/T} + (-1.67 + 0.52j)e^{-(0.11-0.98j)t/T}$$
$$- (1.67 + 0.52j) \, e^{-(0.11+0.98j)t/T}$$

which reduces to

$$f^*(t) = 7.5 - 3.18e^{-0.693t/T} - 2e^{-0.11t/T}\{1.67 \cos(0.98t/T)$$
$$+ 0.52 \sin(0.98t/T)\}$$

Several values of the function $f(t)$ at the sampling instants are:

t	0	T	$2T$	$3T$	$4T$	$5T$	$6T$
$f(t)$	1	3.5	7.0	9.3	9.3	7.6	6.1

Another method of inverting $F(Z)$ is by *long division*. By this method $F(Z)$ is expanded into a power series of Z^{-1} by long division. The coefficient of the Z^{-n} term corresponds to the value of the function $f(t)$ at the nth sampling instant. This can be seen easily if we write the equation that defines the Z-transform of a function as

$$F(Z) = Z[f(t)] = \sum_{n=0}^{\infty} f(nT) \, Z^{-n}$$
$$F(Z) = f(0) + f(T) \, Z^{-1} + f(2T) \, Z^{-2} + f(3T) \, Z^{-3} + \cdots$$

If $F(Z)$ is expanded into a power series, we have $F(Z) = a_0 + a_1 Z^{-1} + a_2 Z^{-2} + a_3 Z^{-3} + \cdots$ By comparison $a_0 = f(0)$; $a_1 = f(T)$ and $a_n = f(nT)$, which is the value of $f(t)$ at the nth sampling instant.

Example 2. Evaluate the inverse of

$$F(Z) = \frac{Z^2(Z^2 + Z + 1)}{(Z - 0.5)(Z - 1)(Z^2 - Z + 0.8)}$$

$$F(Z) = \frac{Z^4 + Z^3 + Z^2}{Z^4 - 2.5Z^3 + 2.8Z^2 - 1.7Z + 0.4}$$

$$Z^4 - 2.5Z^3 + 2.8Z^2 - 1.7Z + 0.4$$

```
                        1 + 3.5Z⁻¹ + 6.95Z⁻² + 9.28Z⁻³ + 9.29Z⁻⁴ + 7.68Z⁻⁵ + 6.19Z⁻⁶
Z⁴ - 2.5Z³ + 2.8Z² - 1.7Z + 0.4 ) Z⁴ + Z³ + Z²
                                   Z⁴ - 2.5Z³ + 2.80Z² - 01.70Z + 00.40
                                        3.5Z³ - 1.80Z² + 01.70Z - 00.40
                                        3.5Z³ - 8.75Z² + 09.80Z - 05.95 + 01.40Z⁻¹
                                                6.95Z² - 08.10Z + 05.55 - 01.40Z⁻¹
                                                6.95Z² - 17.38Z + 19.46 - 11.82Z⁻¹ + 02.78Z⁻²
                                                         9.28Z - 13.91 + 10.42Z⁻¹ - 02.78Z⁻²
                                                         9.28Z - 23.20 + 25.98Z⁻¹ - 15.78Z⁻² + 03.71Z⁻³
                                                                  9.29 - 15.56Z⁻¹ + 13.00Z⁻² - 03.71Z⁻³
                                                                  9.29 - 23.24Z⁻¹ + 26.01Z⁻² - 15.79Z⁻³ + 03.72Z⁻⁴
                                                                         7.68Z⁻¹ - 13.01Z⁻² + 12.08Z⁻³ - 03.72Z⁻⁴
                                                                         7.68Z⁻¹ - 19.20Z⁻² + 25.50Z⁻³ - 13.06Z⁻⁴ + 3.07Z⁻⁵
                                                                                  6.19Z⁻² - 13.42Z⁻³ + 09.34Z⁻⁴ - 3.07Z⁻⁵
```

Thus

$$F(Z) = 1 + 3.5Z^{-1} + 7Z^{-2} + 9.3Z^{-3} + 9.3Z^{-4} + 7.7Z^{-5} + 6.2Z^{-6} + \cdots$$

$f(t) =$	1	3.5	7	9.3	9.3	7.7	6.2
t	0	T	$2T$	$3T$	$4T$	$5T$	$6T$

In summary, the stepwise procedure to determine the Z-transform of a function $f(t)$ is:
1. Substitute nT for t in $f(t)$ to get $f(nT)$.
2. Evaluate the sum $F(Z) = \sum_{n=0}^{\infty} f(nT) Z^{-n}$.
3. Express the result in closed form.

To invert the Z-transform $F(Z)$ to get $f^*(t)$, the procedure is
1. Divide $F(Z)$ by Z to get $F_1(Z)$.
2. Expand $F_1(Z)$ in partial fractions.
3. Multiply $F_1(Z)$ by Z to obtain $F(Z)$.
4. Look up the inverse transform of the terms in $F(Z)$ in the tables of Z-transforms.

Or
1. Obtain a power series of Z^{-n} from $F(Z)$ by long division.
2. Note that the coefficient in front of Z^{-n} is the value of $f(t)$ at the nth sampling instant.

Remember that the inversion procedure gives correct information about $f(t)$ only at the sampling instants. We should also keep in mind that because Z-transforms are derived from Laplace transforms, their use is restricted to the solution of linear-control problems. This in turn means that we can seek only information about the control system in the vicinity of a nominal steady-state operating level. From our work in conventional control this concept of linearizing the process equations around some steady state should be familiar to us. The Laplace-transform methods could then be applied to the linearized equations so as to design the control system. This procedure works well so long as the range of linearity is not exceeded. In studying the subsequent material we should remember that the restrictions of linearity apply to the solution of sampled-data control systems as well.

The problem of Z-transform inversion is conveniently handled by a digital computer. A Fortran-based computer program to perform the inversion is shown in Appendix C1. To use this program, arrange the function to be inverted in the form

$$C(Z) = \frac{P_0 + P_1Z^{-1} + P_2Z^{-2} + P_3Z^{-3} + \cdots}{Q_0 + Q_1Z^{-1} + Q_2Z^{-2} + Q_3Z^{-3} + \cdots} \tag{5.28}$$

where

$$P_0, P_1, \ldots, P_n \text{ and } Q_0, Q_1, \ldots, Q_n \text{ are constants}$$

When provided with these constants as inputs, the program outputs $C(nT)$ according to the equation

$$C(Z) = C(0) + C(T)Z^{-1} + C(ZT)Z^{-2} \qquad (5.29)$$
$$+ C(3T)Z^{-3} + \cdots$$

where $C(nT)$ represents the value of $C(t)$ at the nth sampling instant.

References

1. Ragazzini, J. R., Zadeh, L. A., The Analysis of Sampled-Data Systems, *Trans. AIEE*, **71**, Pt II, (1952) 225–234.
2. Barker, R. H., The Pulse Transfer Function and Its Application to Sampling Servo Systems, *Proc. IEE, London*, **99**, Pt IV, (December 1952) 302–317.

Short Bibliography on Mathematics In Sampled Data Control Systems

1. Kuo, B. C., *Analysis and Synthesis of Sampled-data Control Systems*, Prentice-Hall, Englewood Cliffs, N.J., 1963.
2. Tou, J. T., *Digital and Sampled-data Control Systems*, McGraw-Hill, New York, 1959.
3. Smith, C. L., *Digital Computer Process Control*, Intex Publishers, Scranton, Pa., 1972.

CHAPTER 6

Pulse Transfer Functions

I n conventional control systems the Laplace transform of an input function $X(s)$ is related to the Laplace transform of the output function $Y(s)$ by the transfer function of the system $G(s)$ according to the Equation

$$\frac{Y(s)}{X(s)} = G(s) \qquad (6.1)$$

Schematically, the transfer function is represented as

$$\text{input } X(s) \rightarrow \boxed{G(s)} \rightarrow \text{output } Y(s)$$

In sampled-data systems we must relate the pulsed input to the pulsed output. We have seen that pulsed signals can be conveniently handled by Z transforms. By analogy with the transfer function representation of the conventional control systems, it is tempting to postulate that the Z transform of the pulsed input can be related to the Z transform of the pulsed output according to the expression

$$\frac{Y(Z)}{X(Z)} = G(Z) \qquad (6.2)$$

It will be shown that Equation (6.2) is indeed valid. The term $G(Z)$ is called the pulse-transfer function[1] or the Z-transfer function of the system. To show that Equation (6.2) is correct we have to derive an alternate expression for the sampling process.

6.1. Complex Series Representation of the Sampler

Recall that the sampling operation gives us the values of the output at each sampling instant and can be represented as

continuous $X(t)$→ | Sampler | →$X^*(t)$ sampled
input   output

We have said that the sampled output is related to the continuous input by the expression

$$X^*(t) = \delta_T(t)\, X(t) \qquad (6.3)$$

where

$$\delta_T(t) = \sum_{n=-\infty}^{\infty} \delta(t - nT)$$

Since $\delta_T(t)$ is a periodic function, it can be expanded into a complex Fourier series:

$$\delta_T(t) = \sum_{n=-\infty}^{\infty} C_n\, e^{jn\,\omega_s t} \qquad (6.4)$$

where

ω_s = sampling frequency, in radians per unit time

$$= \frac{2\pi}{T}$$

and the Fourier coefficients are given by

$$C_n = \frac{1}{T} \int_{-T/2}^{+T/2} \delta_T(t)\, e^{-jn\,\omega_s t}\, dt \qquad (6.5)$$

Substituting for $\delta_T(t)$ from Equation (6.3) into Equation (6.5) gives

$$C_n = \frac{1}{T} \int_{-T/2}^{T/2} e^{-jn\omega_s t} \left(\sum_{n=-\infty}^{\infty} \delta(t - nT) \right) dt$$

$$= \frac{1}{T} \int_{-T/2}^{T/2} e^{-jn\omega_s t} \sum_{n=-\infty}^{-1} \delta(t - nT) \, dt \; + \; 0$$

$$\frac{1}{T} \int_{-T/2}^{T/2} e^{-jn\omega_s t} \; \delta(t) \, dt$$

$$+ \frac{1}{T} \int_{-T/2}^{+T/2} e^{-jn\omega_s t} \left(\sum_{n=1}^{\infty} \delta(t - nT) \right) dt \; + \; 0 \tag{6.6}$$

Therefore,

$$C_n = \frac{1}{T} \int_{-T/2}^{+T/2} e^{-jn\omega_s t} \, \delta(t) \, dt \tag{6.7}$$

Now, since

$$\int_{-T/2}^{+T/2} \delta(t) e^{-jn\omega_s t} dt = \int_{-T/2}^{+T/2} \delta(t) dt = 1 \tag{6.8}$$

it follows that

$$C_n = \frac{1}{T} \tag{6.9}$$

Substituting this result in Equation (6.4) gives

$$\delta_T(t) = \frac{1}{T} \sum_{n=-\infty}^{\infty} e^{jn\omega_s t}$$

and

$$X^*(t) = \frac{1}{T} \sum_{n=-\infty}^{\infty} X(t) e^{jn\omega_s t} \tag{6.10}$$

Taking the Laplace transform of this equation we have

$$X^*(s) = l\{X^*(t)\} = \frac{1}{T} \sum_{n=-\infty}^{\infty} X(s + jn\omega_s) \tag{6.11}$$

6.2. Development of the Pulse Transfer Function

We wish to relate the sampled input $X^*(t)$ to the sampled output $Y^*(t)$ as shown in the following figure:

From this figure,

$$Y(s) = X^*(s)\, G(s) \tag{6.12}$$

By analogy to Equation (6.11)

$$Y^*(s) = \frac{1}{T} \sum_{n=-\infty}^{\infty} Y(s + jn\omega_s) \tag{6.13}$$

In view of Equation (6.12),

$$Y(s + jn\omega_s) = G(s + jn\omega_s)\, X^*(s + jn\omega_s) \tag{6.14}$$

Substitution in Equation (6.14) yields

$$Y^*(s) = \frac{1}{T} \sum_{n=-\infty}^{\infty} Y(s + jn\omega_s)$$

$$\tag{6.15}$$

$$= \frac{1}{T} \sum_{n=-\infty}^{\infty} G(s + jn\omega_s)\, X^*(s + jn\omega_s)$$

But because

$$X^*(s + jn\omega_s) = l\left\{ X^*(t)\, e^{jn\omega_s t} \right\}$$

$$= l\left\{ X^*(t)\, e^{jn\,2\pi t/T} \right\} \tag{6.16}$$

substituting $t = nT$ gives

$$X^*(s + jn\omega_s) = l\left\{ X^*(t)\, e^{j2\pi n^2} \right\}$$

Since

$$e^{-j2\pi n^2} = \cos 2\pi n^2 - j \sin 2\pi n^2 = 1 \text{ for all } n,$$

$$X^*(s + jn\omega_s) = l\left\{ X^*(t) \right\} = X^*(s)$$

Equation (6.15) becomes

$$Y^*(s) = \frac{1}{T} \sum_{n=-\infty}^{\infty} G(s + jn\omega_s) X^*(s)$$

$$= \frac{1}{T} X^*(s) \sum_{n=-\infty}^{\infty} G(s + jn\omega_s) \qquad (6.17)$$

By the same reasoning which led to Equation (6.11), this can be written as

$$Y^*(s) = X^*(s) G^*(s) \qquad (6.18)$$

Taking the Z transform of this equation gives

$$Y(Z) = X(Z) G(Z)$$

or $\qquad (6.19)$

$$G(Z) = \frac{Y(Z)}{X(Z)}$$

Note the similarity of Equation (6.19) to the familiar transfer function $G(s)$ given in Equation (6.1), which would be the transfer function of the system if the samplers were removed. This important similarity enables us to treat the Z transforms and the pulse transfer function in much the same manner as the Laplace transforms and the conventional transfer functions. A step-by-step procedure for developing pulse transfer functions is given below.[2]

1. Derive the transfer function $G(s)$ of the system by conventional techniques.
2. For $X(t) = \delta(t)$ (i.e., the impulse function)

$$G(s) = \frac{Y(s)}{X(s)} = Y(s) \qquad (6.20)$$

since $l\{\delta(t)\} = 1$.

3. Invert $G(s)$ to get $g(t)$ which is numerically equal to the impulse response $y(t)$ according to Equation (6.20)
4. Note that for $X(t) = \delta(t)$, $X(Z) = Z\{\delta(t)\} = 1$. Therefore,

$$G(Z) = \frac{Y(Z)}{X(Z)} = Y(Z)$$

5. Thus $G(Z)$ can be obtained from $y(t)$ [or $g(t)$] by substituting nT for t and evaluating the sum of the series as

$$G(Z) = \sum_{n=0}^{\infty} y(nT)\, Z^{-n} \qquad (6.21)$$

For physically realizable systems, convergence of this series is assured. Frequently, Z-transfer functions, with $G(S)$ available as the starting point, are desired, in which case steps 3 and 4 suffice. When $G(s)$ is a complicated function, it may be split into partial fractions; then steps 2 through 5 can be applied to the individual partial fractions.

Example

Find the pulse transfer function of the system shown below:

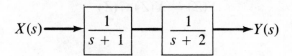

Solution:

$$G(s) = \frac{1}{(s+1)(s+2)}$$

$$= \frac{1}{s+1} - \frac{1}{s+2}$$

Therefore,

$$g(t) = e^{-t} - e^{-2t} = y(t)$$

$$g(nT) = e^{-nT} - e^{-2nT} = y(nT)$$

Now,

$$G(Z) = \sum_{n=0}^{\infty} y(nT)\, Z^{-n} = \sum_{n=0}^{\infty} (e^{-nT} - e^{-2nT})\, Z^{-n}$$

$$= \frac{1}{1 - e^{-T} Z^{-1}} - \frac{1}{1 - e^{-2T} Z^{-1}}$$

After simplification we get

$$G(Z) = \frac{Z(e^{-T} - e^{-2T})}{(Z - e^{-T})(Z - e^{-2T})}$$

References

1. Barker, R. H., The Pulse Transfer Function and Its Application to Sampling Servo Systems, *Proc. IEE, London*, **99**, (4), 302–317, December 1952.
2. Tou, J. T., *Digital and Sampled-data Control Systems*, McGraw-Hill, New York, 1959.

CHAPTER 7

Data Holds

In the sampled-data control loop the continuous signal representing the measured value of the controlled variable is sampled by means of a sampler (i.e., an A/D converter) and compared to the discrete form of the set point to produce an error. An appropriate computer program is executed to produce a control action which consists of discrete data (i.e., a train of impulses of varying strengths). The function of the *holding device* is to reconstruct the continuous signal from the discrete data. This device is called a *digital-to-analog converter*. In this chapter we develop mathematical expressions that describe the operation of some of the common holding devices.

The schematic of the holding device is shown in Figure 7.1. The holding device converts the pulsed-input signal into a continuous signal by interpolation or extrapolation of the input pulses so that the input signal can be approximately reproduced. The smoothing of the pulsed data by a holding device is essentially an extrapolation problem. The extrapolated time function between two consecutive sampling instants nT *and* $(n + 1)T$ depends on its values at the preceding sampling instants $nT, (n - 1)T, (n - 2)T, \ldots$, and can generally be described by a power series expansion of the output between the interval $t = NT$ and $t = (n + 1)T$ as shown in Figure 7.1.

Let $Y(t)$ be the output function and $Y_n(t)$ be its value between sampling instants nT and $(n + 1)T$, that is,

$$Y_n(t) = Y(t) \qquad nT \le t \le (n + 1)T. \tag{7.1}$$

Then,

$$Y_n(t) = Y(nT) + Y^{(1)}(nT)(t - nT) + \frac{Y^{(2)}(nT)}{2!}(t - nT)^2 + \tag{7.2}$$

$$\cdots + \frac{Y^k(nT)}{k!}(t - nT)^k$$

where

$Y(nT)$ = value of $Y(t)$ at $t = nT$

$Y^k(nT)$ = value of the kth-order derivative of $Y(t)$ evaluated at $t = nT$

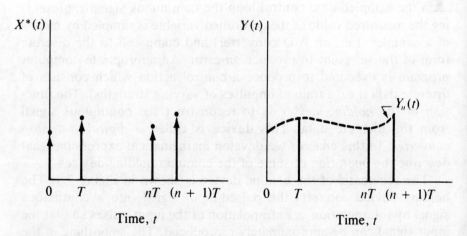

Figure 7.1
Representation of Holding Devices

When a holding device approximates the time function between two consecutive sampling instants by a zero-order polynomial, it is referred to as a *zero-order hold*. The equation for the zero-order hold is

$$Y_n(t) = Y(nT) \qquad nT < t < (n+1)T \tag{7.3}$$

When the device approximates the time function between two successive sampling instants by a first-order polynomial

$$Y_n(t) = Y(nT) + Y^{(1)}(nT)(t - nT) \tag{7.4}$$

it forms a first-order holding device. Similarly, when the time function is described by a kth-order polynomial, the device is referred to as a kth-order holding device. In general, a higher-order holding device produces a better approximation of the desired time function from the input-data pulses.

Since the holding device receives information only at the sampling instants, the values of these derivatives [see Equation (7.2)] can only be estimated from the sampled-input data. These estimates are given by the following expressions:
First-order derivative:

$$Y^{(1)}(nT) = \frac{dY}{dt}\bigg|_{nT} \cong \frac{Y(nT) - Y[(n-1)T]}{T} \tag{7.5}$$

Second-order derivative:

$$Y^{(2)}(nT) = \frac{d^2 Y}{dt^2}\bigg|_{T=nT} = \frac{Y^{(1)}(nT) - Y^{(1)}[(n-1)T]}{T} \tag{7.6}$$

and for the kth-order derivative, we have

$$Y^k(nT) = \frac{Y^{k-1}(nT) - Y^{k-1}[(n-1)T]}{T} \tag{7.7}$$

These derivatives can be expressed in terms of $Y(nT)$ as

$$Y^k(nT) \cong \frac{1}{T^k}\left\{ Y(nT) - kY[(n-1)T] + \frac{k(k-1)}{2!} Y[(n-2)T] \right.$$

$$\left. - \cdots + (-1)^k Y[(n-k)T] \right\} \tag{7.8}$$

From these equations it can be seen that, to obtain an estimated value of a derivative of $Y(t)$, the minimum number of data pulses required is equal to the order of the derivative plus one. Thus first-order derivative calculations require two pulses. To estimate a second-order derivative at a sampling instant nT requires three consecutive pulses at nT, $(n - 1)T$, and $(n - 2)T$. Thus to estimate a higher-order derivative requires a greater delay before a reliable value of that derivative can be obtained. The delay introduced by a higher-order holding device may be detrimental to system stability. On the other hand, to obtain a better reproduction of the desired time function from the input data pulses and to reduce the ripple content it is advisable to employ a higher-order holding device. Consequently, to design a holding device for a sampled-data control system we must compromise between the tolerable ripple content and the specified system stability and dynamic performance. Because of the relatively high cost and constructional complexity of higher-order holding devices and the amount of phase lag introduced by them, the most common holding devices used in computer control systems are the zero-order holding devices, although the first-order holding devices are occasionally used.

7.1 Transfer Function of the Zero-Order Hold

The schematic of a zero-order hold is shown in Figure 7.2. The input to the holding device is a train of impulses of varying strengths. For example, at the first time instant $t = 0$ the input of strength k_0 to the holding device can be expressed as

$$X^*_0(t) = k_0 \, \delta(t) \tag{7.9}$$

During the first time interval the output from the zero-order holding device can be represented as (see Figure 7.3)

$$Y(t) = k_0 \left[u(t) - u(t - T) \right] \tag{7.10}$$

Similar expressions can be written for input and output over N time intervals. Thus the input train can be represented by

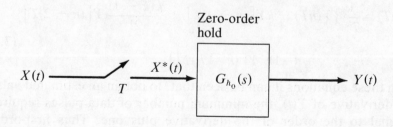

Figure 7.2
Zero-Order Holding Device

$$X^*(t) = \sum_{n=0}^{N} k_n \, \delta(t - nT) \qquad (7.11)$$

and the output by

$$Y(t) = k_0 \left[u(t) - u(t - T) \right] + k_1 \left[u(t - T) - u(t - 2T) \right]$$
$$+ \cdots + k_N \left[u(t - (NT)) - u(t - (N + 1)T) \right] \qquad (7.12)$$
$$= \sum_{n=0}^{N} k_n \left[u(t - nT) - u(t - (n + 1)T) \right]$$

Taking the Laplace transform of the input train gives

$$X^*(s) = \sum_{n=0}^{N} k_n \, e^{-snT} \qquad (7.13)$$

The Laplace transform of the output is

$$Y(s) = \sum_{n=0}^{N} k_n \left\{ \frac{e^{-snT}}{s} - \frac{e^{-(n+1)sT}}{s} \right\} = \frac{1}{s} \sum_{n=0}^{N} k_n e^{-nsT} (1 - e^{-sT})$$
$$\qquad (7.14)$$
$$= \frac{1 - e^{-sT}}{s} \sum_{n=0}^{N} k_n e^{-nsT}$$

Therefore, the transfer function of the zero-order hold (which is the ratio of the Laplace transform of the output to that of the input) is

$$\frac{Y(s)}{X(s)} = G_{h_0}(s) = \frac{\left(\dfrac{1 - e^{-sT}}{s} \right) \sum\limits_{n=0}^{N} k_n e^{-nsT}}{\sum\limits_{n=0}^{N} k_n e^{-nsT}}$$

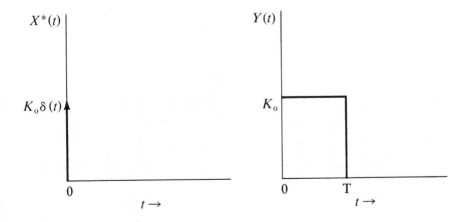

Figure 7.3
Input/Output from Zero-Order Hold during the First Interval

$$G_{h_0}(s) = \frac{1 - e^{-sT}}{s} \tag{7.15}$$

The input-output relationship of the zero-order hold for a typical input function is shown in Figure 7.4.

7.2. Transfer Function of the First-Order Hold

The operation of the first-order hold is shown in Figure 7.5. The extrapolated time function $Y_n(t)$ between two successive sampling instants nT and $(n + 1)T$ is assumed to be a linear function given by

$$Y_n(t) = Y(nT) + Y^{(1)}(nT)(t - nT) \tag{7.16}$$

where

$$Y^{(1)}(nT) = \frac{Y(nT) - Y[(n - 1)T]}{T}$$

or

$$Y_n(t) = Y(nT) + \frac{1}{T}\left[Y(nT) - Y(n - 1)T\right](t - nT) \tag{7.17}$$

Equation (7.17) gives the value of the output between $nT \leqslant t \leqslant (n + 1)T$. Thus if at $t = (n - 1)T$, $X_{n-1}^* = k_{n-1}\delta(t - (n - 1)T)$ and at

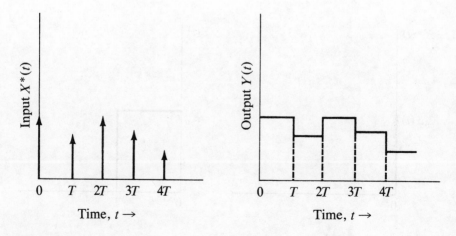

Figure 7.4
Input/Output of Zero-Order Hold

$t = nT$, $X_n^* = k_n \, \delta(t - nT)$, then, $Y(nT)$ will be equal to k_n, and during the time interval $nT \leq t \leq (n + 1)T$ the output will be given by

$$Y_n(t) = k_n + \frac{k_n - k_{n-1}}{T}(t - nT) \tag{7.18}$$

Now, let us derive the transfer function of the first-order hold. Recall that the conventional transfer function can be developed from the impulse response of the system as follows:

$$\text{Transfer function } G(s) = \frac{Y(s)}{X(s)} \tag{7.19}$$

If the input to the system is an impulse function $k_0 \, \delta(t)$ then

$$X(s) = l[k_0 \delta(t)] = k_0 \tag{7.20}$$

Therefore, from Equation (7.19),

$$G(s) = \frac{1}{k_0} Y(s) \tag{7.21}$$

This suggests that if the response of the system to an impulse function can be obtained in the time domain, the Laplace transform of the response gives the transfer function of the system according to Equation (7.21). For the first-order hold, if the input is $k_0 \, \delta(t)$ (i.e., an im-

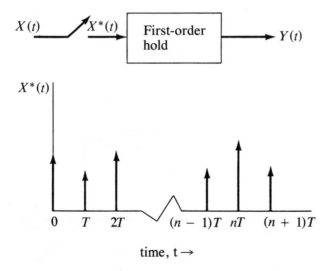

Figure 7.5
First-Order Hold Operation

pulse of strength k_0 at the origin), the output of the holding device will be, from Equation (7.17),

$$Y_0(t) = Y(0) + \frac{Y(0) - Y(-T)}{T} t \qquad 0 < t < T \qquad (7.22)$$

$$= k_0 + \frac{k_0 - 0}{T} t = k_0 \left(1 + \frac{t}{T}\right)$$

$$Y_1(t) = Y(T) + \frac{Y(T) - Y(0)}{T} (t - T) \qquad T < t < 2T \quad (7.23)$$

$$0 + \frac{0 - k_0}{T} (t - T) = k_0 \left(1 - \frac{t}{T}\right)$$

$$Y_2(t) = Y(2T) + \frac{Y(2T) - Y(T)}{T} (t - 2T) \qquad 2T < t < 3T$$
$$(7.24)$$

$$0 + \frac{0 - 0}{T} (t - 2T) = 0$$

The schematic of the input-output representation for the first-order hold is shown in Figure 7.6. Also, from Equation (7.17), $Y_n(t) = 0$ for $n > 2$. Thus the impulse response of the system can be expressed as

$$Y(t) = k_0 \left(1 + \frac{t}{T}\right) u(t) - k_0 \left(1 + \frac{t}{T}\right) u(t - T)$$

$$+ k_0 \left(1 - \frac{t}{T}\right) u(t - T) - k_0 \left(1 - \frac{t}{T}\right) u(t - 2T)$$
$$(7.25)$$

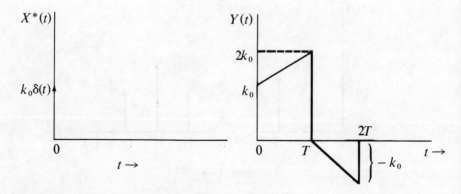

Figure 7.6
Input-Output from First-Order Hold

The reader should verify that this equation gives the results of Equations (7.22) through (7.24). Recall that

$$u(t - nT) = \begin{cases} 1 & t \geqslant nT \\ 0 & \text{otherwise} \end{cases} \tag{7.26}$$

Equation (7.25) can be simplified to give

$$Y(t) = k_0\left(1 + \frac{t}{T}\right)u(t) - 2k_0\frac{t}{T}u(t - T)$$

$$\tag{7.27}$$

$$- k_0\left(1 - \frac{t}{T}\right)u(t - 2T)$$

This equation may be rearranged so that we may readily look up its Laplace transform. Thus

$$Y(t) = k_0\left(1 + \frac{t}{T}\right)u(t) - 2k_0\left(1 + \frac{t-T}{T}\right)u(t - T)$$

$$\tag{7.28}$$

$$+ k_0\left(1 + \frac{t-2T}{T}\right)u(t - 2T)$$

Taking the Laplace transform of Equation (7.28) and dividing by k_0 gives the transfer function of the first-order hold.

$$G_{h_1}(s) = \frac{1}{k_0}l\{Y(t)\} = \frac{1}{s} + \frac{1}{s^2 T} - \frac{2}{s}e^{-Ts} - \frac{2}{s^2 T}e^{-Ts}$$

$$\tag{7.29}$$

$$+ \frac{1}{s}e^{-2Ts} + \frac{e^{-2Ts}}{s^2 T}$$

which can be simplified to give

$$G_{h_1}(s) = \left(\frac{1 + Ts}{T}\right)\left(\frac{1 - e^{-Ts}}{s}\right)^2 \tag{7.30}$$

7.3. Sampling Frequency Considerations[1]

This is perhaps an appropriate place to develop the necessary background that can answer the questions about the minimum sampling frequency in sampled-data control systems. It should be clear that if the sampling frequency approaches infinity, the performance of the sampled data system approaches that of the continuous or analog control system. On the other hand, if we do not sample at all or sample very infrequently, the control-loop performance will be unacceptably poor. Somewhere between these extremes is a minimum sampling rate required for proper signal recovery and an optimum sampling frequency based on economics.

To illustrate the need for the minimum rate, let us begin with the schematic of the sampler followed by a hold circuit as shown in Figure 7.7. The continuous signal $x(t)$ is sampled every T seconds or minutes. The hold circuit reconstructs the continuous signal $y(t)$ from the sampled train $X^*(t)$. The question is: Is there a minimum sampling rate that is required to ensure that adequate information about x will be present in y? The answer to this question may be developed by considering the signals x, X^*, and y in the frequency domain, as described in the following paragraphs.

Consider a signal $x(t)$ that has an amplitude spectrum* as shown in Figure 7.8a. Such a signal is said to be band limited, meaning that $|X(j\omega)|$ is zero for all $\omega > \omega_c$ and $\omega < -\omega_c$. The knowledge of $X(j\omega)$ at each value of ω is equivalent to knowing $x(t)$ for each value of t. Now recall that the sampled train $X^*(t)$ can be expressed in the frequency domain as

$$X^*(j\omega) = \frac{1}{T} \sum_{n=-\infty}^{\infty} X[j(\omega + n\omega_s)] \tag{7.31}$$

where

$$\omega_s = \text{sampling frequency}, 2\pi/T$$

The frequency spectrum of the sampled train is shown in Figure 7.8b. Note from Figure 7.8 that $\omega_c < \omega_s/2$ [i.e., the sampling fre-

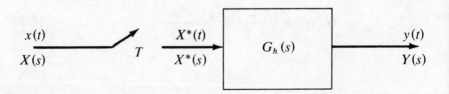

Figure 7.7
Sampler and Hold Operation

---

*The frequency spectrum of a signal $f(t)$ is a plot of its Fourier transform $F(j\omega)$ in the form of amplitude $|F(j\omega)|$ versus frequency and phase angle of $F(j\omega)$ versus frequency.

quency is greater than twice the highest significant frequency in the input signal $x(t)$].

The amplitude spectrum of the sampled train $|X^*(j\omega)|$ consists of an infinite number of spectra, each identical to that of the continuous signal $X(j\omega)$ but reduced in amplitude by a factor $1/T$. The central band in Figure 7.8b is called the primary band, and each of the displaced amplitude spectra is called a side band. From these figures it should be clear that if we are to successfully reconstruct $y(t)$ from $X^*(t)$ such that $y(t)$ has adequate information originally contained in $x(t)$, we must be able to eliminate the side bands (the multiplication by T can be easily accomplished by an operational amplifier).

Now let us see what happens to X^* as it passes through the hold circuit. From Figure 7.7

$$Y(j\omega) = G_h(j\omega) X^*(j\omega) \qquad (7.32)$$

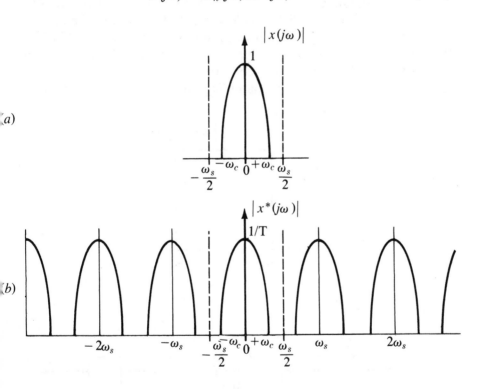

(a)

(b)

Figure 7.8
Amplitude Spectra of Input and Output Signals from the Sampler
(a) Pertains to the Input Function and (b) to the Sampled
Output: Note that the Sampling Frequency is Higher than the
Maximum Frequency of the Input Signal (i.e., $\omega_s > 2\omega_c$)
(Reprinted by Permission from Ref. 1)

If we select the hold circuit having amplitude spectra as shown in Figure 7.9a, it is easy to see that this circuit will act like a filter. It will allow the frequency content of the spectra to pass through, without attenuation, for $|\omega| < |\omega_c|$ but will filter out completely the frequency content for $|\omega| > |\omega_c|$ in accordance with Equation 7.32. This description, shown in Figure 7.9b, depicts idealized filter operation, in that it is not possible to construct a physical hardware which has the amplitude spectra shown in Figure 7.9a. Practical filters can only approach the performance of the ideal filter.

Now let us see what happens if the sampling frequency is less than twice the highest significant frequency of the continuous signal (i.e., $\omega_s < 2\omega_c$), as shown in Figure 7.10a. Again, when we sample this signal we will get an infinite number of spectra, as shown in Figure 7.10b. However, in this case, since $\omega_c > \omega_s/2$, a portion of the spectrum in the interval $\omega_s/2 < \omega < \omega_c$ overlaps onto the spectrum of the adjacent side band. It should be clear that even if we use the ideal filter, we will not be able to reconstruct the signal of Figure 7.10a from the spectra shown in Figure 7.10b. In this case, it will be impossible to recover all the information contained in the original signal. This discussion leads to Shannon's sampling theorem,[2] which states that if

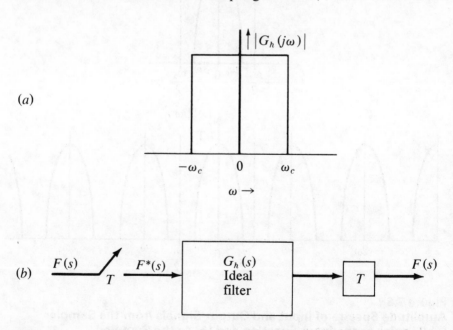

Figure 7.9
Sampler with an Ideal Filter: (a) Amplitude Spectra of Ideal Filter and (b) Schematic for Complete Signal Recovery

a signal contains no frequency higher than ω_c radians per second, it is completely characterized by the values of the signal measured at instants of time separated by $T = \frac{1}{2}(2\pi/\omega_c)$ second.

Recall that we began this discussion by considering the frequency spectra of a band-limited signal. All physical signals contain components over a wide range of frequencies, although the magnitude of the high-frequency components is generally small. Therefore, sampling will result in a certain amount of overlapping, even if an ideal filter were employed. Further, the amplitude spectra of the ideal filter cannot be reproduced exactly by a practical filter (for example, see Figure 7.11 for amplitude spectra of a zero-order hold). Thus the exact reproduction of the continuous signal from the sampled signal is impossible. However, the sampling theorem gives us a useful guide

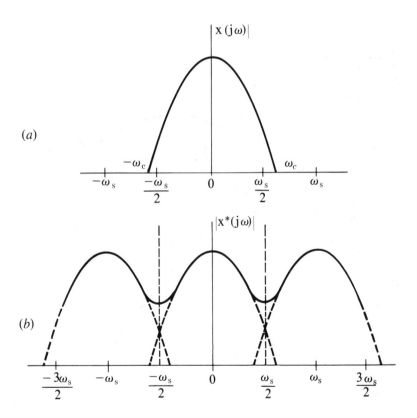

Figure 7.10
Amplitude Spectra of Input and Output Signals when $\omega_s < 2\omega_c$:
(*a*) Input Spectra (*b*) Output Spectra
Reprinted by Permission from C. L. Smith, *Digital Computer Process Control*, Intext Educational Publishers, 1972.

on the minimum sampling frequency in sampled-data control systems.

7.4. Selection of Optimum Sampling Period

The sampling theorem establishes the lower limit on sampling frequency. Somewhere above this minimum is a value for the sam-

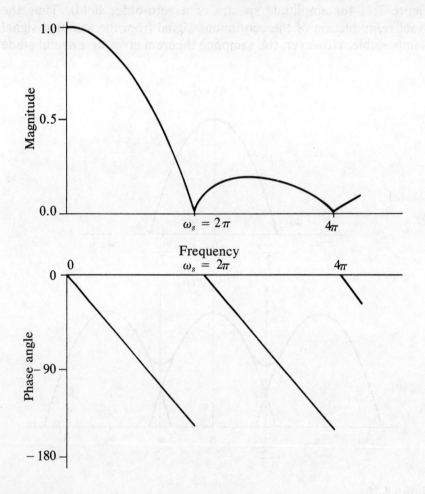

Figure 7.11
Frequency Characteristics of Zero-Order Hold (T = 1)
Reprinted by Permission from C. L. Smith, *Digital Computer Process Control*, Intext Educational Publishers, 1972

pling period that is economically optimum. In this section we present a discussion of how this optimum might be selected.

The simplest method for selecting the sampling period is the one recommended in the DDC guidelines established by the 1963 Users Conference.[3] These guidelines recommended the sampling period of 1 sec for flow loops, 5 sec for level and pressure loops, and 20 sec for temperature and composition loops. Although this procedure is simple to use, the designer may wish to undertake a more detailed analysis to ensure that the sampling period selected is as large as possible, consistent with good control.

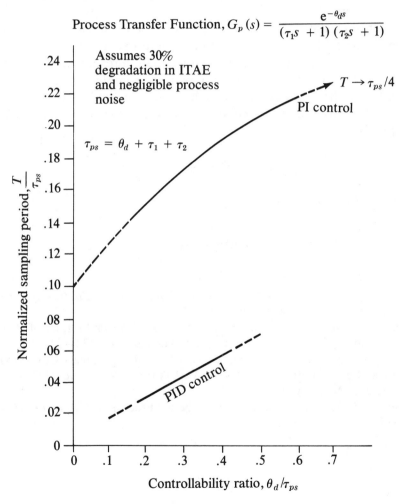

Process Transfer Function, $G_p(s) = \dfrac{e^{-\theta_d s}}{(\tau_1 s + 1)(\tau_2 s + 1)}$

Assumes 30% degradation in ITAE and negligible process noise

$\tau_{ps} = \theta_d + \tau_1 + \tau_2$

$T \to \tau_{ps}/4$

PI control

PID control

Normalized sampling period, $\dfrac{T}{\tau_{ps}}$

Controllability ratio, θ_d/τ_{ps}

Figure 7.12
Selection of Sampling Period For PI Controllers
(Reprinted by Permission of Ref. 4)

A somewhat more refined estimate of the required sampling period may be obtained if the dynamic parameters of the process model are known. The basis of the approach here is the fact that the transient closed-loop response of a sampled-data control system is inferior as compared to that of the equivalent continuous control system because of the dead-time effects of sampling. By fixing the upper limit on the acceptable degradation, the maximum allowable sampling period can be back calculated through simulation. Figure 7.12 shows a plot of normalized sampling period versus controllability for PI and PID control of overdamped second-order processes with dead time.[4]

When a digital computer is used to execute the two- and three-mode (PI and PID) control equations, there exists a lower limit on the sampling period, because if T is too small, a reset deadband may result when a fixed point calculation is used to implement the control algorithm. With correct binary point selection and a 16-bit word length, the lower limit on the sampling period for both PI and PID control is[4]

$$T > \frac{\tau_I}{100}$$

where τ_I is integral time.

References

1. Kuo, B. C., *Analysis and Synthesis of Sampled-Data Control Systems*, Prentice-Hall, Englewood Cliffs, N.J. 1963.
2. Oliver, R. M., Pierce, J. R., Shannon, C. E., The Philosophy of Pulse Code Modulation, *Proc. IRE*, **36**, 11, November 1948, 1324–31.
3. Guidelines and General Information on User Requirements Concerning Direct Digital Control, First Users Workshop on Direct Digital Control, Princeton, N.J., April 3–4, 1963.
4. Fertik, H. A., Tuning Controllers for Noisy Processes, *ISA Trans.* **14**, 4, 1975.

CHAPTER **8**

Open-Loop Response of Sampled-Data Systems

The open-loop system consists of the final control element, the process, and the measuring element in series and can be represented as shown in Figure 8.1. We have studied Z-transforms and their use in representing the sampled data; we have derived the transfer function of the zero-order hold, and in fact we now have the necessary background for evaluating the open-loop response of sampled-data control systems. Referring to Figure 8.1, let us first combine the four blocks into a single block, $G(s)$, which is merely the product of the transfer functions $G_{h_0}(s)$, $G_1(s)$, $G_p(s)$, and $H(s)$. Thus the open-loop system simplifies to

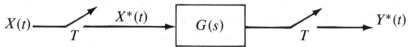

$$X(t)\underset{T}{\nearrow}\quad X^*(t) \rightarrow \boxed{G(s)} \underset{T}{\nearrow}\quad Y^*(t)$$

The Z-transform representation of this block diagram is

$$X(Z) \rightarrow \boxed{G(Z)} \rightarrow Y(Z)$$

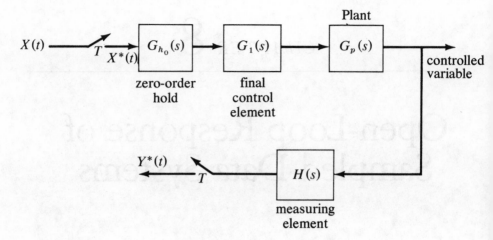

Figure 8.1
Open-Loop Sampled-Data System

that is,

$$\frac{Y(Z)}{X(Z)} = G(Z) \qquad (8.1)$$

where $G(Z)$ is the Z transform of the function whose Laplace transform is $G(S)$.

To obtain the open-loop response of the system to a specified input $X(t)$ we take the Z transform of $X(t)$ to get $X(Z)$ and then combine it with $G(Z)$ to get $Y(Z)$. The inverse Z transform of $Y(Z)$ then gives us the open-loop response at the various sampling instants. The procedure is best illustrated by an example.

8.1. Example of Open-Loop Response

Determine the open-loop response of the sampled-data system shown below to a unit step change in input $X(t)$.

$$X(t) \xrightarrow[\ T\ ]{\ X^*(t)\ } \boxed{G(s) = \frac{9}{s(s^2 + 9)}} \xrightarrow[\ T\ ]{\ Y^*(t)}$$

The sampling period T is 1 sec.

Solution

The block diagram is represented as follows:

$$X(Z) \longrightarrow \boxed{G(Z)} \longrightarrow Y(Z)$$

For a unit step change in $X(t)$

$$X(Z) = \frac{Z}{Z - 1} \tag{8.2}$$

Now let us expand $G(s)$ in partial fractions to get

$$G(s) = \frac{9}{s(s^2 + 9)} = \frac{1}{s} - \frac{s}{s^2 + 9} \tag{8.3}$$

the inverse of which is

$$G(t) = u(t) - \cos 3t \tag{8.4}$$

Now,

$$
\begin{aligned}
G(Z) &= \sum_{n=0}^{\infty} G(nT) Z^{-n} \\
&= \sum_{n=0}^{\infty} \left[u(nT) - \cos (3nT) \right] Z^{-n} \\
&= \frac{Z}{Z - 1} - \frac{1 - Z^{-1} \cos 3T}{1 - 2Z^{-1} \cos 3T + Z^{-2}} \tag{8.5} \\
&= \frac{Z}{Z - 1} - \frac{Z^2 - Z \cos 3T}{Z^2 - 2Z \cos 3T + 1} \\
&= \frac{Z^3 - 2Z^2 \cos 3T + Z - Z^3 + Z^2 \cos 3T + Z^2 - Z \cos 3T}{(Z - 1)(Z^2 - 2Z \cos 3T + 1)} \\
&= \frac{Z^2 + Z - Z^2 \cos 3T - Z \cos 3T}{(Z - 1)(Z^2 - 2Z \cos 3T + 1)} \\
&= \frac{Z(Z + 1)(1 - \cos 3T)}{(Z - 1)(Z^2 - 2Z \cos 3T + 1)}
\end{aligned}
$$

Thus

$$
\begin{aligned}
Y(Z) &= X(Z) \, G(Z) \\
&= \frac{Z}{(Z - 1)} \cdot \frac{Z(Z + 1)(1 - \cos 3T)}{(Z - 1)(Z^2 - 2Z \cos 3T + 1)} \tag{8.6} \\
&= \frac{Z^2 (Z + 1)(1 - \cos 3T)}{(Z - 1)^2 (Z^2 - 2Z \cos 3T + 1)}
\end{aligned}
$$

This can be expanded in partial fractions to give

$$Y(Z) = Z \left\{ \frac{1}{(Z-1)^2} + \frac{1}{2}\frac{1}{Z-1} - \frac{1}{2}\frac{Z+1}{Z^2-2Z\cos 3T+1} \right\} \quad (8.7)$$

So that we can look up the inverse, let us rearrange the above equation as

$$Y(Z) = \frac{Z}{(Z-1)^2} + \frac{1}{2}\frac{Z}{Z-1} - \frac{1}{2}\left(\frac{Z^2 - Z\cos 3T}{Z^2 - 2Z\cos 3T + 1}\right) \quad (8.8)$$

$$- \frac{1}{2}\frac{1+\cos 3T}{\sin 3T}\left(\frac{Z\sin 3T}{Z^2 - 2Z\cos 3T + 1}\right)$$

Now, the inverse can be found from tables of Z-transforms. Thus

$$Y^*(t) = \frac{t}{T} + \frac{1}{2} - \frac{1}{2}\cos 3t - \frac{1}{2}\frac{1+\cos 3T}{\sin 3T}\sin 3t \quad (8.9)$$

For $T = 1.0$ sec, this simplifies to

$$Y^*(t) = t + 0.5 - 0.5\cos 3t - 0.0354 \sin 3t \quad (8.10)$$

The response of $Y^*(t)$ is shown in Figure 8.2. It must be remembered that the output values of $Y^*(t)$ are correct only at the sampling instants. Several forms of the time function $Y(t)$ may be obtained from a given $Y(Z)$ through inverse transformation, yet all of them will indicate correct values only at the sampling instants. The complete time response $Y(t)$ can be obtained by the modified Z-transform method, which is discussed later.

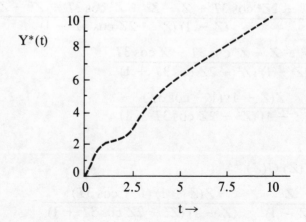

Figure 8.2
Open-Loop Response of Sampled-Data System

8.2 Pulse Transfer Functions in Terms of Impulse Response Coefficients

In recent years powerful control methodologies have been developed that utilize impulse response (IR)-type process models. The IR-type process models can be easily derived from step response data. The advantages of IR-type models are that no parametric modeling is required (that is, the process reaction curve does not have to be fitted to a dead time plus first- or second-order or some other type model), and the order of the process is not at all important. Furthermore, the model can be used to predict the future values of inputs and outputs, a feature that becomes important when operating constraints are encountered.

The open-loop transfer function of a process can be represented by the expression

$$G(Z) = \frac{Y(Z)}{U(Z)} \tag{8.11}$$

For a single unit step function,

$$U(Z) = \frac{1}{1 - Z^{-1}}$$

Equation (8.11) gives

$$G(Z) = (1 - Z^{-1})Y(Z) \Big|_{\text{step}} \tag{8.12}$$

If, on the other hand, the input were a unit impulse function,

$$U(Z) = 1$$

then Equation (8.11) gives

$$G(Z) = Y(Z) \Big|_{\text{impulse}} \tag{8.13}$$

Equating Equations (8.12) and (8.13) we find that

$$Y(Z) \Big|_{\text{impulse}} = (1 - Z^{-1})Y(Z) \Big|_{\text{step}} \tag{8.14}$$

The response of a process to a single unit step change in input shown in Figure 8.3 may be represented as

$$Y(Z)\Big|_{\text{step}} = a_1 Z^{-1} + a_2 Z^{-2} + \dots \tag{8.15}$$

From Equations (8.13), (8.14), and (8.15) we obtain

$$G(Z) = Y(Z)\Big|_{\text{impulse}} = (1 - Z^{-1})(a_1 Z^{-1} + a_2 Z^{-2} + \dots)$$

$$= a_1 Z^{-1} + (a_2 - a_1)Z^{-2} + \dots$$

$$= h_1 Z^{-1} + h_2 Z^{-2} + \dots \tag{8.16}$$

The numbers in the sequence $(h_i, i = 1,2,\dots)$ are called "impulse response coefficients" and a system represented in the form of Equation (8.16) is said to have a model of the impulse-response (IR) type. Similarly, the numbers in the sequence $(a_i, i = 1,2,\dots)$ are called "step-response coefficients," and a system represented in terms of this sequence is said to have a model of the 'step-response' type.

The open-loop response given in Equations (8.15) or (8.16) is based on a single step input. In actual practice, what is needed is a representation that gives the values of the process output in response to a train of step changes, one occurring at each sampling instant. The need for this representation will be clear if it is remembered that the input to the process is the output of a digital-to-analog converter, which is typically represented by a zero-order hold, and, therefore, the D/A converter output is constant between sampling instants. Thus, the input to the process is a train of step changes occurring at the sampling instants. We can obtain the response of a process to a train of step changes simply by the superposition principle, which states that the response to a sum of several functions is equal to the sum of the responses to the individual functions acting separately. A train of step changes can be represented as

$$U(Z) = u_0 Z^0 + u_1 Z^{-1} + u_2 Z^{-2} + \dots \tag{8.17}$$

where u_0 is the magnitude of the step input occurring at $t = 0$, u_1 is the magnitude of the step input occurring at $t = T$, etc., where T is the sampling period. Now let us rearrange Equation (8.11) and write it as

$$Y(Z) = G(Z) \cdot U(Z) \tag{8.17a}$$

Substituting for $G(Z)$ and $U(Z)$ from Equations (8.16) and (8.17), respectively, into Equation (8.17a), we get

$$Y(Z) = (h_1 Z^{-1} + h_2 Z^{-2} + \dots)$$
$$(u_0 Z^0 + u_1 Z^{-1} + u_2 Z^{-2} + \dots)$$
$$= (h_1 u_0)Z^{-1} + (h_2 u_0 + h_1 u_1)Z^{-2} + \dots$$
$$+ (h_n u_0 + h_{n-1} u_1 + \dots h_1 u_{n-1})Z^{-n} + \dots \tag{8.18}$$

$Y(Z)$ can equivalently be written as

$$\hat{Y}(Z) = \hat{y}_1 Z^{-1} + \hat{y}_2 Z^{-2} + \hat{y}_3 Z^{-3} + \ldots \tag{8.19}$$

where the symbol $\char94$ indicates a predicted value.

Comparing Equations (8.18) and (8.19) reveals that

$$\hat{y}_{k+1} = h_1 u_k + h_2 u_{k-1} + \ldots + h_k u_1 + h_{k+1} u_0 \tag{8.20a}$$

or, in alternate form

$$\hat{y}_{k+1} = \sum_{i=1}^{N} h_i u_{k+1-i} \tag{8.20b}$$

Note that $u_i = 0$ for $i < 0$. Equation (8.20b) enables us to predict the response of y to a train of past step inputs.

CHAPTER 9

Closed-Loop Response of Sampled-Data Control Systems

L et us now learn how to evaluate the transient closed-loop response of computer-control systems to set point and load changes. For the purpose of discussion in this chapter, we assume that the transfer functions of all the elements of the closed loop are known. The first step in evaluating closed-loop response is to obtain the closed-loop pulse transfer function of the system.

9.1. Closed-Loop Pulse Transfer Functions

The block diagram of a single-loop, computer-based, closed-loop control system is shown in Figure 9.1. This is the most general description of the feedback control system in that the dynamics of all the elements are assumed to be significant and therefore are included

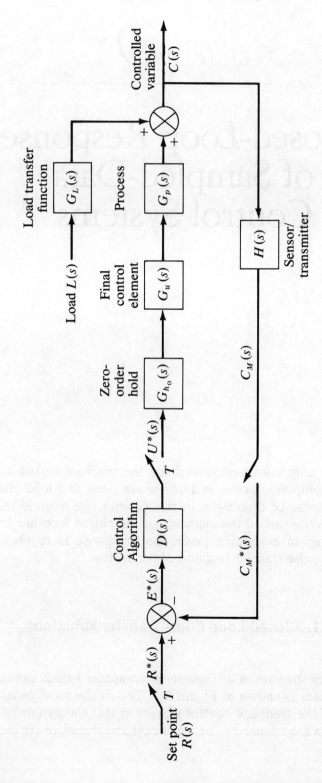

Figure 9.1
Closed-Loop Sampled-Data Control System

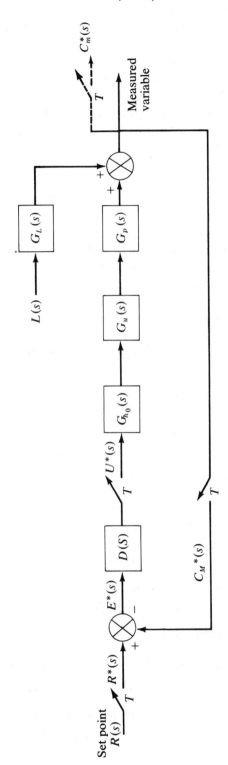

Figure 9.2
Closed-Loop Sampled-Data System with (a) Negligible Sensor/Trans-mitter Lag or (b) Sensor/Transmitter Lag Included in G_p (s) and G_L (s) (The Dashed Line Indicates a Fictitious Sampler, Meaning that the Z-Transform Analysis Will Give Information about an Output Variable Only at the Sampling Instants)

in the block diagram. Several simplifications of this block diagram are possible: If the sensor/transmitter lag is either negligible or is included in the transfer functions $G_p(s)$ and $G_L(s)$, the block diagram can be simplified as shown in Figure 9.2. On the other hand, if the dynamics are determined from experimental tests, the process transfer function $G_p(s)$ will include the dynamics of the final control element, and the sensor/transmitter and the load transfer function will include the dynamics of the sensor/transmitter. In this instance the block diagram of Figure 9.1 reduces to the one shown in Figure 9.3. Finally, if the load dynamics and the process dynamics are equal, that is, $G_L(s) = G_p(s)$, the block diagram of Figure 9.1 can be simplified as shown in Figure 9.4.

The derivation of the closed-loop pulse transfer function proceeds in a manner analogous to that of the continuous control system. However, note that the sampled-data system has a combination of continuous signals and sampled signals, therefore, we must be very careful in deciding which blocks can be combined into a single block. As an illustration, let us derive the closed-loop pulse transfer function of the block diagram of Figure 9.2.

First, we recall from our work in conventional control that the three blocks $G_{h_0}(s)$, $G_u(s)$, and $G_p(s)$ can be combined into a single block (note that there are no samplers between these blocks). Thus, let

$$G_{h_0}(s)\, G_u(s)\, G_p(s) = G(s)$$

Now, from Figure 9.2 we write

$$C_M(s) = U^*(s)\, G(s) + G_L(s)\, L(s) \tag{9.2}$$

$$R^*(s) - C_M{}^*(s) = E^*(s) \tag{9.3}$$

$$U^*(s) = D(s)\, E^*(s) \tag{9.4}$$

First, we take the starred transform of Equation (9.4) to get

$$U^*(s) = D^*(s)\, E^*(s) \tag{9.5}$$

Now, we substitute this equation for $U^*(s)$ into Equation (9.2) to get

$$C_M(s) = D^*(s)\, E^*(s)\, G(s) + G_L(s)\, L(s) \tag{9.6}$$

Substitution of $E^*(s)$ from Equation (9.3) into this equation gives

$$C_M(s) = D^*(s)\{R^*(s) - C^*{}_M(s)\}\, G(s) + G_L(s)\, L(s) \tag{9.7}$$

Next, we take the starred transform of this equation to get

$$C_M{}^*(s) = D^*(s)\, R^*(s)\, G^*(s) - D^*(s)\, C_M{}^*(s)\, G^*(s) + \overline{G_L(s)\, L(s)}^* \tag{9.8}$$

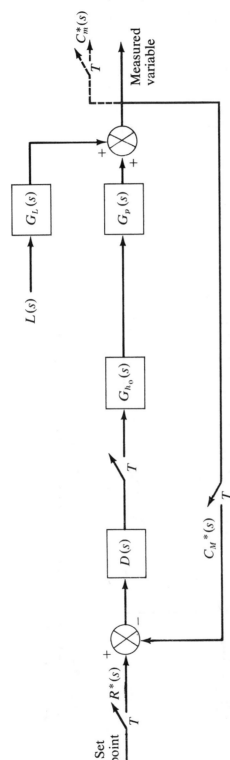

Figure 9.3
Closed-Loop Sampled-Data System where $G_p(s)$ Includes Final Control
Element Dynamics and Sensor/Transmitter Dynamics and $G_L(s)$
Includes Sensor/Transmitter Dynamics

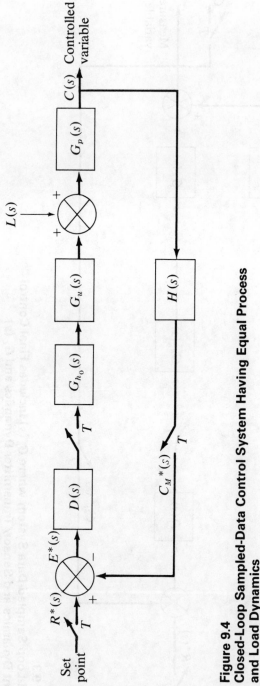

Figure 9.4
Closed-Loop Sampled-Data Control System Having Equal Process and Load Dynamics

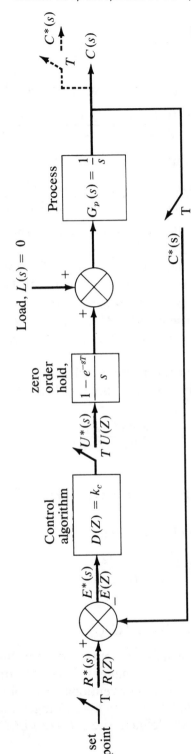

Figure 9.5
Proportional Control of a Sampled-Data System

where

$$G^*(s) = \overline{G_{ho}(s)\, G_u(s)\, G_p(s)}^{\,*}$$

The terms containing $C_M^*(s)$ can now be combined to give

$$C_M^*(s) = \frac{D^*(s)\, R^*(s)\, G^*(s)}{1 + D^*(s)\, G^*(s)} + \frac{\overline{G_L(s)\, L(s)}^{\,*}}{1 + D^*(s)\, G^*(s)} \qquad (9.9)$$

The Z transform of this equation may be taken to give

$$C_M(Z) = \frac{D(Z)\, G(Z)}{1 + D(Z)\, G(Z)}\, R(Z) + \frac{G_L\, L(Z)}{1 + D(Z)\, G(Z)} \qquad (9.10)$$

In taking the starred transform of terms such as those in Equation (9.7) it must be cautioned that (1) a function that is already starred is unaffected and (2) whenever a product of transfer functions appears, we must first multiply the terms and then take the starred transform. Thus the starred transform of $G_L(s)\, L(s)$ is $\overline{G_L(s)\, L(s)}^{\,*}$ and not $G_L(s)^*\, L(s)^*$. Consequently, $Z\{G_L(s)\, L(s)\} \neq G_L(Z)\, L(Z)$ in general. The operation of taking the Z-transform of $G_L(s)\, L(s)$ is denoted as $G_L L(Z)$.

Equation (9.10) relates the response of $C_M^*(t)$ to set point and load changes. For set point changes only it reduces to

$$\frac{C_M(Z)}{R(Z)} = \frac{D(Z)\, G(Z)}{1 + D(Z)\, G(Z)} \qquad (9.11)$$

where

$$G(Z) = Z\{G_{ho}(s)\, G_u(s)\, G_p(s)\}$$

Similarly, we get for load changes alone

$$C_M(Z) = \frac{G_L\, L(Z)}{1 + D(Z)\, G(Z)} \qquad (9.12)$$

where

$$G_L\, L(Z) = Z\{G_L(s)\, L(s)\}$$

Equation (9.11) is the closed-loop pulse transfer function of the sampled-data system to set-point changes. Note the similarity of this equation with its counterpart in conventional control systems. Also observe from Equation (9.12) that for load changes the pulse transfer function as classically defined (i.e., $C_M(Z)/L(Z)$) cannot be

explicitly obtained. This is because the load enters the process without first being sampled.

For complicated loops (for example, the double cascade systems),
this procedure of writing out the equations and solving for $C_M(s)$ to
eventually obtain $C_M(Z)$ can get very tedious. In these instances the
reader is referred to the method of signal flow graphs[1,2] for arriving at
the pulse transfer functions. For single-loop situations the method
described in this chapter is adequate.

9.2. Example to Determine Closed-Loop Transient Response

(a). Determine the pulse-transfer function $C(Z)/R(Z)$ for the
sampled-data control system of Figure 9.5 with $L(S) = 0$.

(b). Evaluate the transient response to a step change in set point.

The closed-loop pulse transfer function for set point changes is

$$\frac{C(Z)}{R(Z)} = \frac{D(Z)\,G_{h_0}G_p(Z)}{1 + D(Z)\,G_{h_0}G_p(Z)}$$

$$G_{h_0}G_p(Z) = Z\{G_{h_0}(s)G_p(s)\} = Z\left[\frac{1}{s}\frac{1 - e^{-sT}}{s}\right]$$

$$= Z\left[\frac{1}{s^2}\right] - Z\left[\frac{e^{-sT}}{s^2}\right]$$

$$= Z\left[\frac{1}{s^2}\right] - Z^{-1}Z\left[\frac{1}{s^2}\right]$$

$$= (1 - Z^{-1})Z\left(\frac{1}{s^2}\right)$$

$$= (1 - Z^{-1})\left(\frac{TZ^{-1}}{(1 - Z^{-1})^2}\right)$$

$$= \frac{TZ^{-1}}{(1 - Z^{-1})}$$

The proportional controller output CO at the nth sampling instant
is related to error by the expression

$$(CO)_n = K_c\, e_n + (CO)_{\text{steady state}}$$

similarly, at $(n - 1)$th sampling instant

$$(CO)_{n-1} = K_c\, e_{n-1} + (CO)_{\text{steady state}}$$

Therefore,

$$(CO)_n - (CO)_{n-1} = K_c\,(e_n - e_{n-1})$$

Taking the Z transform of this equation gives

$$CO(Z) - Z^{-1}\, CO(Z) = K_c E(Z) - Z^{-1} K_c E(Z)$$

$$CO(Z)\left[1 - Z^{-1}\right] = K_c E(Z)\left[1 - Z^{-1}\right]$$

Therefore,

$$D(Z) = \frac{CO(Z)}{E(Z)} = K_c$$

Substitution of the expressions for $G_{h_0} G_p$ (Z) and $D(Z)$ into the pulse-transfer function will give

$$\frac{C(Z)}{R(Z)} = \frac{K_c T Z^{-1}/(1 - Z^{-1})}{1 + K_c T Z^{-1}/(1 - Z^{-1})}$$

$$= \frac{K_c T}{Z + (K_c T - 1)}$$

b. Evaluation of Transient Closed-Loop Response

For a unit step change in set point

$$R(t) = u(t)$$

Therefore,

$$R(Z) = \frac{1}{1 - Z^{-1}} = \frac{Z}{Z - 1}$$

Substitution of $R(Z)$ in the closed-loop pulse transfer function gives

$$C(Z) = \frac{K_c T Z}{(Z - 1)\left[Z + (K_c T - 1)\right]} = ZC_1\,(Z)$$

Therefore,

$$C_1(Z) = \frac{K_c T}{(Z - 1)\left[Z + (K_c T - 1)\right]}$$

Now we expand $C_1(Z)$ in partial fractions. Thus

$$\frac{K_c T}{(Z - 1)\left[Z + (K_c T - 1)\right]} = \frac{A}{Z - 1} + \frac{B}{Z - (1 - K_c T)}$$

Multiply by $Z - 1$ and set $Z = 1$

Then,

$$A = 1$$

Multiply by

$$Z - (1 - K_c T)$$

and set

$$Z = 1 - K_c T$$

Then,

$$B = -1$$

Therefore,

$$C_1(Z) = \frac{K_c T}{(Z - 1)\left[Z + (K_c T - 1)\right]}$$

$$= \frac{1}{Z - 1} - \frac{1}{Z - (1 - K_c T)}$$

and

$$C(Z) = Z\, C_1(Z)$$

$$= \frac{Z}{Z - 1} - \frac{Z}{Z - (1 - K_c T)}$$

$$= \frac{1}{1 - Z^{-1}} - \frac{1}{1 - e^{-\ln 1/(1 - K_c T)}\, Z^{-1}}$$

The second term is of the form

$$\frac{1}{1 - e^{-aT}\, Z^{-1}}$$

the inverse of which is

$$e^{-at}$$

where

$$-aT = -\ln\left[1/(1 - K_cT)\right]$$

$$a = \frac{1}{T} \ln\left[1/(1 - K_cT)\right]$$

Therefore,

$$C(t) = u(t) - e^{-t/T \ln [1/(1 - K_cT)]}$$

$$= u(t) - e^{\ln (1 - K_cT)t/T}$$

$$= u(t) - (1 - K_cT)^{t/T}$$

Remember that this equation is valid only at the sampling instants.

References

1. Ash, R., Kim, W.H., Kranc, G.M., A general Flow Graph Technique for the Solution of Multiloop Sampled Problems, *Trans. A. S. M. E., Journal of Basic Engineering*, June 1960. pp. 360–370.
2. Kuo, B. C., *Analysis and Synthesis of Sampled-data Control Systems*, Prentice Hall, Englewood Cliffs, N.J., 1963, pp. 112–142.

CHAPTER 10

Stability of Sampled-Data Control Systems

$\mathbf{I}$n Chapter 1 we reviewed the stability aspects of conventional control systems. We noted that the stability of continuous-control systems was determined by the location of the roots of the characteristic equation of the system in the s plane. For a stable system all the roots had to be located in the left half of the s plane. Let us now derive the stability criteria for the sampled-data control systems with the aid of Z transforms. Consider the sampled-data control system shown in Figure 10.1. In accordance with our discussion in Chapter 9, the closed-loop pulse transfer function of this system to changes in set point is

$$\frac{C(Z)}{R(Z)} = \frac{D(Z)G_{h_0}G_p(Z)}{1 + D(Z)G_{h_0}G_p(Z)} \tag{10.1}$$

where

$$G_{h_0}G_p(Z) = Z[G_{h_0}(s)G_p(s)] \tag{145}$$

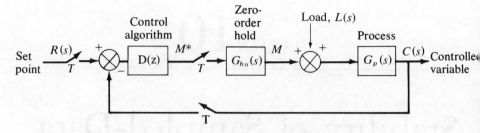

Figure 10.1
Typical Sampled-Data Control System

so that the open-loop pulse-transfer function of the system is

$$A(Z) = D(Z)G_{h_0}G_p(Z) = Z[G_{h_0}(s)G_p(s)]D(Z) \qquad (10.2)$$

The characteristic equation of the sampled-data system is, therefore,

$$1 + A(Z) = 0 \qquad (10.3)$$

The nature of the roots of Equation (10.3) determines the stability and transient behavior of the sampled-data control system. Recall that a stable region in the s domain is the region to the left of the imaginary axis, as shown in Figure 10.2 and that the relationship between Z and s is

$$Z = e^{Ts} \qquad (10.4)$$

We would like to find the corresponding stable region in the Z plane. To do this we would vary s and examine its effect on Z, as described in the following paragraphs.

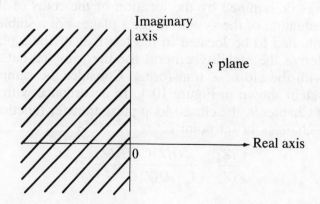

Figure 10.2
Stable Region in *s* plane

For values of s on the imaginary axis, $s = j\omega$. Therefore, in accordance with Equation 10.4

$$Z = e^{j\omega T} \tag{10.5}$$

As ω is varied from 0 to $\omega_s/4 \ (= 2\pi/4T)$, Z varies from $e^{j0T} = 1$ to $e^{j(2\pi/4T)T} = e^{j\pi/2} = \cos \pi/2 + j \sin \pi/2 = j$ that is, along the unit circle in the first quadrant of the Z plane, as shown in Figure 10.3a As ω increases from $\omega_s/4$ to $\omega_s/2$, Z varies from j to -1 along the unit circle in the second quadrant. As ω increases from $\omega_s/2$ to $3\omega_s/4$, Z moves from -1 to $-j$ along the unit circle in the third quadrant. Finally, as ω moves from $3\omega_s/4$ to ω_s, Z traverses from $-j$ to $+1$ along the unit circle in the fourth quadrant. These movements are shown in Figure 10.3b and c. As ω is increased from ω_s to $2\omega_s$, the values of Z trace the unit circle once more. The process is repeated when ω increases or decreases through a range of ω_s. Now let us consider how the region to the left of the

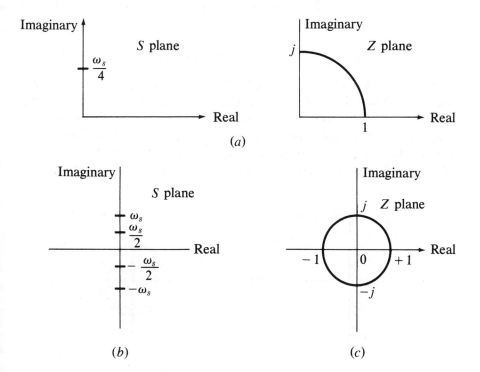

Figure 10.3
Transformations between s and Z Planes

imaginary axis in the s plane maps on to the Z plane. Consider a general point in the s plane whose coordinates are

$$s = \sigma + j\omega \tag{10.6}$$

Therefore,

$$Z = e^{sT} = e^{\sigma T}\,(e^{j\omega T})$$

When $\sigma = -\infty$ and $\omega = 0$ in the s plane, the value of Z is equal to zero. This implies that a point at infinity on the negative real axis of the s plane is mapped into the origin of the Z plane. Also, for $\sigma \le 0$ and $\omega = 0$, $Z = e^{\sigma t}$ describes a line segment between 0 and 1. Thus the negative real axis of the s plane is mapped into that section of the positive real axis of the Z plane that falls inside the unit circle.

Finally, the magnitude of Z is given by

$$|Z| = |e^{sT}| \tag{10.7}$$

For a general point $s = \sigma + j\omega$, Equation (11.7) gives

$$|Z| = |e^{\sigma T}\,e^{j\omega T}| = e^{\sigma T} \tag{10.8}$$

since

$$|e^{j\omega T}| = |\cos \omega T + j \sin \omega T| = \sqrt{\cos^2 \omega T + \sin^2 \omega T} = 1$$

For $\sigma < 0$ (i.e., any point in the left half of the s plane), $e^{\sigma T}$ is less than one. Therefore, $|Z| < 1$, which represents the region covering the interior of the unit circle. Hence, the left half of the s plane is mapped into the area inside the unit circle of the Z plane. For $\sigma > 0$ on the other hand, $|Z|$ is greater than 1, and therefore, the right half of s plane is mapped into the area outside the unit circle. Thus we can

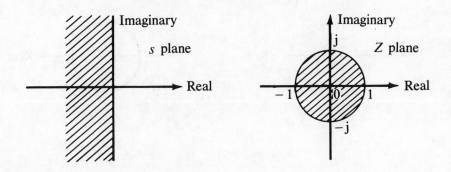

Figure 10.4
Stable Regions in s and Z Plane

state that a *sampled-data control system is stable if all the roots of the characteristic equation 1 + A(Z) = 0 lie inside the unit circle about the origin of the Z plane.* The stable regions are sketched in Figure 10.4.

When the roots of the characteristic equation lie inside the unit circle, the response of output C of the system in Figure 10.1 to a bounded input R or L will be stable. Furthermore, the output of the controller block, with the loop closed, will also be stable. This is because the closed-loop transfer function relating M to R (or L) contains the same characteristic equation, $1 + A(Z) = 0$. Thus, a traditional sampled-data control system of the type shown in Figure 10.1 containing an open-loop stable or unstable plant and/or an open-loop stable controller (e.g., proportional controller) or an open-loop unstable controller (e.g., PI controller) do not pose stability problems for the control loop as long as the roots of $1 + A(Z) = 0$ lie inside the unit circle about the origin in the Z-plane. However, even when the roots of $1 + A(Z)$ are inside the unit circle, it is possible that the response of the controlled variable at the sampling instants is good but the controller output exhibits excessive movements (ringing). The control algorithm having such problems should, of course, be avoided.

In contrast, when model-based control is employed, the closed-loop can, under certain simplifying assumptions, function as an open-loop system. Then, closed-loop stability no longer remains an issue. However, each block between R and C must be open-loop stable if the overall response of C is to be stable. Thus, model-based control works directly only for plants that are open-loop stable and when the controller is also open-loop stable.

10.1 Schur-Cohn Stability Criterion[1]

This method provides an analytical method for determining the absolute stability of sampled-data control systems. The characteristic equation for the sampled-data system is

$$1 + A(Z) = 0 \qquad (10.9)$$

Generally, $A(Z)$ will be in the form of a ratio of two polynomials. If we put the left-hand side of Equation (10.9) over a common denominator and denote the resulting numerator as $F(Z)$, then we can write

$$F(Z) = a_0 + a_1 Z + a_2 Z^2 + \ldots + a_n Z^n = 0 \qquad (10.9a)$$

The first step in determining stability is to write the coefficients of a_k in determinant form as

$$
\Delta_k =
\begin{vmatrix}
a_0 & 0 & 0 & \cdots & 0 & a_n & a_{n-1} & a_{n-k+1} \\
a_1 & a_0 & 0 & \cdots & 0 & 0 & a_n & a_{n-k+2} \\
\cdots & \cdots & a_0 & \cdots & \cdots & \cdots & \cdots & \cdots \\
\cdots & \cdots & \cdots & \cdots & \cdots & \cdots & \cdots & \cdots \\
a_{k-1} & a_{k-2} & a_{k-3} & \cdots & a_0 & 0 & 0 & a_n \\
\bar{a}_n & 0 & 0 & \cdots & 0 & \bar{a}_0 & \bar{a}_1 & \bar{a}_{k-1} \\
\bar{a}_{n-1} & \bar{a}_n & 0 & \cdots & 0 & 0 & \bar{a}_0 & \bar{a}_{k-2} \\
\cdots & \cdots & \bar{a}_n & \cdots & \cdots & \cdots & \cdots & \cdots \\
\cdots & \cdots & \cdots & \cdots & \cdots & \cdots & \cdots & \cdots \\
\bar{a}_{n-k+1} & \bar{a}_{n-k+2} & \bar{a}_{n-k+3} & \bar{a}_n & 0 & \cdots & 0 & \bar{a}_0
\end{vmatrix}
\qquad (10.10)
$$

where $\bar{a}_n$ is the conjugate of a_n. Δ_k $(k = 1, 2, 3, \cdots, n)$ is a determinant that has $2k$ rows and $2k$ columns. The Schur–Cohn criteria states that all the roots of the characteristic equation lie inside the unit circle (i.e., the system is stable) if the following conditions are met:

$$
\Delta_k < 0 \qquad \text{for } k \text{ odd}
$$

$$
\Delta_k > 0 \qquad \text{for } k \text{ even} \qquad (10.11)
$$

Now, as an illustration, let us develop the determinant for a few values of k.

$k = 1$: Δ_1 is as 2×2 determinant

$$
\Delta_1 =
\begin{vmatrix}
a_0 & a_n \\
\bar{a}_n & \bar{a}_0
\end{vmatrix}
\qquad (10.12)
$$

$k = 2$:

$$
\Delta_2 =
\begin{vmatrix}
a_0 & 0 & a_n & a_{n-1} \\
a_1 & a_0 & 0 & a_n \\
\bar{a}_n & 0 & \bar{a}_0 & \bar{a}_1 \\
\bar{a}_{n-1} & \bar{a}_n & 0 & \bar{a}_0
\end{vmatrix}
\qquad (10.13)
$$

k = 3:

$$\Delta_3 = \begin{vmatrix} a_0 & 0 & 0 & a_3 & a_2 & a_1 \\ a_1 & a_0 & 0 & 0 & a_3 & a_2 \\ a_2 & a_1 & a_0 & 0 & 0 & a_3 \\ \overline{a_3} & 0 & 0 & \overline{a_0} & \overline{a_1} & \overline{a_2} \\ \overline{a_2} & \overline{a_3} & 0 & 0 & \overline{a_0} & \overline{a_1} \\ \overline{a_1} & \overline{a_2} & \overline{a_2} & 0 & 0 & \overline{a_0} \end{vmatrix}$$

Let us take an example to illustrate the method.

Example 1. Determine the stability of the sampled-data control system whose open-loop pulse-transfer function is given by

$$A(Z) = \frac{Z}{(2.45Z + 1)(2.45Z - 1)}$$

The characteristic equation is given by

$$1 + A(Z) = 0$$

or

$$F(Z) = (2.45Z)^2 - 1 + Z = 0$$
$$= 6Z^2 + Z - 1 = 0$$

Here n, the order of the characteristic equation, is 2. Therefore, two determinants Δ_1 and Δ_2 must be evaluated to determine stability. From Equations (10.12) and (10.13) we will get

$$\Delta_1 = \begin{vmatrix} -1 & 6 \\ 6 & -1 \end{vmatrix} = -35$$

$$\Delta_2 = \begin{vmatrix} -1 & 0 & 6 & 1 \\ 1 & -1 & 0 & 6 \\ 6 & 0 & -1 & 1 \\ 1 & 6 & 0 & -1 \end{vmatrix} = 1176$$

since

$$\Delta_1 < 0 \text{ and } \Delta_2 > 0$$

according to Equation (10.11) the system is stable.

If $F(Z)$ is a quadratic polynomial with real coefficients and the coefficient of Z^2 is unity, the Schur–Cohn criterion can be simplified. The necessary and sufficient conditions that the roots of the characteristic equation lie inside the unit circle in the Z plane are

$$|F(0)| < 1$$
$$F(1) \;\; > 0 \qquad\qquad (10.14)$$
$$F(-1) > 0$$

Reference

1. Tou, J. T., *Digital and Sampled-data Control Systems,* McGraw-Hill, New York, 1959, p. 238.

CHAPTER 11

Design of Sampled-Data Control Systems

In this chapter we consider the design of digital control algorithms via Z transforms. In Chapter 3 we briefly considered the use of conventional (P, PI, or PID) algorithms in computer control applications. In this chapter we discuss the conventional control algorithms in greater detail. We will shortly see that the algorithms designed by the Z-transform method enable us to specify the desired response characteristics, but their development requires the knowledge of the process transfer function. On the other hand, the development of the conventional control algorithms does not require the knowledge of Z transforms.

Conventional controllers came into existence because the necessary hardware was available. Hardware could be built to produce the desired proportional, integral, and derivative relationships. The availability of digital computers for control applications spurred much research to produce better designs without regard to hardware. Early designs were primarily concerned with response characteristics. Recent research has shown

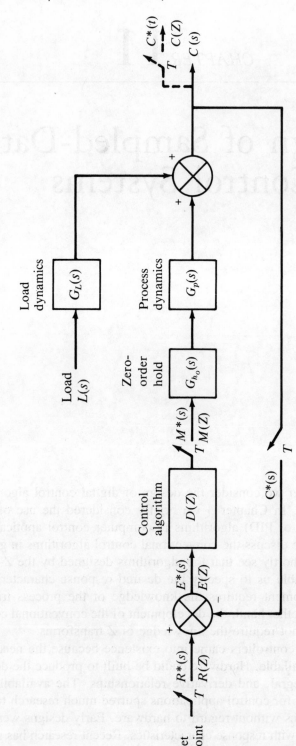

Figure 11.1
Closed-Loop Sampled-Data Control System

that there are many other important considerations. The following is a list of desirable theoretical properties of digital control algorithms.*

1. *Open-Loop Characteristics.* An open-loop stable or unstable control algorithm can be used to control an open-loop stable or unstable process when the control structure employed is the usual sampled-data control system (see Figure 11.1). However, all other things being equal, an open-loop stable control algorithm would probably be preferred. When the control system is expressed in IMC (for Internal Model Control) form, the controller and the process must be open-loop stable as we will see.

2. *Reset Problem.* The algorithm must guarantee offset-free performance in the presence of modeling errors. Modeling errors arise when the nominal model upon which the algorithm is based is different from the real process.

3. *Controller Tuning.* It is desirable that the algorithm contain a small number of independent (adjustable) tuning parameters. The larger the number, the more difficult will be the task of keeping the controller tuned in the field.

4. *Robustness.* A digital control algorithm is based on a nominal plant model that is developed from first principles or obtained from experimental data. Modeling errors will invariably be present. Furthermore, new modeling errors will creep in due to equipment fouling, changing production strategies, etc. The algorithm must maintain stable process operation in the presence of such modeling errors. Robustness issues, having to do with the ability of the control system to maintain stability in the presence of a plant-model mismatch, have received considerable attention in recent years.

5. *Constraint Handling.* Industrial processes must often be operated in the neighborhood of operating constraints. There are two types of constraints: (1) *Stationary* constraints are always present. An example might be a reflux valve of a distillation column that must not be more than $X\%$ closed to avoid weeping on the trays or more than $Y\%$ open to avoid flooding the column. (2) *Moving* constraints come on and disappear during different phases of process operation. Constraints are generally encountered on inputs, but output constraints may be present in specific applications. A digital control algorithm must be designed such that it does not violate operational constraints.

6. *Dead Time Compensations.* In Chapter 1 we examined the detrimental effects of dead time on the response of first-order systems. The

---

*The first author is indebted to Jacques Richalet of Adersa, France, for providing him industrial perspective on the subject.

presence of dead time necessitates lowering of controller gain to maintain stability. We observed that when the apparent dead time θ_d exceeded the dominant time constant of the system τ, the peak offsets, following a step change in load, could approach those of the uncontrolled situation even with best PID controller tuning. Under these conditions, the settling time approaches $9\theta_d$. Since many chemical engineering processes exhibit apparent dead-time characteristics and since dead time is detrimental to control, there is considerable incentive to develop control algorithms that can compensate for such time delays. The concept of dead-time compensation involves the use of a mathematical model of the process to, in effect, remove the dead time from the feedback signal so that the performance of the system is improved.

7. *Working with RHP Zeroes.* When the roots of the numerator polynomial of the Laplace domain transfer function contain one that lies to the right of the imaginary axis in the s-plane, the system is said to have an RHP (right half plane) zero. Such a system gives rise to inverse response; the open-loop response of the system containing an RHP zero begins in a direction that is opposite to the direction in which it eventually settles out. Inverse response occurs in boiler level systems and in some distillation base level systems. It should be clear that such systems will pose difficulties for the PID-type controller in that the controller will take wrong action initially. With digital control, a facility to accommodate inverse response can be built into the algorithm, and this should be considered wherever applicable.

8. *Manipulated Variable Movements.* Excessive movement of the manipulated variable (ringing) should be avoided to reduce actuator wear.

9. *Shaping of Closed-Loop Dynamics.* In the absence of modeling errors, the algorithm should yield a closed-loop response having desired dynamics.

10. *Reset Windup.* If the algorithm contains integral mode, the phenomenon of reset windup may occur. Anti-reset windup features should be provided wherever appropriate.[1]

11. *Bumpless Transfer.* The facility for bumpless transfer from manual to automatic should be provided.

12. *Noisy Processes.* Industrial processes often contain noise. We consider the treatment of noisy process signals later in the chapter. Suffice it to say at this point that the algorithm must function in the presence of noise. Recall that derivative action tends to amplify noise, and, therefore, it should not be used whenever noise is a problem.

To begin the development of digital control algorithms by Z transforms, consider the block diagram of a typical sampled-data control system shown in Figure 11.1. The process transfer function $G_p(s)$ is assumed to include the dynamics of the sensor and the final control element. The objective is to synthesize a control algorithm $D(Z)$ such that the desired loop performance is achieved. Once an expression for $D(Z)$ has been developed, it may be inverted into the time domain to give an equation suitable for computer programming. It is possible to develop the algorithm for set point changes or load changes.

To develop the algorithm for set point changes, recall from our study of Chapter 9 that the closed-loop pulse transfer function of the system of Figure 11.1 for set point changes is

$$\frac{C(Z)}{R(Z)} = \frac{D(Z)G(Z)}{1 + D(Z)G(Z)} \tag{11.1}$$

where

$$G(Z) = Z\{G_{h_0}(s)G_p(s)\}$$

The solution of Equation (11.1) for $D(Z)$ is

$$D(Z) = \frac{1}{G(z)} \cdot \frac{C(Z)/R(Z)}{1 - C(Z)/R(Z)} \tag{11.2}$$

The design procedure is to specify the desired response characteristic, $C(Z)/R(Z)$; for example, the controlled variable shall reach the new set point in one sampling period. Then Equation (11.2) may be solved for $D(Z)$, provided the process transfer function $G_p(s)$ is known.

The transfer function of the process can be developed from a dynamic mathematical model based on first principles (mass, momentum, or energy balances) or it can be determined from experimental tests in the plant. (We consider the experimental evaluation of transfer functions later in the text). It is important to note that the control algorithm $D(Z)$ is developed on the basis of a *model* of the plant $\tilde{G}_p(s)$. In industrial practice, accurate plant models are seldom available; consequently, a plant-model mismatch will invariably be present. A good control algorithm must give acceptable responses in the presence of such a mismatch. Thus, the control algorithm is really determined from the equation

$$D(Z) = \frac{1}{\tilde{G}(Z)} \cdot \frac{C(Z)/R(Z)}{1 - C(Z)/R(Z)} \tag{11.3}$$

where

$$\tilde{G}(Z) = Z\{G_{h_0}(s)\tilde{G}_p(s)\}$$

Some comments on the specification of $C(Z)/R(Z)$ are in order.

1. In sampled-data control systems, $C(Z)/R(Z)$ can never equal 1,

meaning that C cannot reach R instantaneously. This is due to the dead time effect of sampling. $C(Z)/R(Z)$ can at best be set equal to Z^{-1}.

2. If the plant transfer function contains nonminimum phase* elements such as dead time or RHP zeroes, these elements must be accommodated in the specified $C(Z)/R(Z)$. The need for this should be intuitively clear. If the plant transfer function contains, say, a dead time of N sampling instants, then it is obvious that C cannot begin to change, following a change in R, until N sampling instants have elapsed no matter what the input to the process (i.e., the output of the controller block) is.

3. Recent research has shown that the specification for $C(Z)/R(Z)$, which includes the terms in (2) along with a first- or second-order lag, results in excellent design having the desirable properties previously discussed.

These comments are meant to emphasize that achievable closed-loop performance depends directly on the type of process involved, and improvements beyond a certain level may not be feasible in the presence of nonminimum phase elements.

In the light of the foregoing discussion, the closed-loop response characteristic may be postulated to have the form

$$\frac{C(Z)}{R(Z)} = F(Z)G_+(Z) \tag{11.4}$$

where $F(Z)$ specifies the desired shape of the output response and $G_+(Z)$ contains the nonminimum phase elements. $F(Z)$ can also be used to enhance the robustness properties of the algorithm and to reduce ringing, as we will see later.

A control algorithm may also be designed for load changes. In this case we must use the closed-loop pulse transfer function for load changes that is given by

$$C(Z) = \frac{G_L L(Z)}{1 + D(Z)G(Z)} \tag{11.5}$$

where

$$G_L L(Z) = Z\{G_L(s)L(s)\}$$

Recall that in general $G_L L(Z) \neq G(Z)L(Z)$ and, therefore, $C(Z)/L(Z)$ (synonymous to its analog counterpart $C(s)/L(s)$) does not exist. The solution of Equation (11.5) for $D(Z)$ is

---

*Nonminimum phase elements are those for which the phase angle increases or decreases without bound for increasing values of frequency.

$$D(Z) = \frac{\tilde{G}_L L(Z) - C(Z)}{\tilde{G}(Z)C(Z)} \tag{11.6}$$

The procedure in this instance is to select a suitable load input L, specify the desired response, C, and take the Z transforms indicated in Equation (11.6) to obtain $D(Z)$.

Whether we can design $D(Z)$ for set point changes or for load changes depends on whether we can measure the load disturbances and whether we can correctly anticipate the form of these disturbances. Based on a thorough understanding of the process dynamics, the designer must decide whether to base the design on set point changes or load changes. Fortunately, set point designs frequently work well for load changes as well.[2]

In this chapter we will consider several control algorithms starting with the classical algorithms, such as deadbeat control and the Dahlin algorithm, and progressing towards the more recent algorithms such as internal model control and simplified model predictive control. Many of these algorithms offer dead time compensation, meaning that they considerably improve the response of processes containing dead time.

11.1 Deadbeat Algorithm[3]

An algorithm that requires the closed-loop response to have finite settling time, minimum rise time, and zero steady-state error is referred to as a deadbeat algorithm. A specification that satisfies these criteria is

$$\frac{C(Z)}{R(Z)} = Z^{-1} \tag{11.7}$$

This specification requires that the controlled variable shall reach the set point at $t = 1T$, i.e., after a delay of one sampling period. Substituting for C/R from Equation (11.7) into Equation (11.3) gives

$$D(Z) = \frac{1}{\tilde{G}(Z)} \cdot \frac{Z^{-1}}{1 - Z^{-1}} \tag{11.8}$$

This is as far as we can proceed with the development of the algorithm without knowing the process transfer function $G_p(s)$. Let us now consider the design of a deadbeat controller for a process whose transfer function is given by

$$G_p(s) = \frac{1}{0.4s + 1} \tag{11.9}$$

Let us assume that the sampling period $T = 1$. For this example

$$G_{h_0}G_p(Z) = Z\{G_{h_0}(s)G_p(s)\}$$

$$= Z\left\{\frac{1 - e^{-sT}}{s} \cdot \frac{1}{0.4s + 1}\right\} \tag{11.10}$$

$$= Z\left\{\frac{1}{s(0.4s + 1)}\right\} - Z\left\{\frac{e^{-sT}}{s(0.4s + 1)}\right\}$$

Applying the theorem on translation of the function to Equation (11.10) gives

$$G_{h_0}G_p(Z) = Z\left\{\frac{1}{s(0.4s + 1)}\right\} - (Z^{-1})Z\left\{\frac{1}{s(0.4s + 1)}\right\} \tag{11.11}$$

$$= (1 - Z^{-1})Z\left\{\frac{1}{s(0.4s + 1)}\right\}$$

This equation should be rearranged slightly so that we may look up the Z transform from tables. Thus

$$G_{h_0}G_p(Z) = (1 - Z^{-1})Z\left\{\frac{1}{s(0.4s + 1)}\right\}$$

$$= (1 - Z^{-1})Z\left\{\frac{2.5}{s(s + 2.5)}\right\} \tag{11.12}$$

$$= \frac{(1 - Z^{-1})Z(1 - e^{-2.5T})}{(Z - 1)(Z - e^{-2.5T})}$$

$$= \frac{1 - e^{-2.5T}}{Z - e^{-2.5T}}$$

The next step is to substitute this expression for $G_{h_0}G_p(Z)$ into Equation (11.8), which gives

$$D(Z) = \frac{Z - e^{-2.5T}}{1 - e^{-2.5T}} \cdot \frac{Z^{-1}}{1 - Z^{-1}} \tag{11.13}$$

For $T = 1$ this equation becomes

$$D(Z) = \frac{(Z - 0.082)Z^{-1}}{0.918(1 - Z^{-1})} \tag{11.14}$$

$$= \frac{1 - 0.082Z^{-1}}{0.918(1 - Z^{-1})}$$

Equation (11.14) is the deadbeat algorithm for this example in the Z domain. Let us now invert this algorithm into the time domain. To accomplish this we note from Figure 11.1 that

$$D(Z) = \frac{M(Z)}{E(Z)} \tag{11.15}$$

where

$$M(Z) = Z \text{ transform of controller output}$$

$$E(Z) = Z \text{ transform of error } R(Z) - C(Z).$$

Thus,

$$D(Z) = \frac{M(Z)}{E(Z)} = \frac{1 - 0.082Z^{-1}}{0.918(1 - Z^{-1})} \tag{11.16}$$

The cross multiplication of terms in Equation (11.16) gives

$$0.918\, M(Z) - 0.918Z^{-1}M(Z) = E(Z) - 0.082Z^{-1}E(Z)$$

This equation can be inverted to give the algorithm for computing the controller output. Thus,

$$M_n = M_{n-1} + 1.09E_n - 0.089\, E_{n-1} \tag{11.17}$$

where

$$M_n = \text{controller output at the } n\text{th sampling instant}$$

$$M_{n-1} = \text{controller output at the } (n-1)\text{th sampling instant}$$

$$E_n = \text{error (set point } - \text{ measurement) at the } n\text{th sampling instant}$$

$$E_{n-1} = \text{error at } (n-1)\text{th sampling instant}$$

The FORTRAN or BASIC statements

$$M = M + 1.09 * E - 0.089 * E1 \tag{11.18}$$
$$E1 = E$$

C RETURN FOR NEW SAMPLE AND REPEAT ALGORITHM

will accomplish the computations indicated in Equation (11.17).

In the above example, the process selected is first order. If the process were to contain a time delay that is greater than the sampling period, then substitution in Equation (11.8) will result in a $D(Z)$ that will require future values of the error to determine the current value of the controller output, which is physically impossible. When this occurs, the controller is said to be physically unrealizable. Thus, if the process contains a delay of NT $(N > 1)$ sampling periods, then, C/R must be set to $Z^{-(N+1)}$, where N is the largest integer sampling periods in dead time, θ_d.

Although the deadbeat control specification is particularly simple, the closed-loop response is likely to deviate from the specification due to the inertia present in industrial-scale processes and modeling errors. Furthermore, the possible existence of ringing should be investigated. For a more thorough discussion of the limitations of deadbeat control strategies, the reader is referred to Kuo.[3]

11.2 Dahlin Algorithm[4]

Dahlin's algorithm specifies that the closed-loop sampled-data control system behave as though it were a first-order process with dead time. Thus, the closed-loop response specification, Equation (11.4) for this case, is

$$\frac{C(Z)}{R(Z)} = F(Z)G_+(Z) = F(Z)Z^{-(N+1)} \tag{11.19}$$

where $F(Z)$ represents the first-order lag and N is the number of sampling periods in the process dead time θ_d. To derive the expression for $F(Z)$, consider the differential equation for a first-order process

$$\tau_f \frac{dY}{dt} + Y(t) = X(t) \tag{11.20}$$

Note that in Equation (11.20) the steady-state gain has been selected to be equal to one since C/R must have a gain of one. That is, C must equal R at the new state. The numerical solution of Equation (11.20) is

$$Y_n = (1 - \alpha_f)X_n + \alpha_f Y_{n-1} \tag{11.21}$$

where

$$\alpha_f = e^{-T/\tau_f} \ (0 < \alpha_f < 1)$$

Taking the Z transform of terms in Equation (11.21) and rearrangement gives

$$\frac{Y(Z)}{X(Z)} = F(Z) = \frac{1 - \alpha_f}{1 - \alpha_f Z^{-1}} \tag{11.22}$$

Substituting for $F(Z)$ from Equation (11.22) into Equation (11.19) gives

$$\frac{C(Z)}{R(Z)} = \frac{1 - \alpha_f}{1 - \alpha_f Z^{-1}} \cdot Z^{-(N+1)} \tag{11.23}$$

Now Equation (11.23) may be substituted in Equation (11.3) to give

$$D(Z) = \frac{(1 - \alpha_f)Z^{-(N+1)}}{1 - \alpha_f Z^{-1} - (1 - \alpha_f)Z^{-(N+1)}} \cdot \frac{1}{\tilde{G}(Z)} \qquad (11.24)$$

The time constant of the closed-loop response τ_f, or, equivalently, the parameter α_f, is an adjustable parameter that is selected by trial and error in the field. As we will see later in the chapter, Dahlin's algorithm is essentially the same as the recently developed internal model control for first-order plus dead time types of processes. It turns out that α_f can be used to enhance robustness of the loop in the presence of modeling errors. A high value of α_f close to 1 gives a high degree of robustness, but the response becomes quite sluggish. Smaller values of α_f close to zero improve dynamic response, but the system becomes more sensitive to modeling errors. Even in the absence of modeling errors, the parameter α_f can be used to shape the output response and to reduce ringing. Although the Dahlin algorithm is likely to work quite well for first-order plus dead time types of processes, it is unsuitable for processes with inverse response due to the particular choice of $G_+(Z) = Z^{-(N+1)}$ employed.

As an illustration, let us consider the design of the Dahlin algorithm for computer control of a process having the model

$$G_p(s) = \frac{e^{-0.8s}}{0.4s + 1} \qquad (11.25)$$

with $T = 0.4$ and, therefore, $N = 2$. Now

$$G(Z) = Z\left\{\frac{1 - e^{-sT}}{s} \cdot \frac{e^{-0.8s}}{0.4s + 1}\right\} \qquad (11.26)$$
$$= \frac{0.6321Z^{-3}}{1 - 0.3679Z^{-1}}$$

For $\tau_f = 0.15$, substitution of Equation (11.26) into Equation (11.24) gives

$$D(Z) = \frac{M(Z)}{E(Z)} = \frac{0.9305 - 0.3423Z^{-1}}{0.6321 - 0.0439Z^{-1} - 0.5882Z^{-3}} \qquad (11.27)$$

Cross multiplying and inverting this equation into the time domain gives the algorithm in a form suitable for programming as

$$M_n = 1.4721E_n - 0.5421E_{n-1} + 0.0695M_{n-1} + 0.9306M_{n-3} \qquad (11.28)$$

The control system of this example was simulated on a digital computer. The resulting closed-loop response to a unit step change in set point is shown in Figure 11.2. An interesting set of exercises based on

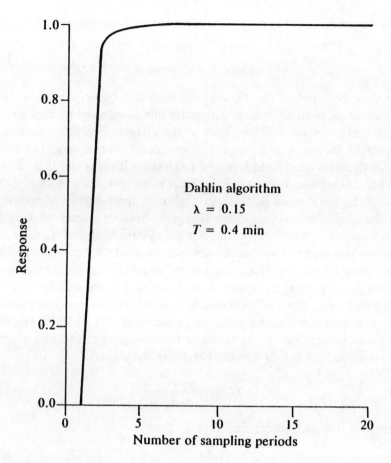

Figure 11.2
Response of the Control System to a Unit Step Change in Set Point

this example would be to use Equation (11.28) for the algorithm but introduce a variety of modeling errors for the process block and repeat the simulation and observe the resulting responses. The parameter α_f can then be varied to study its effect on closed-loop responses in the presence of modeling errors.

Let us consider another example.[2] The plant has the transfer function

$$G_P(s) = \frac{1}{(0.5s + 1)(s + 1)^2(2s + 1)} \tag{11.29}$$

The plant is modeled as first order with dead time having the transfer function

$$\tilde{G}_p(s) = \frac{e^{-1.46s}}{3.34s + 1} \tag{11.30}$$

If $T = 1$, then the pulse transfer function for the plant model is

$$\tilde{G}(Z) = \left\{ \frac{1 - e^{-sT}}{s} \cdot \frac{e^{-1.46s}}{3.34s + 1} \right\} \tag{11.31}$$

$$= \frac{Z^{-2}(0.1493 + 0.1095Z^{-1})}{1 - 0.7413Z^{-1}}$$

Note that in this instance the dead time θ_d is not an integer multiple of the sampling period T. Therefore, the evaluation of Z transform of the expression in Equation (11.31) will require the use of modified Z transforms. We will study modified Z transforms in a later chapter.

Now, substituting Equation (11.31) into Equation (11.24) gives

$$D(Z) = \frac{(1 - \alpha_f) Z^{-2}}{1 - \alpha_f Z^{-1} - (1 - \alpha_f)Z^{-2}} \cdot \frac{1 - 0.7413 Z^{-1}}{Z^{-2} (0.1493 + 0.1095Z^{-1})} \tag{11.32}$$

The performance of the Dahlin algorithm utilizing the fourth-order transfer function for the process for several different values of τ_f was evaluated through simulation. The resulting set point responses are shown in Figure 11.3(a). The response corresponding to $\tau_f = 2$ would probably be preferred by control engineers. For $\tau_f = 2$ (or $\alpha_f = e^{-T/\tau_f} = 0.606$), the Dahlin algorithm, Equation (11.32), becomes

$$D(Z) = \frac{0.392(1 - 0.7413Z^{-1})}{(1 - 0.606Z^{-1} - 0.392Z^{-2})(0.1493 + 0.1095Z^{-1})} \tag{11.33a}$$

$$= \frac{2.63(1 - 0.7413Z^{-1})}{(1 - Z^{-1})(1 + 0.392Z^{-1})(1 + 0.738Z^{-1})} \tag{11.33b}$$

or

$$D(Z) = \frac{2.63Z^2(Z - 0.7413)}{(Z - 1)(Z + 0.392)(Z + 0.738)} \tag{11.33c}$$

For this choice of α_f, the controller $D(Z)$ exhibits considerable ringing as seen from Figure 11.3b, although the response of the controlled variable appears quite acceptable.

Dahlin has suggested a procedure to reduce ringing based on pole-zero locations of $D(Z)$. In this connection, a *zero* represents a root of the numerator polynomial and a *pole* is a root of the denominator polynominal of $D(Z)$. The point $Z = -1$ has been called the *ringing*

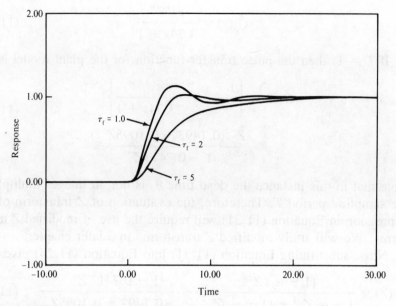

Figure 11.3a
Performance of Dahlin's Algorithm for Various Values of τ_f
(Reproduced with permission from Ref. 1)

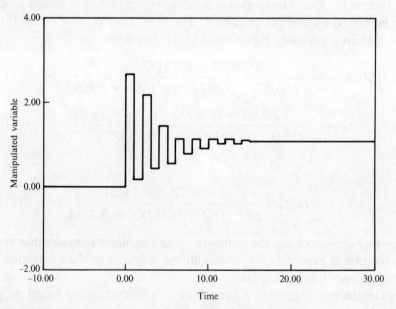

Figure 11.3b
Valve Action for Response in Fig. 6-13 Corresponding to $\tau_f = 2$
(Reproduced with permission from Ref. 1)

node and the pole at $Z = -1$ is the *ringing pole*. Moving the pole away from the ringing node towards zero reduces ringing. A pole in the right half Z-plane ($0 < Z < 1$) is suggested to reduce ringing amplitude, while zeroes in the right half plane aggravate it. Dahlin suggests that the ringing pole may simply be eliminated and gain adjusted to reduce ringing. Removing ringing poles is synonymous to using a higher-order filter for $F(Z)$.

In the example under scrutiny, $D(Z)$ has two poles in the left half Z-plane at $Z = -0.738$ and at $Z = -0.392$ (see Equation (11.33)) that may be suspected to cause ringing. Ringing may be reduced to acceptable levels by eliminating the pole $Z = -0.738$, which is closest to the ringing node. When this is done, the Dahlin algorithm, Equation (11.33b), becomes

$$D(Z) = \frac{(2.63/1.738)(1 - 0.7413Z^{-1})}{(1 - Z^{-1})(Z + 0.392Z^{-1})} \qquad (11.34)$$

or

$$D(Z) = \frac{1.51Z(Z - 0.7413)}{(Z - 1)(1 + 0.392)} \qquad (11.35)$$

Note that although we have eliminated the term $1 + 0.738Z^{-1}$ from Equation (11.33b), its gain 1.738, which is found by letting Z approach one, has been retained in Equation (11.34). This is necessary since the gain of C/R is desired to be unity. Note also that it is not sufficient to remove the term $(Z + 0.738)$ from Equation (11.33c) since that renders the powers of Z in the numerator to be greater than that in the denominator, making the controller unrealizable.

The use of Equation (11.34) for the algorithm and the fourth-order transfer function for the process gives the set point response, and the associated manipulated variable moves as shown in Figure 11.4. The algorithm is now essentially free of ringing.

A better insight into the problem of ringing may be obtained by examining, instead of $D(Z)$, the following closed-loop transfer function relating M to R.

$$\frac{M}{R} = \left(\frac{C}{R}\right)\frac{1}{G} \qquad (11.36)$$

where

$$\frac{C}{R} = \frac{DG}{1 + DG} = FG_+ \qquad (11.37)$$

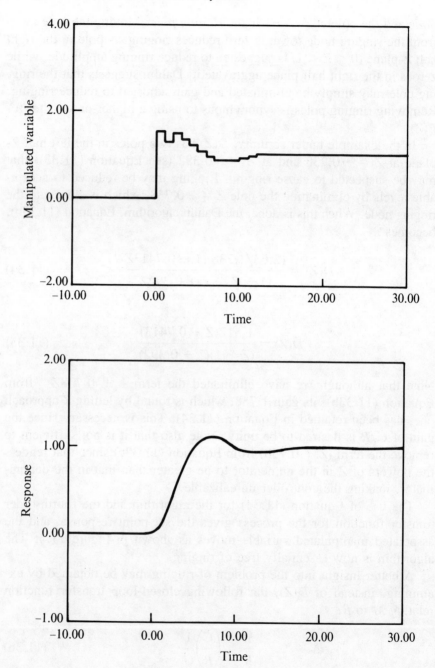

Figure 11.4
Response and Value Action for Dahlin's Algorithm with Ringing
Pole Removed ($\tau_f = 2$; $T = 1$)
(Reproduced with permission from Ref. 1)

The Z transform variable has been omitted for brevity. Now, as long as the roots of $1 + DG$ lie inside the unit circle, the response of C and M will be stable. We wish to find out if there is ringing in M. The term C/R in Equation (11.36) refers to the specification FG_+ and, therefore, it cannot contribute to ringing in M. However, the term $1/G$ can, as we will see from the following discussion.

1. If the process is first order with dead time and modeling errors are absent, then $1/\tilde{G}$ can lead to ringing only if the dead time is not an integer multiple of the sampling period. In such a case a simple alternative would be to increase the dead time in the model just enough so that it equals an integer number of sampling periods in computing $D(Z)$. The effect of the modeling error so introduced can be accommodated by adjusting α_f.

2. If the process is first order with dead time and modeling errors are present, M can exhibit ringing, leading possibly to rippling behavior of C between sampling instants. In such a case, one may adjust α_f to reduce ringing.

3. If $\tilde{G}$ involves a second-order process with dead time, then, $1/\tilde{G}$ will lead to ringing even in the absence of modeling errors. In such a case, ringing can be reduced by a higher-order $F(Z)$, although the approach will no longer be the Dahlin algorithm as it was originally proposed.

11.3. Digital Equivalent of a Conventional Controller

The operation of an ideal PID controller is described by

$$v = v_0 + K_c \left(e + \frac{1}{\tau_I} \int_0^t e \, dt + \tau_D \frac{de}{dt} \right) \qquad (11.38)$$

In conventional control applications, a controller whose output approximates the right side of Equation (11.38) can be built through the use of pneumatic components or operational amplifiers, integrators, and summers. In computer control applications a discrete equivalent to Equation (11.38) is employed. In the development of algorithms that are based on Z transforms we specify the nature of the response to be achieved, whereas in the digital equivalent to the PID controller we "adjust" the constants K_c, τ_I, and τ_D so as to achieve desired response. The computer control system containing the PID control algorithm can be simulated and the constants adjusted so as to minimize the value of an integral of the following types:

1. Integral of the square error,

$$\text{ISE} = \int_0^\infty [e]^2 \, dt \qquad (11.39a)$$

2. Integral of the absolute value of the error,

$$\text{IAE} = \int_0^\infty |e| \, dt \qquad (11.39b)$$

3. Integral of the time multiplied by the absolute value of the error,

$$\text{ITAE} = \int_0^\infty t|e| \, dt \qquad (11.39c)$$

Just which criterion to choose depends on the type of response desired.[1]

For example, large errors contribute more heavily to ISE than to IAE, which means that ISE will favor responses with smaller overshoots for load changes, as shown in Figure 11.5. Note that ISE gives a longer settling time. In the ITAE integral, time appears as a factor, and, therefore, this criterion heavily penalizes errors that occur late in time but virtually ignores errors that occur early in time. Figure 11.5 shows that ITAE criterion gives shortest settling time but has the largest overshoot among the three criteria considered.

To obtain the digital equivalent of the PID controller, the derivative and the integral terms of Equation (11.38) are numerically approximated to give an expression for the output of the algorithm at the nth sampling instant. Thus,

$$v_n = v_0 + K_c \left[e_n + \frac{T}{\tau_I} \sum_{i=0}^n e_i + \frac{\tau_D}{T} (e_n - e_{n-1}) \right] \qquad (11.40)$$

where

v_n = controller output at nth sampling instant

e_n = error (set point − measurement) at the nth sampling instant

v_0 = steady-state output of the control algorithm that gives zero error

Equation (11.40) is referred to as the "position" form of the control algorithm, since the actual controller output is computed. To derive an alternate form of the algorithm we write the expression for controller output at the $(n - 1)$th sampling instant as

$$v_{n-1} = v_0 + K_c \left[e_{n-1} + \frac{T}{\tau_I} \sum_{i=0}^{n-1} e_i + \frac{\tau_D}{T} (e_{n-1} - e_{n-2}) \right] \qquad (11.41)$$

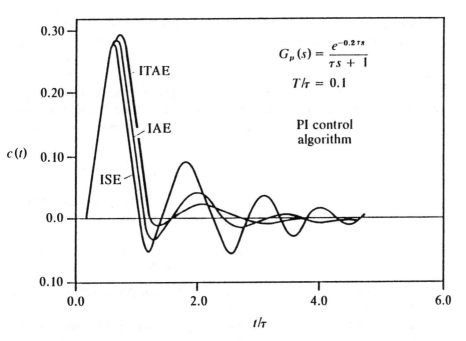

Figure 11.5
Closed-Loop Response of a First-Order Process with Dead Time to a
Unit Step Change in Load with Different Performance Criteria
(Reproduced with Permission from A. M. Lopez, et al., Tuning PI &
PID Digital Controllers, *Instruments and Control Systems*, 42, 2,
February 1969. p. 89)

Then we subtract Equation (11.41) from Equation (11.40) to obtain

$$v_n = v_{n-1} + K_c \left[(e_n - e_{n-1}) + \frac{T}{\tau_I} e_n \right.$$

$$\left. + \frac{\tau_D}{T} (e_n - 2e_{n-1} + e_{n-2}) \right] \qquad (11.42)$$

Equation (11.42) is referred to as the velocity form of the PID algorithm, because it computes the incremental output instead of the actual output of the controller. The velocity form of the algorithm also provides some protection against reset windup, because it does not incorporate sums of error sequences.

Perhaps this is an appropriate place to point out that the performance of the computer control algorithm depends not only on the tuning constants but also on the sampling period. Indeed, it can be shown[5] that although a second-order conventional control system is stable for all values of the proportional gain constant, computer control of the same system can give unstable response for some specific combinations of the proportional gain and the sampling period. Also, we should remember from our discussion in Chapter 7 that the technical requirements of the sampling theorem must be satisfied.

Using simulation techniques and the IAE performance criteria, Fertik[6] developed controller parameter charts for conventional PI and PID controllers. The tuning constants are based on a process model of the form

$$G_p(s) = \frac{K_p \, e^{-\theta_d s}}{(\tau_1 s + 1)(\tau_2 s + 1)} \tag{11.43}$$

The charts are shown in Figures 11.6 and 11.7. The abscissa for the charts is the process controllability parameter, which is defined as

$$\text{process controllability} = \frac{\theta_d}{\tau_1 + \tau_2 + \theta_d} \tag{11.44}$$

Thus the values of the parameter range from 0 for processes containing no dead time to 1 for pure dead time processes.

For computer control applications the same charts are used, except that the process controllability is calculated according to the equation

$$\text{process controllability} = \frac{\theta_d + T/2}{\tau_1 + \tau_2 + \theta_d + T/2} \tag{11.45}$$

where T = sampling period.

The above calculation corrects for the dead time effects of sampling. Thus the same controller parameter charts, Figures 11.6 and 11.7, can be used for digital PI or PID controllers.

We have included a computer program in Appendix C2 that determines the transient closed-loop response of sampled-data control systems to set point or load changes. The program is based on one or two time constants plus a dead time model. If the parameters of the open-loop process model are available, the reader may enter trial values of the tuning constants, execute the program, plot the load or set point response, and adjust the tuning constants until satisfactory performance (at least by a visual check) is achieved.

The use of impulse response models significantly simplifies the determination of closed-loop responses. A block diagram and a flow chart of a computer program to simulate the control system based on impulse response representation for the process is shown in Figure 11.8. The reader is encouraged to develop a computer program based on this flow chart, execute it for a sample problem, and compare the resulting closed-loop responses with those obtained from the program in Appendix C2.

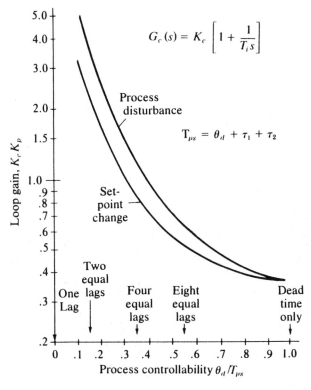

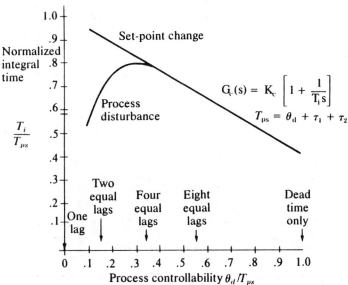

Figure 11.6
Controller Parameter Charts for PI Controllers
(By Permission from Ref. 6)

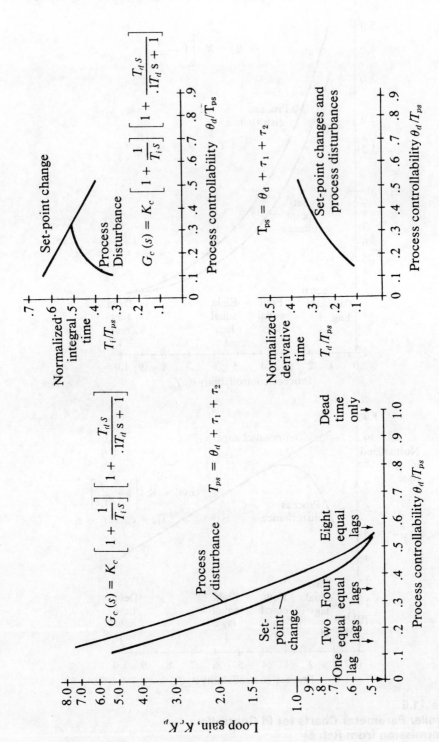

Figure 11.7
Controller Parameter Charts for PID Controllers
(By Permission from Ref. 6)

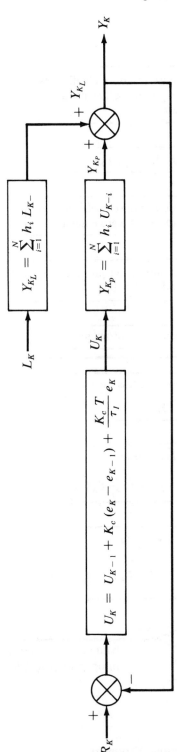

$$Y_{K_L} = \sum_{i=1}^{N} h_i L_{K-}$$

$$Y_{K_p} = \sum_{i=1}^{N} h_i U_{K-i}$$

$$U_K = U_{K-1} + K_c(e_K - e_{K-1}) + \frac{K_c T}{\tau_I} e_K$$

Notes:

1. R_K, L_K are changed at $T = 0$. Thus at $T = 0$, $Y_o = 0$; R_K and/or $L_K \neq 0$, and $U_o \neq 0$.
2. The above representation assumes that $G_L L(z) = G_L(z) L(z)$. And, therefore, it is applicable only for step changes in L.

Figure 11.8a
Block Diagram of Sampled Data Control System with Impulse Response Representation for the Process

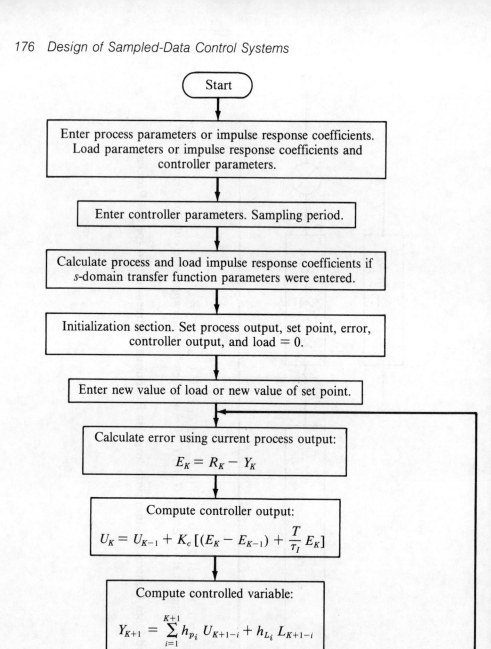

Figure 11.8*b*
Flow Chart of Closed-loop Response Program

11.4 Smith Predictor Algorithm[7]

The Smith Predictor algorithm was developed by O. J. M. Smith[7] in 1957. The technique is a model-based approach to better control of systems with long dead times and has come to be known as the Smith predictor. This was one of the advanced control strategies developed years ago that was shelved because of a lack of practical hardware to implement it (it requires a pure dead-time element having a delay equal to that of the process—difficult if not impossible with analog hardware). When digital computers for on-line control applications appeared on the market, the Smith Predictor was "rediscovered" and tried in many applications.

It is possible to apply the direct synthesis methodology discussed in the previous sections to develop the Smith Predictor algorithm. Rather than doing that, we present the original derivation to provide a historical perspective. Furthermore, the IMC algorithm, which could also have been developed by the direct synthesis method, utilized the Smith Predictor as a basis for development. Once the reader has studied the derivation of the Smith Predictor and the IMC algorithm, the derivation of either by the direct synthesis method should pose no difficulty.

Let us develop the Smith Predictor algorithm for a process that can be represented by a first-order lag plus dead time model. The block diagram of a conventional control system for this process is shown in Figure 11.9.

As shown in Figure 11.9, the process is conceptually split into a pure lag and a pure dead time. If the fictitious variable B could be measured somehow, we could connect it to the controller, as shown in Figure 11.10. This would move the dead time outside the loop. The controlled variable C would repeat whatever B did after a delay of θ_d. Since there is no delay in the feedback signal B, the response of the system would be greatly improved. This scheme, of course, cannot be implemented, because B is an unmeasurable (fictitious) signal.

Now, suppose we develop a model of the process and apply the manipulated variable M to the model as shown in Figure 11.11. If the model were perfect and $L = 0$ (i.e., no load disturbances are present), then C will equal C_M and $E_M = C - C_M = 0$. The arrangement shown in Figure 11.11 reveals that, although the fictitious process variable B is unavailable, we can get at B_M in the model. The B_M will be equal to B unless modeling errors or load upsets are present. We use B_M as the feedback signal, as shown in Figure 11.12. The difference, $C - C_M$, is the error E_M, which arises because of modeling errors or load upsets.

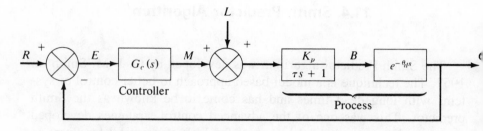

Figure 11.9
Conventional Feedback Loop Having Dead Time

The arrangement shown in Figure 11.12 will control the model well but perhaps not the process, if load upsets occur or if our model is inaccurate. To compensate for these errors, a second feedback loop is implemented using E_M, as shown in Figure 11.13. This is the Smith Predictor control strategy. The $G_c(s)$ is a conventional PI or PID controller, which can be tuned much more tightly because of the elimination of dead time from the loop. A block diagram of the Smith Predictor drawn for computer control applications is shown in Figure 11.14.

The closed-loop transfer function of the system shown in Figure 11.13 for $L = 0$ is

$$\frac{C(s)}{R(s)} = \frac{G_c(s)G_p'(s)e^{-\theta_d s}}{1 + G_c(s)G_M'(s) - G_c(s)G_M'(s)e^{-\theta_M s} + G_c(s)G_p'(s)e^{-\theta_d s}} \tag{11.46}$$

If $G_M'(s) = G_p'(s)$ and $\theta_M = \theta_d$, Equation (11.46) reduces to

$$\frac{C(s)}{R(s)} = \frac{G_c(s)G_p'(s)}{1 + G_c(s)G_p'(s)} e^{\theta_d s} \tag{11.47}$$

Equation 11.47 is the closed-loop transfer function of the desired configuration, based on the fictitious signal B, which was shown earlier in

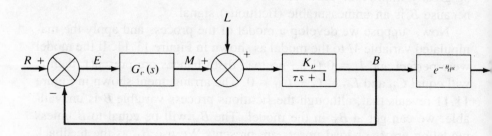

Figure 11.10
Desired Configuration of the Feedback Loop

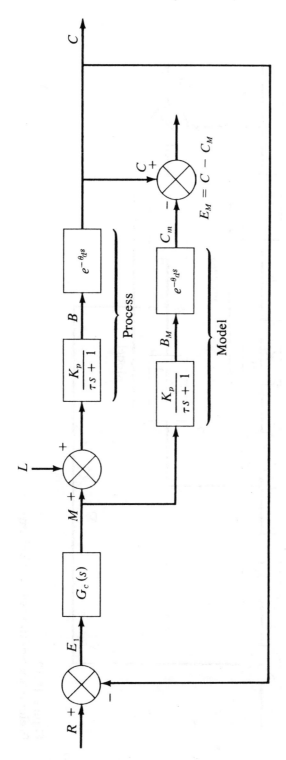

Figure 11.11
Feedback Arrangement Incorporating a Process Model

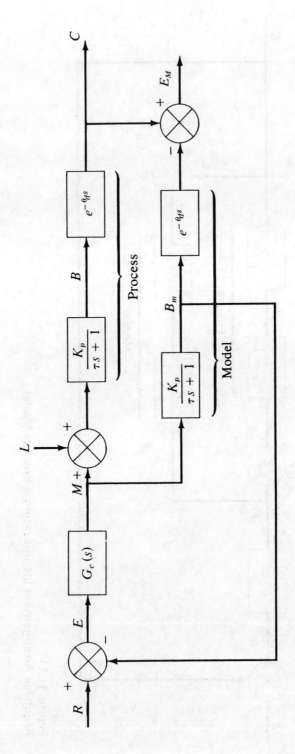

Figure 11.12
Preliminary Smith Predictor Scheme

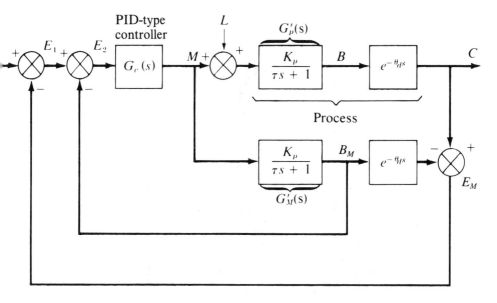

Figure 11.13
Final Smith Predictor Control System

Figure 11.10. To be assured of success, the model parameters must be known to a high degree of accuracy.

A comparison of Equation (11.47) with the closed-loop transfer function of the system of Figure 11.9 will show that the Smith Predictor strategy has removed the dead time from the characteristic equation. Thus, stability is improved and the controller can be tuned more tightly.

As an example, let us consider the design of a Smith Predictor algorithm for computer control of a first-order process having dead time whose transfer function is given by

$$G_p(s) = \frac{K_p e^{-\theta_d s}}{\tau s + 1} \tag{11.48}$$

where

$$K_p = \text{process gain}$$

$$\theta_d = \text{dead time}$$

$$\tau = \text{process time constant}$$

The Smith Predictor scheme for this computer control application is shown in Figure 11.14. The implementation of this scheme would require that we develop equations for B_M, C_M, and U_K.

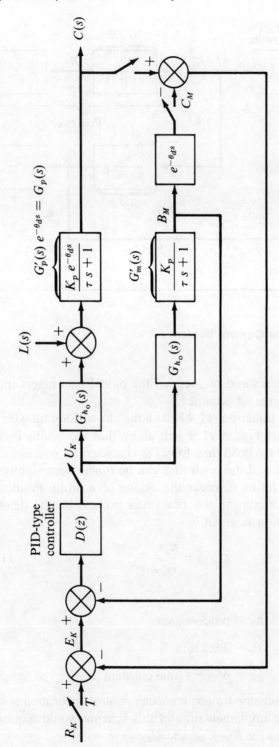

Note: The diagram assumes perfect process modeling.

Figure 11.14
Smith Predictor Block Diagram for Computer Control Applications

From Figure 11.14 observe that the model output C_M is related to the input U by

$$\frac{C_M(Z)}{U(Z)} = Z\{G_{h_0}(s)G_p(s)\} \tag{11.49}$$

where

$C_M(Z) = Z$ transform of model output

$U(Z) = Z$ transform of input

$G_{h_0}(s) = $ transfer function of zero-order hold, $\dfrac{1 - e^{-sT}}{s}$

$G_p(s) = $ transfer function of process, $\dfrac{K_p e^{-\theta_d s}}{\tau s + 1}$

Substituting the transfer functions in Equation (11.49) we get

$$\frac{C_M(Z)}{U(Z)} = Z\left\{\frac{1 - e^{-sT}}{s} \cdot \frac{K_p e^{-\theta_d s}}{\tau s + 1}\right\} \tag{11.50}$$

If we denote the integral number of sampling periods in dead time as N, then the following equality holds:

$$\theta_d = (N + \beta)T \tag{11.51}$$

where $\beta = $ a fraction between 0 and 1. With this expression for θ_d, Equation (11.50) can be written as

$$\frac{C_M(Z)}{U(Z)} = Z\left\{\frac{1 - e^{-sT}}{s} \cdot \frac{K_p e^{-(N + \beta)T/s}}{\tau s + 1}\right\} \tag{11.52}$$

Taking Z transform of Equation (11.52) gives

$$\frac{C_M(Z)}{U(Z)} = K_p Z^{-N}(1 - Z^{-1})Z^{-1}\left(\frac{1}{1 - Z^{-1}} - \frac{e^{-(1-\beta)T/\tau}}{1 - e^{-T/\tau}Z^{-1}}\right) \tag{11.53}$$

Cross multiplying and inverting gives an equation for the model output containing time delay. Thus,

$$C_{M,K} = A_2 C_{M,(K-1)} + A_2 K_p\left(\frac{1}{A_3} - 1\right) U_{K-(N+2)} + K_p\left(1 - \frac{A_2}{A_3}\right) U_{K-(N+1)} \tag{11.54}$$

where

$$A_2 = e^{-T/\tau}$$

$$A_3 = e^{-\beta T/\tau}$$

Now, we need to get the model output containing no dead time. For this case,

$$\frac{B_M(Z)}{U(Z)} = Z\left\{ \frac{1 - e^{-sT}}{s} \cdot \frac{K_p}{\tau s + 1} \right\}$$

$$= K_p(1 - Z^{-1})\left[\frac{1}{1 - Z^{-1}} - \frac{1}{1 - e^{-T/\tau}Z^{-1}} \right]$$

(11.55)

Cross multiplying and inverting gives

$$B_{M,K} = K_p(1 - A_2)U_{K-1} + A_2 B_{M,K-1}$$

(11.56)

From Figure 11.14 the error is computed from the equation

$$E_K = R_K - C_K - (B_{M,K} - C_{M,K})$$

(11.57)

Note that if the model is perfect and L is zero,

$$C_K = C_{M,K}$$

(11.58)

and the input to the controller will be based on

$$E_K = R_K - B_{M,K}$$

(11.59)

Thus, dead time has been compensated. The digital controller must be a conventional control algorithm (i.e., P, PI, or PID). If the controller is based on Z transforms, it incorporates the effect of dead time, and the Smith Predictor algorithm is not applicable.

Experimental Applications

A few applications of the Smith Predictor algorithm have appeared in the literature. Meyer, et al.[8] have listed these references in their paper. They have described an application of the Smith Predictor algorithm for automatic control of top product composition in a distillation column. A schematic of the pilot-scale distillation control system is shown in Figure 11.15. The process transfer function relating the top product composition X_D to feed rate F and reflux R_e is

$$X_D(s) = \frac{e^{-60s}}{1002s + 1} R_e(s) + \frac{0.167\, e^{-486s}}{895s + 1} F(s)$$

(11.60)

The delay and time constants are given in seconds. The parameters of the digital PI controller were tuned according to the relations

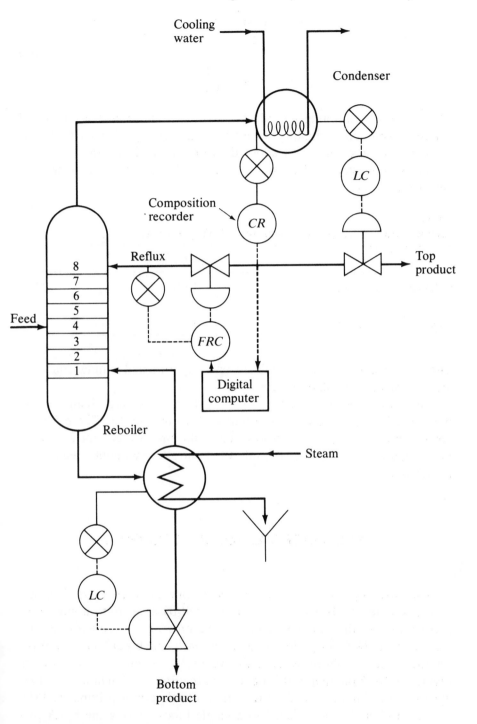

Figure 11.15
Schematic Diagram of the Distillation Column

$$K_c = \frac{0.984}{K_p}\left(\frac{T}{\tau}\right)^{-0.986}$$

$$\tau_I = 1.644\tau\left(\frac{T}{\tau}\right)^{0.707}$$

(11.61)

where T is the sampling period. The authors found that the controlled variable exhibited oscillatory behavior, and, hence, a further trial-and-error tuning procedure was adopted. Table 11.1 shows the controller constants suggested by Equation (11.61) and those adopted in the experimental work.

Table 11.1.
Comparison of Controller Settings of Meyer et al.

	K_c, grams/s/%	τ_1, sec
Calculated according to Equation 14.16	15.8	225
Experimentally tuned	10.0	250

The differences between calculated and experimental tuning constants are indicative of modeling errors.

The experimental responses of the system to a step change in set point and load are sketched in Figures 11.16 and 11.17. These figures clearly show the superior performance of the Smith Predictor algorithm in comparison with the digital PI controller without dead time compensation.

11.5 INTERNAL MODEL CONTROL

The IMC algorithm can be derived in one of three ways. The approach adopted by Brosilow[10] was motivated by a desire to improve the performance of the Smith Predictor in the presence of modeling errors. To derive the IMC algorithm by this approach, the block diagram of the Smith Predictor control system, Figure 11.14, is redrawn as shown in Figure 11.18. Note that in this depiction, the process dynamics and load dynamics need not be equal. Now, the feedback loop involving the PID-type controller can be reduced to a single block, giving the block diagram in Figure 11.19. It is emphasized that Figures 11.14, 11.18, and

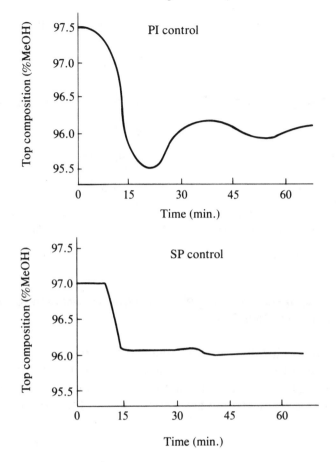

Figure 11.16
Experimental Response Using PI and SP Control System to 1%
Decrease in Composition Set Point

11.19 are entirely equivalent. In the classical Smith Predictor strategy
the controller G_I of Figure 11.19 is given by the transfer function

$$G_I(Z) = \frac{G_c(Z)}{1 + G_c(Z)\,G_M'(Z)} \qquad (11.62)$$

where $G_c(Z)$ represents the PID-type controller and $G_M'(Z)$ is the pulse
transfer function of the process model without the dead time. Due to
this particular choice of $G_I(Z)$ given in Equation (11.62) involving a
PID-type controller, the system response would not be good in the presence of modeling errors. In the following paragraphs it will be shown
that a better choice for $G_I(Z)$ is available. The block diagram shown in
Figure 11.19 depicts the IMC structure.

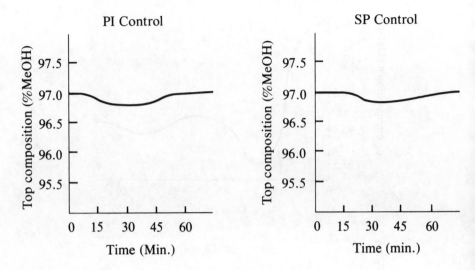

Figure 11.17
Experimental Response Using PI and SP Control Scheme to 22%
Decrease in Feed Flow Rate

In the approach followed by Garcia and Morari[11] one begins with the block diagram of a typical SISO sampled-data control system as shown in Figure 11.20. The transfer function, $G(Z)$, as shown in Figure 11.20, includes all the elements that are associated with the process block in a

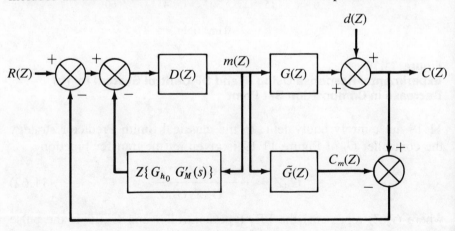

Note: Samplers are not shown for brevity, $G(Z) = Z\{G_{h_0}(s)\ G_\rho(s)\}$ and ~ indicate that the term pertains to a model.

Figure 11.18
Alternate Depiction of the Smith Predictor Control Strategy

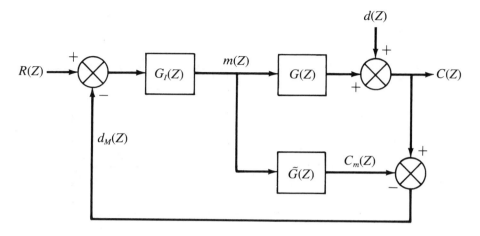

Figure 11.19
Basic IMC Structure

sampled-data control loop. The algorithm $D(Z)$ may be of the PID type
or it may be one of the Z-transform-based control algorithms.

Now, let us make use of the mathematical model of the process $\tilde{G}(Z)$
and add the effect of input m upon it to yield a new system, shown in
Figure 11.21. It is obvious that the systems in Figures 11.20 and 11.21
are entirely equivalent. The feedback loop around $D(Z)$ can be simplified
by using the familiar closed-loop transfer function relationship

$$G_I(Z) = \frac{D(Z)}{1 + D(Z)\,\tilde{G}(Z)} \tag{11.63}$$

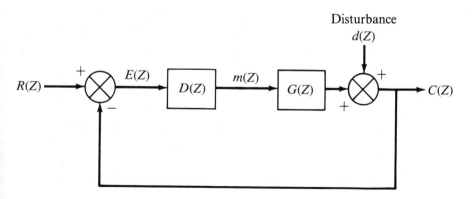

Figure 11.20
Typical Sampled-data Control System

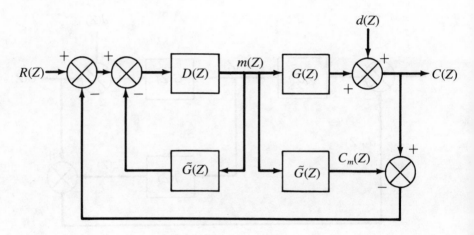

Figure 11.21
Equivalent Model-based Control System

The resulting block diagram is shown in Figure 11.22. Figure 11.22 depicts the basic IMC structure. One can naturally ask, if Figures 11.20 and 11.22 are entirely equivalent, then what is the advantage of the IMC representation? Garcia and Morari and Brosilow have shown that the IMC algorithm, $G_I(Z)$ is much easier to design than $D(Z)$ and that the IMC structure permits us to include robustness as a design objective in a very explicit manner. This can be seen by examining the feedback signal, $d_M(Z)$ in Figure 11.22, which is

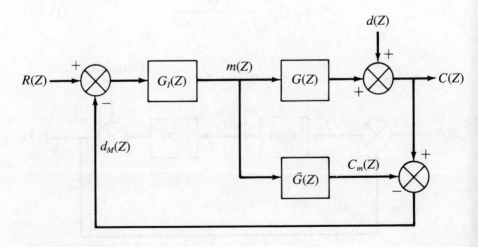

Figure 11.22
Basic IMC Structure

$$d_M(Z) = [1 + \{G(Z) - \tilde{G}(Z)\} G_I(Z)]^{-1} d(Z) \qquad (11.64)$$

If $G(Z) = \tilde{G}(Z)$ (that is, the model is perfect), then the feedback signal is simply $d(Z)$. Consequently, the control loop is effectively an open-loop system and, therefore, stability is not an issue. In the presence of plant-model mismatch, $d_M(Z)$ may be suitably modified to obtain robustness.

To design the IMC controller, one begins with the following transfer functions derived from Figure 11.22:

$$M(Z) = \frac{G_I(Z)}{1 + G_I(Z)[G(Z) - \tilde{G}(Z)]} [R(Z) - d(Z)] \qquad (11.65)$$

and

$$C(Z) = d(Z) + \frac{G(Z) G_I(Z)}{1 + G_I(Z)[G(Z) - \tilde{G}(Z)]} [R(Z) - d(Z)] \qquad (11.66)$$

For stability the necessary and sufficient condition is that the roots of the following two characteristic equations must lie inside the unit circle in the Z-plane.

$$\frac{1}{G_I(Z)} + [G(Z) - \tilde{G}(Z)] = 0 \qquad (11.67)$$

and

$$\frac{1}{G_I(Z) G(Z)} + \frac{1}{G(Z)} [G(Z) - \tilde{G}(Z)] = 0 \qquad (11.68)$$

In the absence of a plant-model mismatch, these equations reduce to

$$\frac{1}{G_I(Z)} = 0 \qquad (11.69)$$

and

$$\frac{1}{G_I(Z) G(Z)} = 0 \qquad (11.70)$$

Thus, when $G(Z) = \tilde{G}(Z)$, the controller and process poles must be inside the unit circle in the Z-plane. For an open-loop stable plant, therefore, stability (but not necessarily robustness) is ensured for any $G_I(Z)$ that is stable.

Now on to the best choice of $G_I(Z)$. Equation (11.66) shows that, in the absence of modeling errors, perfect set point control can be achieved by setting

$$G_I(Z) = \frac{1}{\tilde{G}(Z)} \tag{11.71}$$

The process transfer function, $\tilde{G}(Z)$, in general, can contain time delay and zeroes outside the unit circle. The inversion indicated in Equation (11.71) in such cases can lead to an unrealizable pure predictor and/or an unstable controller that would violate the stability requirements, Equations (11.69) and (11.70). Therefore, it is generally not possible to base the design of $G_I(Z)$ on the exact process inverse. Instead, the next best thing is to split the model into three parts, one containing the time delay, another containing the zeroes outside the unit circle, and a third containing the remaining terms of $\tilde{G}(Z)$, giving a factorization

$$\tilde{G}(Z) = \tilde{G}_+(Z)\,\tilde{G}_{+1}(Z)\,\tilde{G}_-(Z) \tag{11.72}$$

where $\tilde{G}_+(Z)$ contains the time delay and $\tilde{G}_{+1}(Z)$ accommodates the zeroes outside the unit circle.

If the process contains a dead time of N sampling instants, then, $\tilde{G}_+(Z)$ should be chosen according to

$$\tilde{G}_+(Z) = Z^{-(N+1)} \tag{11.73}$$

In the absence process dead time,

$$\tilde{G}_+(Z) = Z^{-1} \tag{11.74}$$

Equation (11.74) reflects the delay inherent in the sampling process.

If the plant transfer function contains zeroes outside the unit circle, $\tilde{G}_{+1}(Z)$ may be selected according to

$$\tilde{G}_{+1}(Z) = \prod_{=1} \left(\frac{Z - v_i}{Z - \bar{v}_i}\right)\left(\frac{1 - \bar{v}_i}{1 - v_i}\right) \tag{11.75}$$

where v_i are p zeroes of $\tilde{G}(Z)$ and

$$\bar{v}_i = v_i \quad \text{for} \quad |v_i| \le 1$$
$$\bar{v}_i = 1/v_i \quad \text{for} \quad |v_i| > 1$$

If the system contains a single zero outside the unit circle, then Equation (11.75) reduces to

$$\tilde{G}_{+1}(Z) = \frac{v - Z}{vZ - 1} \tag{11.76}$$

If the system contains no zeroes, then

$$\tilde{G}_{+1}(Z) = 1 \tag{11.77}$$

Once the factorization process is completed, the IMC controller is designed according to

$$G_I(Z) = \frac{1}{\tilde{G}_-(Z)} \qquad (11.78)$$

Equation (11.78) gives the so-called perfect IMC controller, which is based on a 100 percent accurate process model. Robustness in the presence of modeling errors may be achieved by inserting a filter in the feedback path as shown in Figure 11.23. With the filter in the line, the characteristic equation (Equation 11.68) becomes

$$G_I(Z)^{-1} + F(Z)[G(Z) - \tilde{G}(Z)] = 0 \qquad (11.79)$$

For a given plant-model mismatch $F(Z)$ is selected such that all the roots of Equation (11.79) lie inside the unit circle in the Z-plane.

Garcia and Morari[11] state that a first-order filter

$$F(Z) = \frac{1 - \alpha_f}{1 - \alpha_f Z^{-1}} \qquad (11.80)$$

with $0 \le \alpha_f < 1$, may be used to stabilize the closed-loop system for a given plant-model mismatch. The filter constant α_f is given by

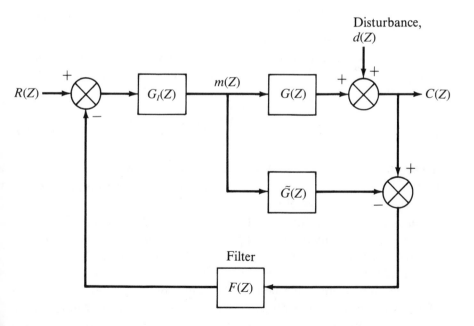

Figure 11.23
IMC with Filter

$$\alpha_f = e^{-T/\tau_f} \tag{11.81}$$

where T is the sampling period and τ_f is the filter time constant. The filter constant can serve as a tuning parameter: very small values of α_f close to zero improve dynamic performance, but the system becomes very sensitive to modeling errors; while large values of α_f close to 1 improve robustness, but the response becomes very sluggish. Thus, a suitable compromise must be found between robustness and dynamic performance. In specific applications, higher-order filters may yield better results (Rivera, et al.).[12]

As an illustration of the filtering concepts, consider a first-order process containing a dead time of one sampling instant, while its model, accurate in terms of gain and time constant, contains no dead time.[11] Then,

$$G(Z) = Z^{-2} G_-(Z)$$
$$\tilde{G}(Z) = Z^{-1} G_-(Z) \tag{11.82}$$

where $G_-(Z)$ is the delay-free model. For this case

$$G_I(Z)^{-1} = G_-(Z)$$

Substitution in Equation (11.68) gives

$$G(Z) [1 + (Z^{-2} - Z^{-1})] = 0 \tag{11.83}$$

Equation (11.83) has two roots on the unit circle at $Z_1, Z_2 = 0.5 \pm 0.866i$ and, consequently, the system is on the verge of instability. Now, if a first-order filter, Equation (11.80), is introduced, Equation (11.79) becomes

$$G_-(Z) \left[1 + \frac{1 - \alpha_f}{1 - \alpha_f Z^{-1}} (Z^{-2} - Z^{-1}) \right] = 0 \tag{11.84}$$

having the roots $Z_1, Z_2 = 0.5 \pm 0.5 \sqrt{4\alpha_f - 3}$. A value of α_f in the range of $0 < \alpha_f < 1$ gives new roots that are inside the unit circle, and the system has been stabilized.

In addition to the filter block one can make the system follow a set point trajectory by inserting an appropriate reference model block, $H^R(Z)$, into the control structure. The complete IMC structure is shown in Figures 11.24(a) and (b). In the latter, the filter block and the reference trajectory block are combined into a single block. Although the filter block is meant to ensure robustness in the presence of modeling errors, it can also compensate for certain types of disturbance dynamics when no modeling errors are present.

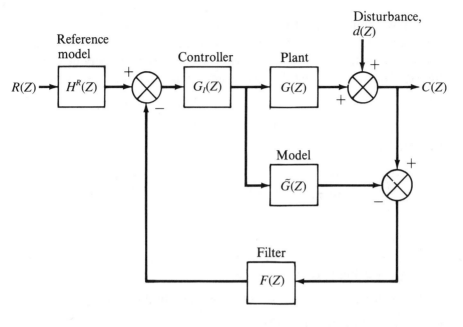

Figure 11.24*a*
Complete IMC Structure with Filter and Reference Trajectory

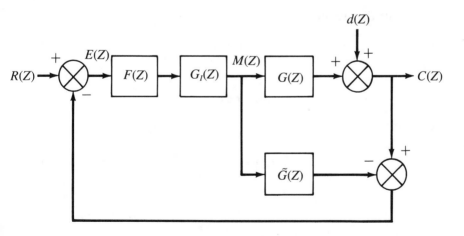

Figure 11.24*b*
Block Diagram of SMPC in IMC Form with Filter

Applications of IMC

A few applications of IMC have appeared in the literature (see, e.g., References 13, 14, 15, 16). The application reported by Arkun et al.,[13] involves a steam-heated heat exchanger. The experimentally determined transfer function relating the exit water temperature to steam flow is

$$\frac{Y(s)}{M(s)} = \tilde{G}_p(s) = \frac{3.5e^{-3s}}{10s + 1} \tag{11.85}$$

For this case

$$\tilde{G}(Z) = \frac{Y(Z)}{M(Z)} = Z\{G_{h_0}(s)\,\tilde{G}_p(s)\} = z^{-(N+1)}\frac{3.5(1 - \bar{\alpha})}{1 - \bar{\alpha}Z^{-1}} \tag{11.86}$$

where

$$\bar{\alpha} = e^{-T/10}$$

T = sampling period, 0.3 time units

N = integer multiple of sampling periods in dead time

Referring to Equation (11.72),

$$\tilde{G}(Z) = \tilde{G}_+(Z)\,\tilde{G}_{+1}(Z)\,\tilde{G}_-(Z) = Z^{-(N+1)}\frac{3.5\,(1 - \bar{\alpha})}{1 - \bar{\alpha}Z^{-1}} \tag{11.87}$$

As per our discussion,

$$\tilde{G}_+(Z) = Z^{-(N+1)} \tag{11.88}$$

and

$$\tilde{G}_{+1}(Z) = 1 \tag{11.89}$$

Therefore,

$$\tilde{G}_-(Z) = \frac{3.5\,(1 - \bar{\alpha})}{1 - \bar{\alpha}Z^{-1}} \tag{11.90}$$

and

$$G_I(Z) = \frac{1 - \bar{\alpha}Z^{-1}}{3.5(1 - \bar{\alpha})} \tag{11.91}$$

In this application $H^R(Z)$ has been selected to be the same as $F(Z)$ and, therefore, the block diagram of the control system appears as shown in Figure 11.24(b). Now,

$$\frac{M(Z)}{E(Z)} = F(Z)\,G_f(Z) \tag{11.92}$$

Using Equation (11.80) for $F(Z)$ we obtain

$$\frac{M(Z)}{E(Z)} = \frac{1 - \alpha_f}{1 - \alpha_f Z^{-1}} \cdot \frac{1 - \bar{\alpha}Z^{-1}}{3.5(1 - \bar{\alpha})} \tag{11.93}$$

Crossmultiplying the terms shown in Equation (11.93) and inverting the result gives the control algorithm in the time domain:

$$M(k) = \frac{1 - \alpha_f}{3.5(1 - \bar{\alpha})}\,[e(k) - \bar{\alpha}e(k - 1)] + \alpha_f M(k - 1) \tag{11.94}$$

where

$$e(k) = y^{\text{set}}(k) - [y(k) - y_M(k)] \tag{11.95}$$

The expression for the model output, $y_M(k)$, is obtained by crossmultiplying the terms in Equation (11.86) as

$$y_M(k) = \bar{\alpha}y_M(k - 1) + 3.5(1 - \bar{\alpha})\,M(k - N - 1) \tag{11.96}$$

It has been emphasized that the input M to the model in Equation (11.96) should be the actual (possibly constrained) input to the system. This feature provides antireset windup automatically.[11]

The results of this application are shown in Figure 11.25. The effects of the filter time constant τ_f for set point and load changes are shown in Figures 11.25(a) and 11.25(b) for both directions. The trade-off between speed of response and stability is apparent. Step changes in water flow rate were used to test the effect of modeling errors. The results for gain mismatch are shown in Figure 11.25(c) and those for time delay mismatch are shown in Figure 11.25(d).

Chawla has shown that the direct synthesis method can be used to derive Equation (11.78) and the IMC structure. In this approach one begins with the closed-loop response characteristic of the form given in Equation (11.4). Substitution in Equation (11.3) gives

$$D(Z) = \frac{1}{\tilde{G}(Z)} \frac{F(Z)\,G_+(Z)}{1 - F(Z)\,G_+(Z)} \tag{11.97}$$

The process pulse transfer function is represented in the form

$$\tilde{G}(Z) = G_-(Z)\,G_+(Z) \tag{11.98}$$

In view of Equation (11.98) the control algorithm, Equation (11.97), can be written as

$$D(Z) = \frac{F(Z)\,G_I(Z)}{1 - F(Z)\,G_I(Z)\,\tilde{G}(Z)} \qquad (11.99)$$

where

$$G_I(Z) = \frac{1}{G_-(Z)}$$

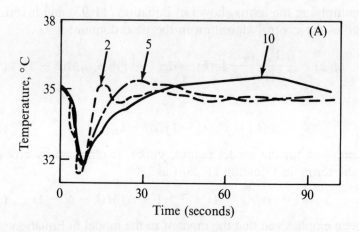

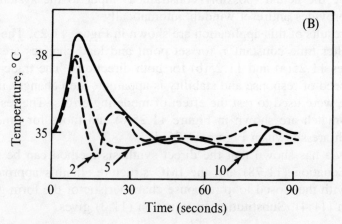

Notes:

A. Water flow changes from 15.6 to 19.8 kg/min
B. Water flow changes from 19.8 top 15.6 kg/min

Figure 11.25*a*
Effect of Filter Time Constants on Load Responses (*T* = 0.3 sec)
(Reprinted by permission from "Experimental Study of Internal
Model Control," *Ind. Eng. Chem. Proc. Des. Dev.,* **vol. 25, no. 1, pp.**
102–108, © **1986, American Chemical Society.)**

The control law is shown in block diagram form in Figure 11.26(a) or equivalently in Figures 11.26(b) or 11.26(c). Figure 11.26(c) represents the IMC structure.

If $F(Z)$ is selected to be a first-order lag, Equation (11.22), it may be observed that the IMC algorithm is the same as the Dahlin algorithm.

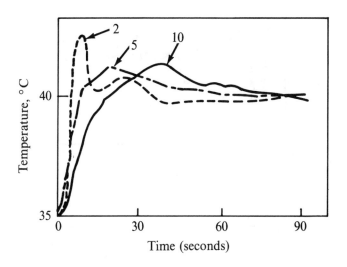

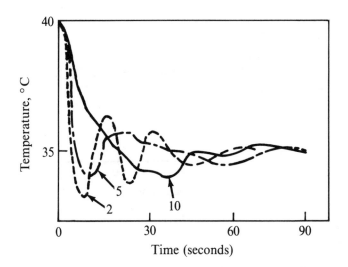

Figure 11.25*b*
Effect of Filter Time Constants for Set Point Changes in Exit Water Temperature (T = 0.3 sec)

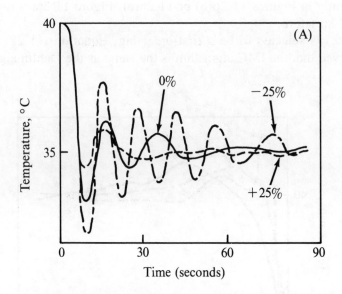

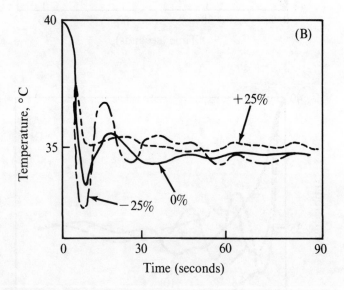

Notes:
A. $\tau_f = 2.25$ sec
B. $\tau_f = 4.5$ sec

Figure 11.25*c*
Effect of Gain Mismatch on Dynamic Performance

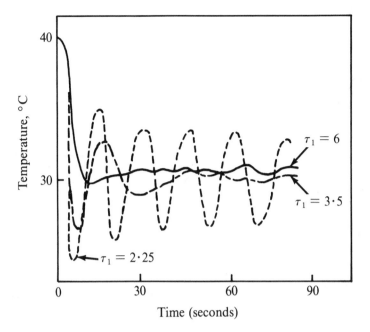

Figure 11.25d
Effect of Time Delay Mismatch on Dynamic Performance
for Different Filter Time Constants (Model Delay is 40% Smaller than
Process Delay)

11.6 Analytical Predictor Algorithm[9]

In the Smith Predictor algorithm the process model was used to compensate for the dead time prior to sending the signal to the controller. An alternate approach would be to use a process model to "predict" the future value of the controlled variable and use the predicted value as the input to the controller. This is the basic notion behind the analytical predictor (AP) algorithm originally proposed by Moore.[9] A block diagram illustrating the analytical predictor concept is shown in Figure 11.27. The analytical predictor predicts the value of the controlled variable T' time units in future from current inputs where T' is the sum of the system dead time plus one-half of the sampling period, that is,

$$T' = \theta_d + \frac{1}{2}T$$
$$T' = (N + \beta)T + 0.5\,T \qquad (11.100a)$$

or

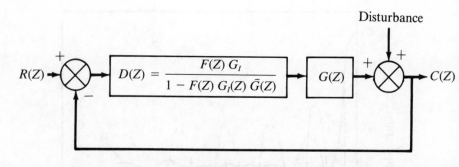

Figure 11.26a
Derivation of IMC Structure by Direct Synthesis Method

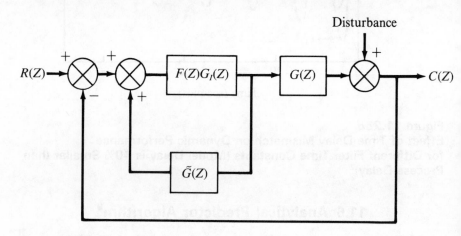

Figure 11.26b
Equivalent Representation of Figure 11.25a

$$\frac{T'}{T} = N + \beta + 0.5 \qquad (11.100b)$$

Control degradation of sampled-data systems occurs because of sampling. In many instances the dynamic effect of sampling is similar to that of pure dead time equal to one-half the sampling period. It is for this reason the analytical predictor predicts the value of the controlled variable over T' units of time, which includes the process dead time and the effect of sampling.

Consider the design of the analytical predictor algorithm for a first-order process with dead time. The differential equation representing this class of models is

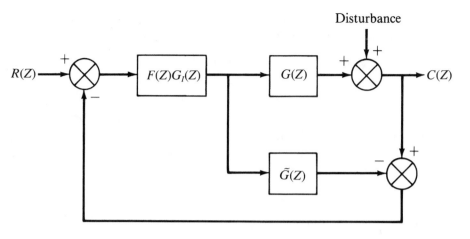

Figure 11.26c
IMC Structure

$$\tau \frac{dC}{dt} + C(t) = K_p U(t - \theta_d) \qquad (11.101)$$

where

τ = process time constant

K_p = process gain

θ_d = total dead time (dead time in process + dead time in measurement line)

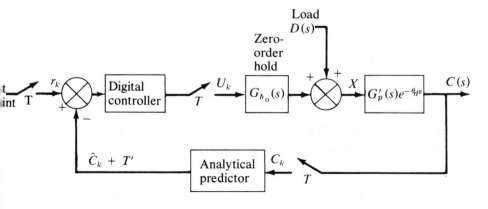

Figure 11.27
Analytical Predictor (AP) Control System

The analytical solution of Equation (11.101) gives, as shown by Moore, the following expression for the predicted output:

$$\hat{C}_{K+T'} = A_1\{A_3 A_2^N C_K + A_2^N K_p(1 - A_3)U_{K-(N+1)} + \tag{11.102}$$

$$K_p(1 - A_2)\sum_{i=1}^{N} A_2^{i-1} U_{K-i}\} + K_p(1 - A_1)U_K$$

where A_2 and A_3 have been defined just below Equation (11.54) and $A_1 = e^{-T/2\tau}$. This predicted output is used to calculate the controller output according to

$$U_K = K_c(r_K - \hat{C}_{K+T'}) \tag{11.103}$$

Equation (11.103) represents a proportional control algorithm and, therefore, an offset

$$\lim_{t \to \infty} R(t) - C(t) = K_c K_p/(K_c K_p + 1)$$

will develop for our system whenever a step change in set point is made. This offset can be eliminated by "calibrating" the set point according to

$$\rho_K = \frac{K_c K_p + 1}{K_c K_p} r_K \tag{11.104}$$

The control algoithm then becomes

$$U_K = K_C(\rho_K - \hat{C}_{K+T'}) \tag{11.105}$$

Moore has also developed algorithms to compensate the analytical predictor algorithm for measurable and unmeasurable disturbances. If the disturbance d_k can be measured, the controller calculations should be done according to

$$U_K = K_c(\rho_K - \hat{C}_{K+T'}) - d_K \tag{11.106}$$

Note from Figure 11.27 that the input to the process is

$$X_K = U_K + d_K \tag{11.107}$$

In the presence of disturbances, the input to the process is X_K. However, if the control calculations are based as shown in Equation (11.106), the effect d_K will be cancelled.

In the more usual situations where disturbances cannot be measured, Moore suggests the following procedure.

If a disturbance is present, it affects the controlled variable C_K. By comparing the predicted value of the controlled variable with C_K, an estimate of the disturbance can be obtained as

$$\hat{d}_K = \hat{d}_{K-1} + \omega T(C_K - \hat{C}_K) \tag{11.108}$$

The constant ω can be used as a tuning parameter to be chosen on line or through simulation. At the instant when the loop is switched to computer control, d_0 is assumed to be zero, and d_1 is calculated as

$$\hat{d}_1 = \omega T(C_K - \hat{C}_K) \tag{11.109}$$

Thereafter, d_K is computed from Equation (11.108). The predicted value $\hat{C}_K$ is given by

$$\hat{C}_K = A_2 C_{K-1} + K_p\left(1 - \frac{A_2}{A_3}\right)(U_{K-(N+1)} + \hat{d}_{K-1}) + \tag{11.110}$$
$$K_p\left(\frac{A_2}{A_3}\right)(1 - A_3)(U_{K-(N+2)} + \hat{d}_{K-1})$$

The estimated disturbance $\hat{d}_K$ can be incorporated into Equation (11.102) to provide a better prediction of the future output. Thus,

$$\hat{C}_{K+T'} = A_1 A_3 A_2^N C_K + A_2^N K_p(1 - A_3)(U_{K-(N+1)} + \hat{d}_K) + $$
$$K_p(1 - A_2) \sum_{i=1}^{N} A_2^{i-1}(U_{K-i} + \hat{d}_K) + K_p(1 - A_1)(U_K + \hat{d}_K) \tag{11.111}$$

The controller equation for this case is

$$U_K = K_c(\rho_K - \hat{C}_{K+T'}) - \hat{d}_K \tag{11.112}$$

Substituting from Equation (11.104) for ρ_K and from Equation (11.111) for $\hat{C}_{K+T'}$, we get

$$U_K = K_c\left\{\frac{K_c K_p + 1}{K_c K_p} r_K - A_1 A_3 A_2^N C_K - \right.$$
$$A_1 A_2^N K_p(1 - A_3)(U_{K-(N+1)} - \hat{d}_K) - $$
$$A_1 K_p(1 - A_2) \sum_{i=1}^{N} A_2^{i-1}(U_{K-i} + \tag{11.113}$$
$$\left. \hat{d}_K) - K_p(1 - A_1)(U_K - \hat{d}_K)\right\} - \hat{d}_K$$

Simplifying this equation gives

$$U_K = C_1 r_K - C_2 C_K - C_3(U_{K-(N+1)} + \hat{d}_K) - \tag{11.114}$$
$$C_4 \sum_{i=1}^{N} A_2^{i-1}(U_{K-i} + \hat{d}_K) - C_5 \hat{d}_K - \hat{d}_K$$

where

$$C_1 = \frac{K_c K_p + 1}{K_p[1 + K_c K_p(1 - A_1)]}; \qquad C_2 = \frac{K_c A_1 A_3 A_2^N}{1 + K_c K_p(1 - A_1)}$$

$$C_3 = \frac{K_c K_p A_1 A_2^N(1 - A_3)}{1 + K_c K_p(1 - A_1)}; \qquad C_4 = \frac{K_c K_p A_1(1 - A_2)}{1 + K_c K_p(1 - A_1)}$$

and

$$C_5 = \frac{K_c K_p(1 - A_1)}{1 + K_c K_p(1 - A_1)}$$

In this final form of the algorithm there are terms involving current measurements, past controller outputs, set point, and disturbance. The term $\hat{d}_k$ is computed via Equations (11.108) and (11.110). Note that in Equation (11.114) the presence of d_k provides a form of integral action, since the value of the estimated disturbance will continue to change with time until the measured value of the controlled variable C_k matches the predicted value $\hat{C}_k$.

Moore suggests the deadbeat tuning procedure for K_c and ω. The equations are

$$K_c = \frac{A_1}{K_p(1 - A_1)}$$

$$\omega = \frac{1}{T K_p(1 - A_2)}$$

(11.115)

Experimental Applications

Meyer et al.[8] have surveyed the experimental applications of the analytical predictor algorithm. They have applied the analytical predictor technique to the automatic control of a pilot-scale distillation column. The transfer function of the process is given in Equation (11.60). In this case as well, the authors found that further on-line tuning of K_c and ω was necessary. Table 11.2 shows the calculated and experimentally determined constants. Again, the differences in tuning constants are indicative of modeling errors. The response of the system to a step change in set point and load is shown in Figures 11.28 and 11.29. These Figures indicate that the AP control scheme performs better than the PI controller.

**Table 11.2
Comparison of Calculated and Experimentally
Determined Controller Constants**

	K_c, *grams/sec/%*	ω, *grams/%/sec²*
Calculated according to Moore	32.9	0.29
Experimental values	10	0.06

11.7 Simplified Model Predictive Control (SMPC)[18,19,20]

In IMC design methodology one begins with the premise of perfect set point control, then backs off and designs the control law that gives achievable performance. In SMPC one begins at the other end of the spectrum with the assumption that the control algorithm must be at least good enough to give open-loop performance. This is tantamount to solving the *minimum energy* problem. Once the control algorithm is derived, a tuning parameter is introduced to speed up the response to such an extent that a user-defined performance measure is satisfied. The derivation refers to a typical sampled-data control system shown in Figure 11.20.

In this instance

$$\frac{C(Z)}{R(Z)} = \frac{1}{K_p} \tilde{G}(Z) \tag{11.116}$$

The gain of the process, K_p, is inserted in the denominator of the right-hand side of Equation (11.116) to insure that C/R has a gain of one. Substituting into Equation (11.3) gives

$$D(Z) = \frac{C(Z)/R(Z)}{1 - C(Z)/R(Z)} \cdot \frac{1}{\tilde{G}(Z)} \tag{11.117}$$

$$= \frac{\tilde{G}(Z)/K_p}{1 - \frac{\tilde{G}(Z)}{K_p} \cdot \tilde{G}(Z)} \cdot \frac{1}{\tilde{G}(Z)}$$

or

$$D(Z) = \frac{M(Z)}{E(Z)} = \frac{1}{K_p - \tilde{G}(Z)} \tag{11.118}$$

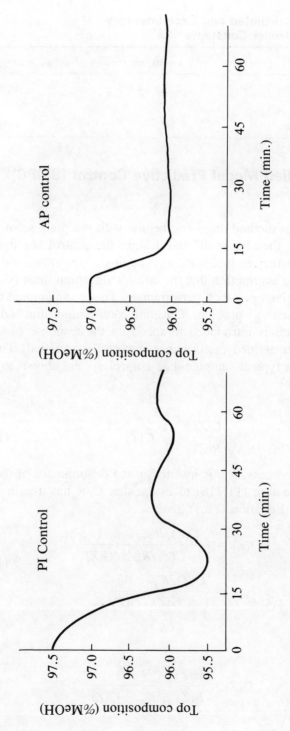

Figure 11.28
**Experimental Response using PI and AP Control System to 1%
Decrease in Composition Set Point**

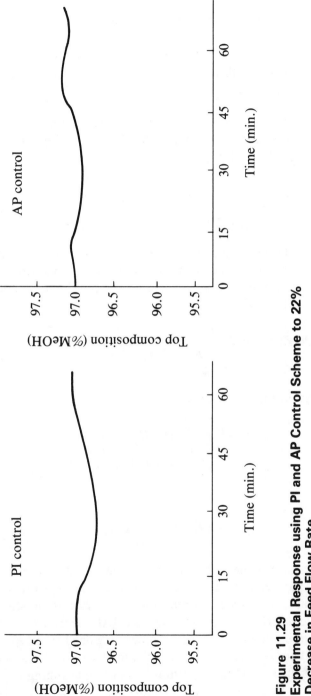

Figure 11.29
**Experimental Response using PI and AP Control Scheme to 22%
Decrease in Feed Flow Rate**

Crossmultiplying the terms in Equation (11.118) gives

$$M(Z) = \frac{1}{K_p} E(Z) + \frac{1}{K_p} \tilde{G}(Z) M(Z) \tag{11.119}$$

As was seen in Chapter 8, the pulse transfer function of the process model can be written with the aid of impulse response coefficients as

$$\tilde{G}(Z) = h_1 Z^{-1} + h_2 Z^{-2} + h_3 Z^{-3} + \ldots + h_n Z^{-n} \tag{11.120}$$

where n represents the number of data points used to represent the open-loop step response curve. Substitution for $\tilde{G}(Z)$ from Equation (11.120) into Equation (11.119) gives

$$M(Z) = \frac{1}{K_p} E(Z) + \frac{1}{K_p} [h_1 Z^{-1} M(Z) + h_2 Z^{-2} M(Z) \\ + h_3 Z^{-3} M(Z) + \ldots + h_n z^{-n} M(Z)] \tag{11.121}$$

Equation (11.121), in the absence of modeling errors, will give a closed-loop set point response that is the same as the normalized open-loop response. The response can be speeded up by introducing a constant in Equation (11.121), giving

$$M(Z) = \alpha E(Z) + \frac{1}{K_p} [h_1 Z^{-1} M(Z) + h_2 Z^{-2} M(Z) \\ + h_3 Z^{-3} M(Z) + \ldots + h_n z^{-n} M(Z)] \tag{11.122a}$$

or equivalently in the time domain

$$M_k = \alpha E_k + \frac{1}{K_p} [h_1 M_{k-1} + h_2 M_{k-2} \\ + h_3 M_{k-3} + \cdot + \cdot + h_n M_{k-n}] \tag{11.122b}$$

or

$$M_k = \alpha E_k + \frac{1}{K_p} \sum_{k=1}^{n} h_i M_{k-i}$$

Equation (11.122b) is the final form of the SMPC algorithm. The term α is a tuning parameter that can be determined through off-line simulation such that a user-selected optimization criterion is satisfied. Some commonly used performance measures are IAE, ISE, and ITAE. We will present tuning guidelines for first-order plus dead time types of processes a little later in this section. The parameter αK_p varies between 1 and ∞. Note that the SMPC algorithm has only one tuning constant as opposed to up to three for PID-type controllers. Furthermore, the memory requirements of this digital control algorithm are modest and the algorithm can be applied to processes with inverse response characteristics. SMPC cannot, however, be directly applied to open-loop unstable processes.

Theoretical Properties of the SMPC Algorithm

The SMPC algorithm has the following important properties:

1. *Open-loop characteristics.* For first-order plus dead time types of process models, the open-loop step response of the algorithm is similar to that of a PI controller: the controller output, following a step change in error, suddenly increases (or decreases), which is followed by a ramp. An illustrative plot of the controller output for one of the simulated processes studied is shown in Figure 11.30.

2. *No reset problem.* It will be shown that the algorithm guarantees offset-free performance in the presence of modeling errors. Consider the closed-loop transfer function of the system to load changes:

$$\frac{C}{d} = \frac{1}{1 + GD} \qquad (11.123)$$

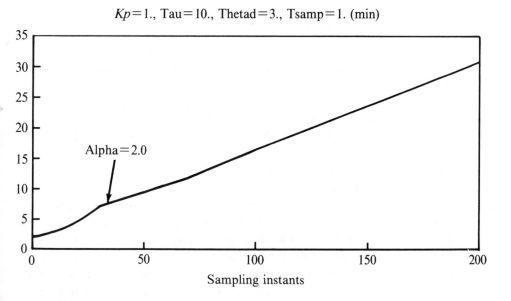

$Kp=1.,\ Tau=10.,\ Thetad=3.,\ Tsamp=1.\ (min)$

Alpha$=2.0$

Sampling instants

Figure 11.30
Open-loop Response of SMPC Controller to a Step Change in Error

The Z-transform operator has been omitted for brevity. For SMPC

$$D = \frac{\alpha K_p}{K_p - \tilde{G}} \tag{11.124}$$

where $\tilde{G}$ is the process model. Upon substitution for D from Equation (11.124) into Equation (11.123) gives

$$\frac{C}{d} = \frac{K_p - \tilde{G}}{K_p - \tilde{G} + \alpha K_p G} \tag{11.125}$$

or introducing the Z-transform operator and rearranging gives

$$C(Z) = \frac{K_p - \tilde{G}(Z)}{K_p - \tilde{G}(Z) + \alpha K_p G(Z)} d(Z) \tag{11.126}$$

Now

$$\lim_{t \to \infty} C(t) = \lim_{Z \to 1} (1 - Z^{-1})C(Z)$$
$$= \lim_{Z \to 1} \frac{(1 - Z^{-1})(K_p - \tilde{G}(Z))}{K_p - \tilde{G}(Z) + \alpha K_p G(Z)} \frac{1}{(1 - Z^{-1})} \tag{11.127}$$

or

$$\lim_{t \to \infty} c(t) = \frac{(K_p - K_p)}{K_p - K_p + \alpha K_p K_p^*} = 0$$

where the steady-state gain of the actual process is K_p^* while that of the model is K_p.

Equation (11.127) shows that the algorithm guarantees offset-free performance in the presence of modeling errors.

3. *Controller tuning.* The SMPC algorithm for single-input single-output systems has only one tuning parameter. As has been pointed out earlier, this parameter is determined such that a user-selected optimization criterion is satisfied.

4. *Robustness of the algorithm.* The SMPC system can be expressed equivalently as an internal model control system. Thus, the robustness tools available to IMC, such as feedback filtering, are available to SMPC. Alternately, the value of the tuning parameter, α, may simply be reduced to enhance the robustness of the system in the presence of modeling errors. As an illustration, the closed-loop set point and load responses of an arbitrary first-order with dead time process in the presence of mod-

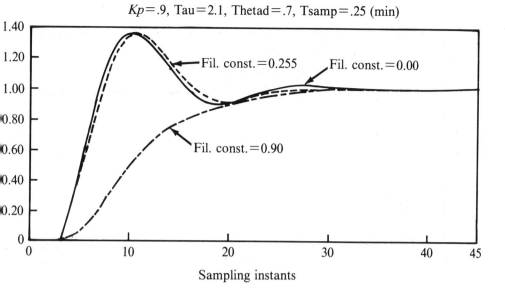

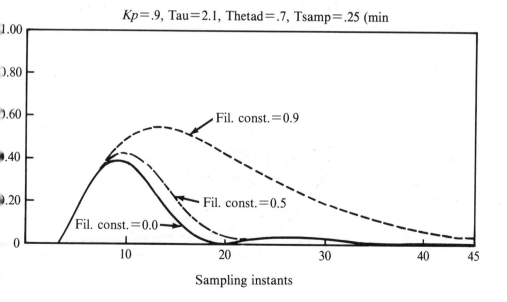

Figure 11.31a
Response of SMPC System in the Presence of +15% Modeling Error
in All Parameters — IMC Form

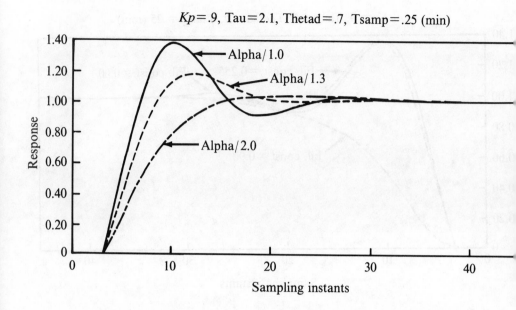

$Kp=.9$, Tau$=2.1$, Thetad$=.7$, Tsamp$=.25$ (min)

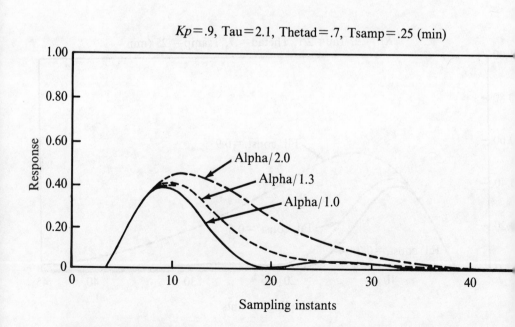

$Kp=.9$, Tau$=2.1$, Thetad$=.7$, Tsamp$=.25$ (min)

Figure 11.31*b*
Response of SMPC System in the Presence of +15% Modeling Error in All Pararmeters — with Scalar Reduction in α

eling errors, with both approaches, are shown in Figures 11.31(a) and 11.31(b).

5. *Constraint handling.* Any constraint on the input and/or output is incorporated by suitably modifying the performance index. The suggested procedure is to solve the unconstrained optimization problem first and examine the resulting output response and the manipulated variable moves. If constraints are violated, the optimization problem must be solved again with the performance index modified to include appropriate weights on inputs and/or outputs. As may be expected, if the value of α (based on minimum ISE) is reduced, the movements of the manipulated variable are reduced and the response becomes more sluggish. In the presence of constraints, the performance index, say Equation (11.39a) written for digital control applications, may be modified according to

$$\text{ISE} = \sum_{i=1}^{n} (\gamma E_i^2 + \beta' M_i^2) \tag{11.128}$$

where

β' = input penalty parameter

γ = weights on output

The objective of the optimization effort would be to determine α subject to the minimization of the function given in Equation (11.128). As an illustration, the closed-loop responses for an arbitrary FODT process with $\gamma = 1$ and varying β' are shown in Figure 11.32.

6. *Reference trajectory.* The process output can be made to follow a desired reference trajectory by including a reference model ahead of the set point, as shown in Figure 11.24.

7. *Dead time compensation.* If SMPC is expressed as an IMC system, it becomes clear that some dead time compensation is inherent in the algorithm. The results in the following section will show that the benefit increases with increasing values of θ_d for first-order with dead time systems.

8. *Working with systems having zeroes in the right half s-plane.* The use of impulse response coefficients in conjunction with a suitable sampling frequency circumvents the usual problems encountered with PID-type controllers in controlling processes that have inverse response. Further research at assessing the benefits of SMPC in comparison with PID control is in process.

9. *Shaping of closed-loop dynamics.* The suggested procedure is to base the initial design of α on minimizing an ISE type of performance index and to examine the resulting closed-loop responses and the as-

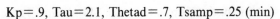

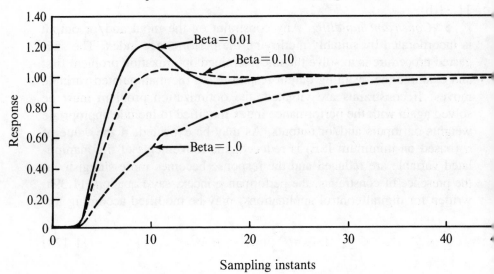

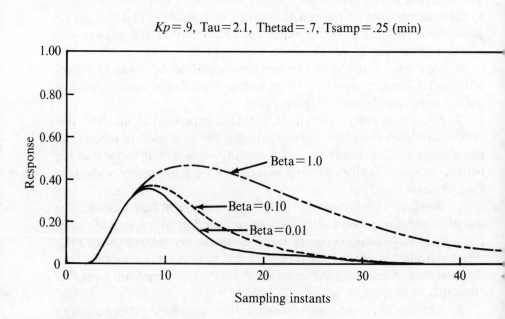

Figure 11.32
Response of SMPC System with Input Constraints

sociated manipulated variable moves for excessive movement. The oscillations in output responses and the movements of the manipulated variables can be reduced, albeit at the expense of ISE (integral of the squared error), by dividing the values of α (ISE$_{min}$) by a scalar. The simulations may then be repeated until satisfactory output and input behavior are achieved.

As an illustration, the set point and load responses of an arbitrary first-order with dead time process for several values of α are shown in Figure 11.33.

Tuning Guidelines for First-Order with Dead Time Processes[20]

Several investigators have proposed tuning guidelines for PID-type controllers. The oldest among them are the Ziegler-Nichols method [21] (a closed-loop method) and the Cohen-Coon method[22] (an open-loop method). Both give one quarter decay in the output response curve if the process is exactly first order with dead time and approximately quarter decay response for more complex processes. When digital computers became popular for process control, the subject of tuning received renewed attention, which resulted in guidelines for selecting digital PID constants.[23,2,6] Recently, Rivera et al.[12] published tuning guidelines for PID controllers in the framework of internal model control.

In this section SMPC tuning guidelines for first-order with dead time types of processes are presented. As we have pointed out in Chapter 1 it is often possible to describe the dynamics of a complex high-order process by an FODT process model. Consequently, industrial processes can often be described by a model of this type, which is the reason for its use here.

FODT processes are characterized by a dynamic parameter, the controllability ratio, which is the normalized ratio of dead time to time constant, $\Theta_d/(\Theta_d + \tau)$. For direct digital control loops the equivalent ratio is $(\Theta_d + T/2)/(\Theta_d + \tau + T/2)$, where T is the sampling period and the term $T/2$ accounts for the dead time effects of sampling.

Recall that for SMPC, the tuning parameter α is determined such that it minimizes a user-selected performance index. In this work, the ISE performance index given by Equation (11.39(a)) has been used.

With the aid of optimization software, the optimum α was obtained for numerous values of $(\Theta_d + T/2)/(\Theta_d + \tau + T/2)$. The effect of sampling frequency on α was also studied. A plot of optimum as a function of $(\Theta_d + T/2)/(\Theta_d + \tau + T/2)$ is shown in Figure 11.34. With this graph, the implementation of SMPC for FODT processes becomes particularly simple.

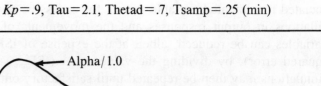

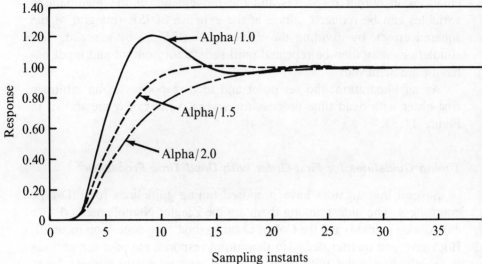

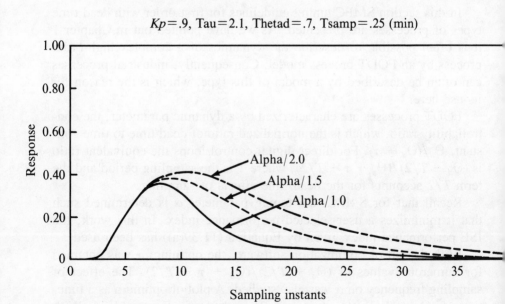

Figure 11.33
Shaping of Closed-Loop Responses in the Absence of Modeling Errors

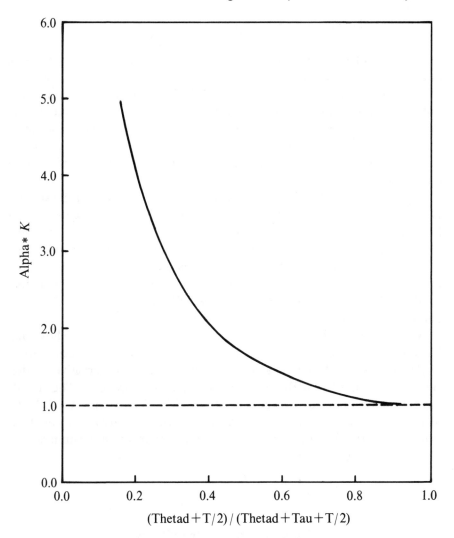

Figure 11.34
SMPC Tuning Plot for First-Order with Dead-Time Processes
(Performance Index: ISE)

Comparison of PID and SMPC[20]

A digital computer study was made to compare the performance of
PI and PID control with SMPC for FODT processes. Both set point
responses and load responses were evaluated. In the first set of runs,
perfect modeling was assumed for SMPC. In the second set of runs, a
modeling error of $\pm 15\%$ in each of the three parameters, K_p, Θ_d, and

τ, was introduced. The ISE performance measure was used for comparison purposes. The results are shown in Tables 11.3 and 11.4. Illustrative plots of set point and load responses for an arbitrary process with and without modeling errors are shown in Figures 11.35 and 11.36. The results show that the performance of SMPC is better than PI control but not as good as PID control. For large Θ_d, SMPC performs better than PI and PID control. These conclusions remain valid in the presence of modeling errors. Furthermore, process noise, which presents problems for PID control, does not appear to affect the SMPC performance adversely.

In the light of the simulation results and the ensuing discussion, it is believed that SMPC will be very useful in industrial applications. For SISO systems the main advantage of SMPC over PID-type controllers is the ease with which the former algorithm can be tuned. Thus, when step response coefficients for SISO systems are available, SMPC would be the preferred algorithm. Furthermore, the robustness tools available to SMPC but not to PID-type controllers will make the former algorithm particularly attractive in field applications.

Chawla et al., in an article in *Hydrocarbon Processing*, show that by replacing the item $1/K_p$ in Equation (11.121) at both places with $(1 - \beta Z^{-1})/K_p (1 - \beta)$, a new algorithm they call the "Conservative Model-Based Controller" (CMBC), results in superior properties. The details have been left out as an exercise for the reader. The extension of CMBC to multivariable systems is straightforward and has given excellent results. The term β is the only tuning parameter in the algorithm whose values lie between 0 and 1.

11.8 Kalman Algorithm

The Kalman algorithm is not to be confused with the Kalman filter that is used in state and parameter estimation. To synthesize a digital control algorithm by Kalman's approach, one places restrictions on C and M, instead of the usual C/R. Thus, suppose we assume the following expressions based on our knowledge of the process

$$C(Z) = C_1 Z^{-1} + Z^{-2} + Z^{3} + \dots \qquad (11.129a)$$

and

$$M(Z) = M_0 + M_1 Z^{-1} + M_f Z^{-2} + M_f Z^{-3} + \dots \qquad (11.129b)$$

Table 11.3(a)
ISE for PID, PI, and SMPC for FODT for a Step Change in Set Point for No Modeling Errors.

Sr. No.	Process			TSAMP = 0.25 min ISE Error			TSAMP = 0.5 min ISE Error		
	K_p	min	Θ_d min	SMPC	PID	PI	SMPC	PID	PI
1.	5.0	10.1	1.7	2.657	2.088	2.741	*	2.291	2.913
2.	2.5	2.9	1.0	1.577	1.345	1.666	1.769	1.605	1.869
3.	1.4	5.1	1.9	2.810	2.281	2.976	2.973	2.519	3.152
4.	5.0	20.1	10.7	14.649	11.713	15.196	14.819	11.954	15.97
5.	6.0	12.1	14.7	18.482	15.977	20.933	18.632	15.942	23.75
6.	3.6	3.7	4.7	5.979	5.236	6.639	6.109	5.431	6.79
7.	2.1	2.1	3.7	4.571	4.146	5.123	4.6927	4.323	5.26
8.	0.8	0.1	0.17	0.269	0.263	0.279	0.500	*	*
9.	4.0	0.2	0.5	0.762	0.787	0.841	1.002	*	*
10.	0.8	0.1	0.19	0.304	0.295	0.316	*	*	*
11.	6.0	0.1	0.2	0.318	0.322	0.342	*	*	*
12.	4.0	1.1	10.9	11.542	11.783	12.844	11.6699	11.985	13.177
13.	5.0	1.0	15.0	15.619	22.03	21.552	15.776	21.632	21.624

Table 11.3(b)
ISE for PID, PI and SMPC for a FODT for a Step Change in Load for No Modeling Errors

Sr. No.	K_p	1	O_d	ISE SMPC	ISE PID	ISE PI
1.	5.0	10.1	1.7	7.243	5.469	9.57
2.	2.5	2.9	1.0	2.225	1.809	2.78
3.	1.4	5.1	1.9	1.253	0.932	1.566
4.	5.0	20.1	10.7	105.181	73.102	116.35
5.	6.0	12.1	14.7	342.642	294.95	409.53
6.	3.6	3.7	4.7	41.824	35.353	50.428
7.	2.1	2.1	3.7	12.807	10.891	15.372
8.	0.8	0.1	0.17	0.104	0.086	0.097
9.	4.0	0.2	0.5	9.985	10.349	11.420
10.	0.8	0.1	0.19	0.121	0.105	0.119
11.	6.0	0.1	0.2	6.816	6.993	7.782
12.	4.0	1.1	10.9	170.938	172.524	192.137
13.	5.0	1.0	15.0	371.393	399.215	501.08

where no restrictions need be placed on the value of C_1, and M_f equals the reciprocal of the process steady-state gain. A plot of the desired response and the associated (assumed) manipulated variable moves is shown in Figure 11.36. In this illustration M is assumed to have two intermediate values. It turns out that the number of intermediate values of M equals the order of the process.

If the control algorithm is to be designed for a unit step change in set point, then, $R(Z) = 1/(1 - Z^{-1})$ and

$$
\begin{aligned}
\frac{C(Z)}{R(Z)} &= (1 - Z^{-1})(C_1Z^{-1} + Z^{-2} + Z^{-3}\ldots) \\
&= C_1Z^{-1} + (1 - C_1)Z^{-2} \\
&= P_1Z^{-1} + P_2Z^{-2} \\
&= P(Z)
\end{aligned}
\tag{11.130a}
$$

and

$$
\begin{aligned}
\frac{M(Z)}{R(Z)} &= (1 - Z^{-1})(M_0 + M_1Z^{-1} + M_fZ^{-2} + M_fZ^{-3} + \ldots +) \\
&= M_0 + (M_1 - M_0)Z^{-1} + (M_f - M_1)Z^{-2} \\
&= q_0 + q_1 Z^{-1} + q_2 Z^{-2} \\
&= Q(Z)
\end{aligned}
\tag{11.130b}
$$

Table 11.4(a)
ISE for Step Change in Set Point for SMPC, PID, and PI with 15% Modeling Errors (Controller Constants Tuned for No Mismatch)

Process No.	ISE for 15% (+ve) mis-match in K_p*			ISE for 15% (+ve) mis-match in O_d			ISE for 15% (+ve) mis-match in Tau 1			ISE for 15% (+ve) mis-match in Tau 1, K_p, O_d		
	SMPC	PID	PI	SMPC	PID	PI	SMPC	PID	PI	SMPC	PID	PI
1.	2.711	2.149	2.807	2.317	1.832	2.407	2.673	2.172	2.789	2.313	1.849	2.439
2.	1.610	1.383	1.709	1.369	1.159	1.462	1.568	1.389	1.683	1.371	1.179	1.491
3.	2.869	2.346	3.048	2.441	1.999	2.612	2.789	2.364	3.005	2.444	2.030	2.661
4.	14.943	12.039	15.632	12.611	10.16	13.461	14.454	12.046	15.414	12.655	10.377	13.831
5.	18.882	17.706	21.338	15.891	14.776	18.212	18.073	16.867	20.455	16.161	14.787	18.301
6.	6.114	5.377	6.828	5.16	4.592	5.827	5.866	5.329	6.589	5.259	4.732	6.036
7.	4.676	4.257	5.279	3.949	3.605	4.502	4.487	4.197	5.069	4.050	3.729	4.684
8.	0.291	0.269	0.287	0.257	0.258	0.259	0.264	0.257	0.268	0.266	0.264	0.264
9.	0.781	0.805	0.866	0.615	0.634	0.698	0.759	0.793	0.834	0.646	0.669	0.738
10.	0.321	0.302	0.324	0.270	0.259	0.275	0.288	0.279	0.299	0.282	0.268	0.284
11.	0.324	0.330	03522	0.265	0.273	0.290	0.298	0.305	0.324	0.271	0.283	0.302
12.	11.811	12.085	13.204	10.891	10.309	11.197	11.469	11.867	12.733	11.100	10.832	11.743
13.	15.983	33.406	21.885	15.203	14.981	18.782	15.553	17.763	21.409	15.307	15.859	19.389

* +ve % mismatch means process has been overestimated

Table 11.4(b)
ISE for Step Change in Load for Mismatch of 15% (Tuning Parameters Optimized Using Set Point Response)

Process No.	ISE for +15% Mismatch in All Parameters			ISE for −15% Mismatch in All Parameters		
	SMPC	*PID*	*PI*	*SMPC*	*PID*	*PI*
1.	6.588	5.064	8.850	8.034	5.919	10.403
2.	1.870	1.559	2.402	2.649	2.076	3.214
3.	1.053	0.806	1.354	1.527	1.089	1.845
4.	81.65	60.058	91.073	136.49	90.897	149.552
5.	241.86	210.393	295.71	463.05	405.396	527.134
6.	30.29	25.044	37.797	62.28	46.984	71.822
7.	8.98	7.833	11.162	20.00	16.024	22.968
8.	0.084	0.073	0.079	0.164	0.145	0.174
9.	5.86	6.197	7.041	12.78	13.004	14.809
10.	0.089	0.078	0.088	0.236	0.254	0.301
11.	4.52	4.872	5.447	20.08	19.795	22.468
12.	117.43	115.138	129.084	288.27	279.459	303.608
13.	253.11	268.705	340.99	568.801	582.173	712.258

+ ve mismatch process overestimated
− ve mismatch process underestimated

Now

$$\tilde{G}(Z) = \frac{C(Z)}{M(Z)} \qquad (11.131)$$
$$= \frac{P(Z)}{Q(Z)}$$

Thus, the coefficients in $\tilde{G}(Z)$ must equal those in $P(Z)$ and $Q(Z)$. Note that

$$\sum_{i=1}^{2} P_i = P_1 + P_2 = 1 \qquad (11.132)$$

and

$$\sum_{i=0}^{2} q_i = q_0 + q_1 + q_2 + 1/K_p \qquad (11.133)$$

These relationships do not generally hold for the pulse transfer functions, but dividing by the sum of the numerator coefficients will ensure that both Equations (11.132) and (11.133) hold. Since $P(Z)$ and $Q(Z)$ are

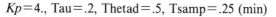

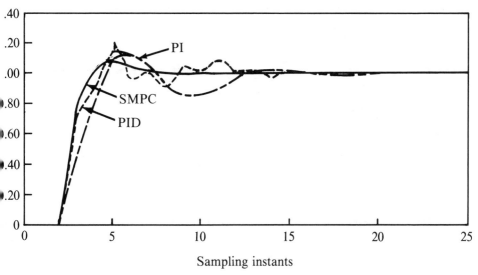

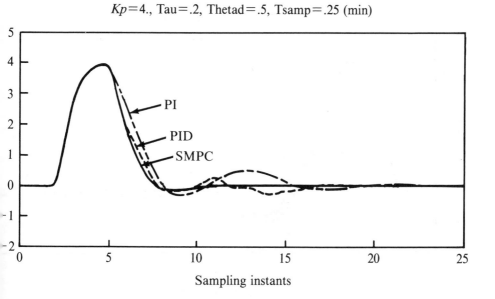

Figure 11.35
Comparison of SMPC with PI and PID Control in the Absence
of Modeling Errors

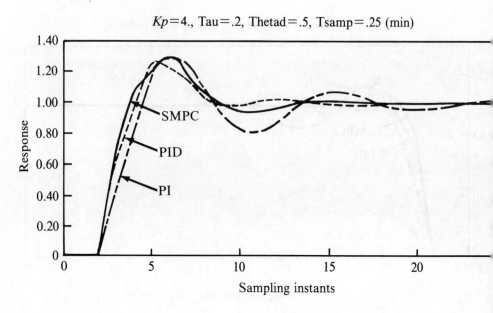

$Kp=4.$, Tau$=.2$, Thetad$=.5$, Tsamp$=.25$ (min)

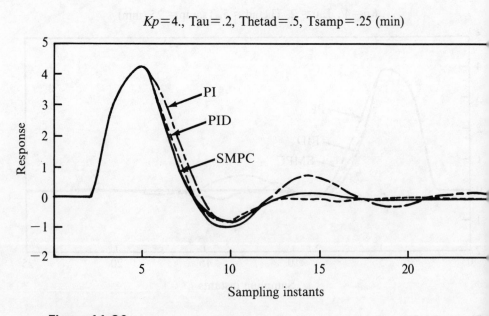

$Kp=4.$, Tau$=.2$, Thetad$=.5$, Tsamp$=.25$ (min)

Figure 11.36
Comparison of SMPC with PI and PID Control in the Presence
of +15% Modeling Error in All Parameters

known, the control algorithm $D(Z)$ can be derived by substitution in Equation (11.3). Thus

$$D(Z) = \frac{C(Z)/R(Z)}{1 - C(Z)/R(Z)} \cdot \frac{1}{\tilde{G}(Z)} \tag{11.134a}$$

$$= \frac{P(Z)}{1 - P(Z)} \cdot \frac{Q(Z)}{P(Z)}$$

or

$$D(Z) = \frac{Q(Z)}{1 - P(Z)} \tag{11.134b}$$

It can be shown that Equation (11.134b) holds even if the process has dead time of N sampling instants. Of course, in such a case the specification for $C(Z)$, Equation (11.129a) must include the appropriate number of zeros. As an illustration,[2] consider the design of Kalman algorithm for the fourth-order process (Equation 11.29) that is described by the model

$$\tilde{G}_p(s) = \frac{1e^{-1.46s}}{3.34s + 1} \tag{11.135}$$

If $T = 1$, then,

$$\tilde{G}(Z) = G_{h_0}\tilde{G}_p(Z) = \frac{Z^{-2}(0.1493 + 0.1095Z^{-1})}{1 - 0.7413Z^{-1}} \tag{11.136}$$

Since the numerator coefficients do not sum to 1, let us divide the numerator and denominator of Equation (11.136) by $(0.1493 + 0.1095)$ to obtain

$$G(Z) = \frac{Z^{-2}(0.577 + 0.423Z^{-1})}{3.86 - 2.86Z^{-1}} = \frac{P(Z)}{Q(Z)} \tag{11.137}$$

Therefore, the control algorithm is

$$D(Z) = \frac{Q(Z)}{1 - P(Z)} \tag{11.138}$$

$$= \frac{3.86 - 2.86Z^{-1}}{1 - Z^{-2}(0.577 + 0.423Z^{-1})}$$

The performance of the algorithm for the fourth-order process (see Equation (11.29)) is depicted in Figure 11.37.

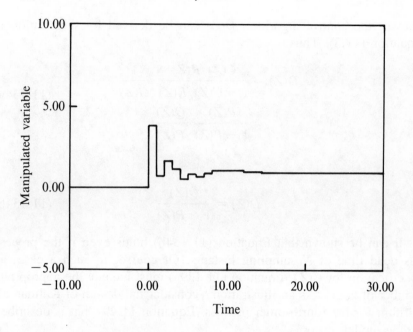

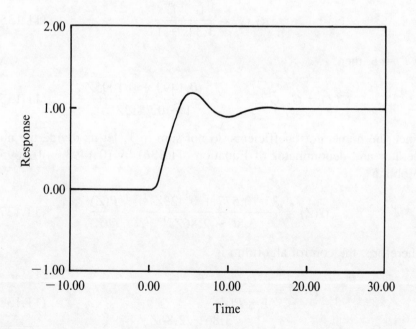

Figure 11.37
Response of the Fourth-order System Using the Kalman Algorithm
(Reprinted with permission from *Digital Computer Process Control*,
© 1972, Cecil L. Smith)

11.9. Algorithm of Gautam and Mutharasan[25]

Computer control loops frequently employ the digitial equivalent of the conventional controller. The loops are often tuned for set point changes. The response of the system is not as good for load changes. Therefore, it is highly desirable to have an algorithm that performs equally well for set point and load changes.

Gautam and Mutharasan[25] have presented a general-purpose control algorithm for a first-order process with dead time that appears to perform well in the presence of set point *and* load changes.

We shall present the algorithm for a general first-order lag plus dead time model. Numerical values of the dead time, gain, and time constant are based on process identification studies on an industrial system. From Figure 11.38 we note that, in the absence of load disturbances, L equals zero and M equals X (or M^* equals X^*). The relationship between Y^* and X^* (or M^*) in the Z domain is

$$\frac{Y(Z)}{X(Z)} = Z\{G_{h_0}(s)G_p(s)\} \qquad (11.139)$$

where

$\quad G_{h_0}(s) =$ transfer function of the zero-order hold,

$$= \frac{1 - e^{-Ts}}{s}$$

$\quad G_p(s) =$ process transfer function, $\dfrac{K_p e^{-\theta_d s}}{\tau s + 1}$

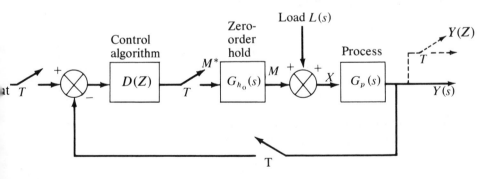

Figure 11.38
Computer-Control System

$$T = \text{sampling period, 0.5 min (chosen arbitrarily)}$$

$$\theta_d = \text{dead time, 0.76 min}$$

$$K_p = \text{process gain, } -1$$

$$\tau = \text{time constant, 0.4 min}$$

Thus,

$$\frac{Y(Z)}{X(Z)} = Z\left\{ \frac{1 - e^{-sT}}{s} \cdot \frac{K_p e^{-\theta_d s}}{\tau s + 1} \right\} \tag{11.140}$$

With modified Z transforms, Equation (11.140) gives

$$\frac{Y(Z)}{X(Z)} = \frac{K_p Z^{-(N+1)} (1 - e^{-mT/\tau}) + K_p Z^{-(N+2)} (e^{-mT/\tau} - e^{-T/\tau})}{1 - e^{-T/\tau} Z^{-1}} \tag{11.141}$$

This equation can be inverted to obtain the input-output relationship for the process. Thus,

$$y_k = e^{-T/\tau} y_{k-1} + K_p\{(1 - e^{-mT/\tau}) x_{k-(N+1)} + (e^{-mT/\tau} - e^{-T/\tau}) x_{k-(N+2)}\} \tag{11.142}$$

where the subscript k refers to the kth sampling instant, N is the largest integral number of sampling periods in θ_d, and m, θ, are given by

$$m = 1 - \frac{\theta}{T} \tag{11.143}$$

$$\theta = \theta_d - NT$$

By introducing A_1, A_2, and A_3 to represent the constants defined earlier, Equation (11.142) can be solved for $x_{k-(n+1)}$. Thus,

$$x_{k-(N+1)} = \frac{A_3}{K_p(A_3 - A_2)} y_k - \frac{A_2 A_3}{K_p(A_3 - A_2)} y_{k-1} - \frac{A_2(1 - A_3)}{(A_3 - A_2)} x_{k-(N+2)} \tag{11.144}$$

But in the presence of load disturbances Figure 11.38 indicates that the input to the process is given by

$$x_{k-(N+1)} = L_{k-(N+1)} + M_{k-(N+1)} \tag{11.145}$$

Therefore, an estimate of the load condition can be obtained by combining Equations (11.144) and (11.145). Thus,

$$L_{k-(N+1)} = \frac{A_3}{K_p(A_3 - A_2)}y_k - \frac{A_2 A_3}{K_p(A_3 - A_2)}y_{k-1} -$$
$$\frac{A_2(1 - A_3)}{(A_3 - A_2)}x_{k-(N+2)} - M_{k-(N+1)} \qquad (11.146)$$

and, therefore, the manipulated variable should be calculated from the equation

$$M_k = \frac{y_{set}}{K_p} - L_{k-(N+1)}$$

$$= \frac{y_{set}}{K_p} - \frac{A_3}{K_p(A_3 - A_2)}y_k + \frac{A_2 A_3}{K_p(A_3 - A_2)}y_{k-1} + \qquad (11.147)$$
$$\frac{A_2(1 - A_3)}{(A_3 - A_2)}x_{k-(N+2)} + M_{k-(N+1)}$$

The value of $x_{k-(N+2)}$ for use in this equation is obtained from the recursive equation, derived from Equation (11.144). Thus,

$$x_{k-(N+2)} = \frac{A_3}{K_p(A_3 - A_2)}y_{k-1} - \frac{A_2 A_3}{K_p(A_3 - A_2)}y_{k-2} -$$
$$\frac{A_2(1 - A_3)}{(A_3 - A_2)}x_{k-(N+3)} \qquad (11.148)$$

In this development, since the sampling period is 0.5 min and dead time is 0.76 min, N equals 1. This makes $\theta = 0.26$ and $m = 0.48$. Substituting these values into Equations (11.147) and (11.148) gives the algorithm in a form that is suitable for implementation on the control computer. Thus,

$$M_k = -y_{set} + 2.216y_k - 0.6349y_{k-1} + 0.5813X_{k-3} + M_{k-2} \quad (11.149)$$
$$+ \omega(y_{set} - y_k)$$

where ω = tuning parameter, which is to be determined by trial and error, and

$$x_{k-3} = -2.216y_{k-1} + 0.6351y_{k-2} - 0.5812x_{k-4} \qquad (11.150)$$

The last term in Equation (11.149) has been added to improve the speed of response of the system. If $K = 1$ denotes the time at which the loop is switched to computer control, M_1 is the first value of the controller output computed from Equation (11.149) and the values of x_{k-3} for the next three sampling periods, that is, x_{-2}, x_{-1}, x_0, are the steady-state values of the process output y_{set}. Thereafter, x_{k-3} is computed from Equation (11.150). Note that the contribution of the last term in Equation

(11.149) is significant only when the process shows a deviation from set point.

Simulation Results. The process model and the control algorithm were implemented on a digital computer. The differential equation representing the process model was numerically solved by the fourth-order Runge-Kutta method. The integration step size was 0.02 min. The responses of the process to set point and load changes are shown in Figures 11.39 and 11.40, respectively. The best value of ω turned out to be -0.1 for both set point changes and load changes. Also shown in these figures is the response of the process using the PI control algorithm given in the following equation:

$$M_k = M_{k-1} - K_c \left[(e_k - e_{k-1}) + \frac{T}{\tau_I} e_k \right] \qquad (11.151)$$

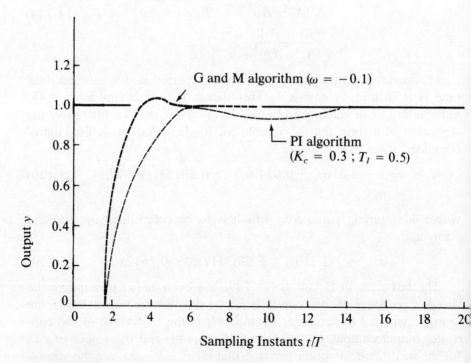

Figure 11.39
Set Point Response of G and M Algorithm

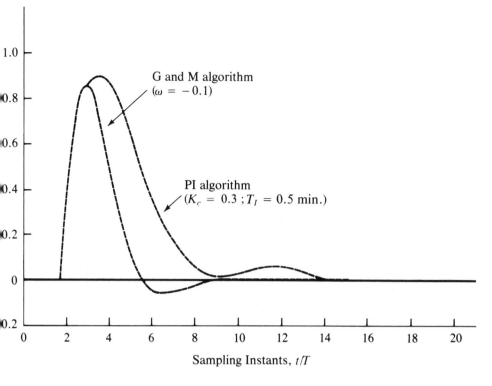

Figure 11.40
Response of G & M Algorithm to a Load Change

where

$$K_c = \text{gain},\ 0.3$$
$$\tau_I = \text{reset},\ 0.5\ \text{min}$$

The best values of proportional gain and reset were also found by trial and error. From Figures 11.39 and 11.40, it can be seen that the present algorithm performs well for both set point and load changes.

We wish to point out that the derivations of the Kalman algorithm and the Gautam-Mutharasan algorithm have a basis somewhat different from the direct synthesis method used elsewhere in this chapter. No comparisons of Kalman and Gautam-Mutharasan algorithms with IMC or SMPC are currently available. A comparative study of the various algorithms reported in this chapter would be an excellent set of exercises for the student.

11.10 Treatment of Noisy Process Signals

We may define the term "noise" as those fluctuations in the process signal that do not contain useful control information. In process control applications noise arises from one or more of the following sources:

Presence of those process disturbances that are too rapid to be reduced significantly by control action.

Turbulence around sensors (e.g., liquid flow or furnace pressure) and instrument noise.

Stray electrical pickup such as from ac power lines.

Process noise resulting from these sources occupies a frequency band that is much wider than that of the controllable process fluctuations. Since control action can only minimize the effects of these low-frequency, controllable fluctuations, process noise always causes a degradation of control performance. Further, the noise can cause aliasing of the signal (sampling-induced low-frequency noise) if the sampling frequency is not carefully chosen (see Figure 11.41). High-frequency noise, which cannot be attenuated by control, also causes excessive actuator wear. It should be clear from this introduction that we must in-

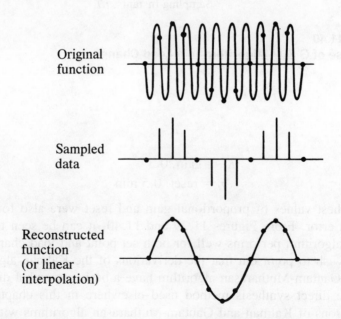

Original function

Sampled data

Reconstructed function (or linear interpolation)

Figure 11.41
Aliasing Error Resulting from Sampling at Rates 8/7 Samples per Cycle or Original Function (Reprinted by Permission from Ref. 7)

corporate noise-reduction techniques in those sampled-data control loops that contain noisy process signals. Some process signals are relatively noise-free. For example, when temperature is measured in a large thermal capacity, such as a thermowell, noise reduction is not required.

The frequency bands occupied by the different noise sources are shown in Figure 11.42. Noise reduction is accomplished via pure analog and/ or digital filtering, as shown in Figure 11.43.

Analog Filters

Stray electrical pickup can be minimized by proper grounding, shielding, and routing of wires. This type of high-frequency noise can be eliminated by an analog (RC) filter, shown schematically in Figure

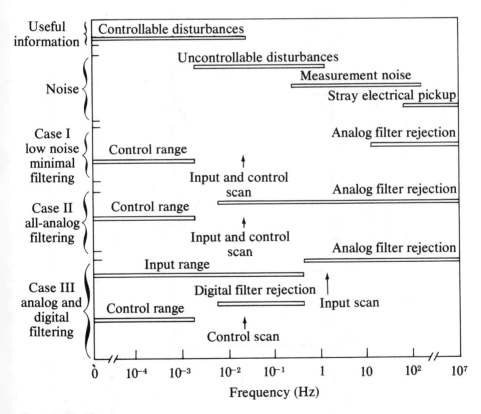

Figure 11.42
Frequency Ranges for Input Signal and Associated Scanning and Filtering Functions for a Typical Process
(Reprinted by Permission from Ref. 7)

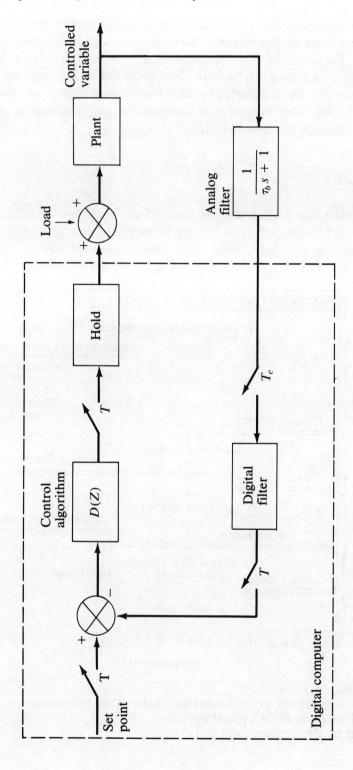

Figure 11.43
Sampled-Data Control System with Noise Reduction Schemes

11.44. The time constant of the analog filter τ_b is usually about $T_c/2$. The equation for the RC filter shows that it behaves as a simple first-order lag. Analog filters that are built from passive elements (resistors and capacitors) give time constants up to a few seconds. For filter time constants greater than a few seconds, active elements using operational amplifiers are needed. Since the filter serves only one input signal, this construction becomes relatively expensive. Therefore, combined analog/digital filtering is usually preferred, as shown in Figure 11.43.

Digital Filtering Algorithms

There are two algorithms for digital filtering: one is based simply on the arithmetic average of N samples; the other is the digital equivalent of the first-order lag.

The mean value of the filter output that should be employed in control calculations according to the first method is

$$y_k = \frac{1}{N} \sum_{i=0}^{N-1} x_{k-i} \tag{11.152}$$

where

x_k = kth input to the filter

y_k = filter output

N = number of input
samples

Since N samples of x_k are required to produce a filter output y_k, the successive time intervals for control action (i.e., the sampling period T) is $T = NT_c$.

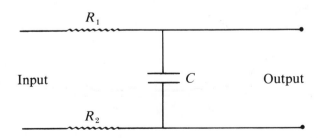

Figure 11.44
Analog RC Filter

The second algorithm is the digital equivalent to the first-order lag. Recall that the differential equation for the first-order lag is

$$\tau_f \frac{dy}{dt} + y = x(t) \qquad (11.153)$$

where x and y are input and output, respectively, and τ_f is the time constant of the first-order system. The numerical approximation of the differential equation is

$$y_N = \frac{1}{Q} x_N + \left(1 - \frac{1}{Q}\right) y_{N-1} \qquad (11.154)$$

where Q is a design parameter that is related to the sampling period T and the filter time constant τ_f as

$$Q = \frac{1}{1 - e^{-T/\tau f}} \qquad (11.155)$$
$$\cong \frac{\tau_f}{T} \text{ for large } \frac{T_f}{T} \text{ (say, greater than 10).}$$

To determine the digital filter time constant, consider the frequency spectrum of a process signal shown in Figure 11.45.[1] If we sample at a frequency of ω_s (equal to $2\pi/T$), the frequency components above $\omega_s/2$ will be folded onto the low-frequency components if no filtering were used. The use of a digital first-order lag filter will greatly attenuate (though not completely eliminate) these components. Since the corner frequency of the first-order lag is simply the reciprocal of its time constant, the filter time constant is given by

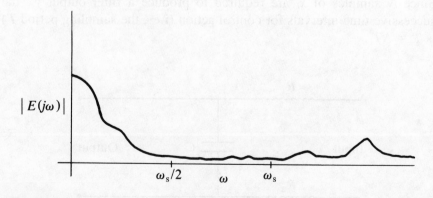

Figure 11.45
Typical Frequency Spectrum of a Process Signal (Reproduced with Permission from Ref. 1)

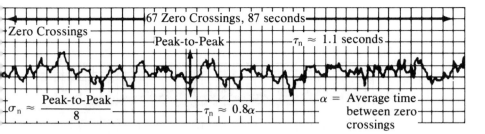

Figure 11.46
Measuring Noise Parameters (By Permission from Ref. 6)

Table 11.5
Selection of Scan and Control Sampling Rates for Processes with Noise

Type of control algorithm	Scan sampling period, T_c		Control Sampling period, T
	$\tau_n < \tau_b$		$\tau_n > 2\tau_b$
PI	$\dfrac{\tau_b}{2} < T_c < \tau_b$	$\dfrac{1}{30}\,\tau_f < T_c < \dfrac{\tau_n}{2}$	$\dfrac{\tau_i}{100} < T < \dfrac{\tau_f}{2}$
PID (with a filter on the derivative mode)	—Normally not used when significant process noise is present. —May be used when noise bandwidth does not extend into the higher frequencies amplified by the rate term. May also be applied to processes with wide-bandwidth noise provided σ_n is small enough so that the standard deviation of the controller output, σ_m, is acceptable, where $$\sigma_m = k_c G \sigma_n$$ $G \cong 10$ for analog control. May be reduced for digital control to bring σ_m down to an acceptable level. If σ_m is still too large as indicated by the final control element activity, use the two-mode controller.		$\dfrac{\tau_d}{10G} < T < \dfrac{\tau_d}{G}$ $G = \dfrac{\text{derivative time, } \tau_d}{\text{derivative filter time constant}}$

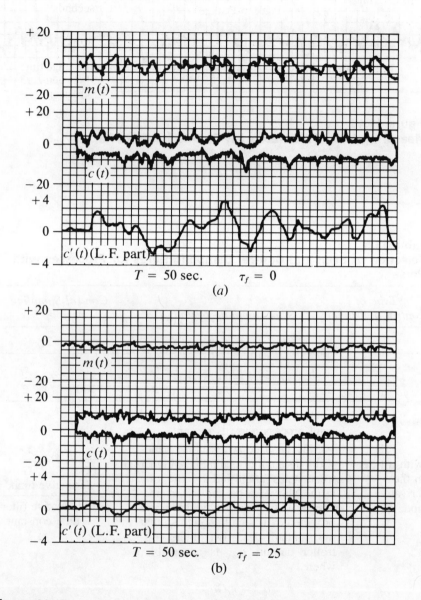

Figure 11.47
Effect of Noise on Process Variables for Control with and without a Filter. Each Time Division is 250 sec (Reprinted by Permission from Ref. 7)

$$\tau_f = \frac{1}{\omega_s/2} = \frac{2T}{2\pi} = \frac{T}{\pi}$$

A rule of thumb is to set the digital filter time constant equal to $T/2$.

Characterization and Measurement of Process Noise Parameters

The process noise characteristics (magnitude and bandwidth) can be estimated from steady-state process data when there are no changes in the manipulated variable. Figure 11.46 shows such data for an illustrative process[26]. It can be assumed that most random noise found in industrial control can be represented by exponentially correlated noise with a dominant time constant τ_n having a magnitude that can be measured by its standard deviation σ_n. From the chart record, Figure 11.46, these parameters are calculated as

$$\tau_n = 0.8\alpha$$

and

$$\sigma_n = \frac{\text{peak-to-peak amplitude}}{8}$$

where α, the average time between zero crossings, equals the total time interval divided by the number of zero crossings in that interval. Zero crossings refer to the intersection of the process variable with the estimated mean as the process variable cuts back and forth across the mean. The estimation of σ_n is based on finding the peak-to-peak amplitude for the measured interval and dividing by 8. For digitally executed PI and PID control algorithms, these noise parameters will assist in the selection of the scan sampling period T_c and the control sampling period T, as shown in Table 11.5.

As an illustration, a chart record of control with and without filtering is shown in Figure 11.47. The beneficial effects of filtering are clearly evident.

NOMENCLATURE

| A_1 | Constant; exp $(-T/2\tau)$ | A_3 | Constant; exp $(-\beta T/\tau)$ |
| A_2 | Constant; exp $(-T/\tau)$ | B, C | Controlled variables |

$D(Z)$	Control algorithm	s	Laplace transform variable
$d(Z)$	Disturbance		
d_k	Disturbance	T	Sampling time
E	Error	T	Sampling period
$F(Z)$	Transfer function of desired closed-loop response	T'	Total dead time; $\theta_d + T$ or $\theta_d + T/2$
$F(s)$	Feed flow	U_k	Input function
G_c	PID controller	v	Controller output
$G_{h_o}(s)$	Zero-order hold	x	Input variable
G_I	IMC controller	x_D	Top composition
$G(Z)$	Process (pulse) transfer function	y	Output variable
		Z	Z transform variable
$G'_M(s)$	Model transfer function without dead time		
$G_p(s)$	Process transfer function	**Greek Letters**	
$G'_p(s)$	Process transfer function without dead time	β	Fractional part of θ_d/T
$G_L(s)$	Load transfer function	ω	Tuning parameter
$G_+(Z)$	Transfer function containing nonminimum-phase elements	ρ_k	Calibrated set point
		θ	Fractional dead time; $\beta \times T$
$G_-(Z)$	Invertible part of $G(Z)$	θ_d	Process dead time
h	Impulse response coefficient	τ	Time constant of the process
$H(s)$	Measuring element transfer function	τ_f	Filter time constant
		τ_D	Derivative time
K_c	Controller gain	τ_I	Integral time
K_L	Gain relating C to L	α_f	Filter constant
K_p	Process gain	γ	Weight on output
K_r	Reset gain	β'	Weight on input
K_T	Integral controller gain	ν	"Zero" of $G(Z)$
L	Load disturbance	α	SMPC tuning parameter
$L(s)$	Load variable		
m	Constant; $1 - \theta/T$	**Superscripts, Subscripts**	
M	Manipulated variable		
N	Integral part of θ_d/T	$\sim$	Pertaining to model
R, r_k	Set point	m	Pertaining to model
R_e	Reflux	p	Pertaining to model

References

1. McMillan, G. K., *Tuning and Control Loop Performance,* ISA, Research Triangle Park, NC, 1983.
2. Smith, C. L., *Digital Computer Process Control,* Intext Educational Publishers, Scranton, PA, 1972.
3. Kuo, B. C., *Analysis and Synthesis of Sampled-Data Control Systems,* Prentice-Hall, Englewood Cliffs, NJ, 1963.
4. Dhalin, E. B., "Designing and Tuning Digital Controllers," *Instruments and Control Systems,* **41, 6,** June 1968.
5. Tou, J. T., *Digital and Sampled-Data Control Systems,* McGraw-Hill, New York, 1959, p. 242.
6. Fertik, H. A., "Tuning Controllers for Noisy Processes," *ISA Trans.,* **14, 4,** 1975.
7. Smith, O. J. M., "Close Control of Loops with Dead Time," *Chemical Engineering Progresses,* **53, 5,** 1957, 217–219.
8. Meyer, C., et al., "An Experimental Application of Time Delay Compensation Techniques to Distilation Column Control," *I&EC Proc. Des. Dev.,* **17, 1,** 1978.
9. Moore, C. F., *Selected Problems in the Design and Implementation of Direct Digital Control,* Ph.D. thesis, Department of Chemical Engineering, Louisiana State University, 1969.
10. Brosilow, C. B., "The Structure and Design of Smith Predictors from the Viewpoint of Inferential Control," Joint American Control Conference, Denver, Colorado, June 1979.
11. Garcia, C. E. and M. Morari, "Internal Model Control 1—A Unifying Review and Some New Results," *Ind. Eng. Chem. Process Des. Dev.,* **21,** 1982, p. 308–323.
12. Rivera, D. E., M. Morari and S. Skogestad, "Internal Model Control 4-PID Controller Design," *Ind. Eng. Chem. Process Des. Dev., 25,* 1986, p. 252–265.
13. Arkun, Y., J. Hollett, W. M. Canney and M. Morari, "Experimental Study of Internal Model Control," *I&EC Proc. Des. Dev.,* **25, 1,** 1986, pp. 102–108.
14. Levien, K. L. and M. Morari, "Internal Model Control of Coupled Distillation Columns," *Proc. Am. Cont. Conf.,* Boston, June 1985.
15. Wassick, J., "Multivariable Internal Model Control for a Full-Scale Distillation Column," Paper Presented at ACC, St. Paul Minnesota, June, 1987.
16. Matsko, T. N., "Internal Model Control for Chemical Recovery," *CEP,* December 1985, pp. 46–51.
17. Chawla, V. K., Private Communication, Department of Chemical Engineering, University of Louisville, January 1987.
18. Tu, F. C. Y. and J. Y. H. Tsing, "Synthesizing a Digital Algorithm for Optimized Control," *In. Tech,* May 1979.

19. Arulalan, G. R. and P. B. Deshpande, "Simplified Model Predictive Control," *I&EC Proc. Des. Dev.*, **25, 2,** 1987.
20. Vaidya, C. M. and P. B. Deshpande, "Single-Loop Simplified Model Predictive Control," *Hydrocarbon Processing,* June, 1988.
21. Ziegler, J. G. and N. B. Nichols, "Optimum Settings for Automatic Controllers," *Trans ASME,* **64, 11,** November 11, 1942 p. 759.
22. Cohen, G. H. and G. A. Coon, "Theoretical Considerations for Retarded Control," Taylor Instrument Company Bulletin, TDS-10A102.
23. Gallier, P. W. and R. E. Otto, "Self-Tuning Computer Adapts DDC Algorithms," *Instrumentation Technology,* February 1968, p. 65–70.
24. Kalman, R. E., Discussion following article "Sampled-Data Processing Techniques for Feedback Control Systems," by A. R. Beregand, J. R. Ragazzini, *Trans. AIEE,* November 1956, pp. 236–247.
25. Gautam, R. and R. Muthrasan, "A General Direct Digital Control Algorithm for a Class of Linear Systems," *AIChE J.,* **24, 2,** 1978, 360–64.
26. Goff, K. W., "Dynamics in Direct Digital Control," Part I *ISA J.,* **13, 11,** November 1966, 44–49; Part II, *ISA J.,* **12,** December 1966, 44–54.

CHAPTER 12

Modified Z Transforms

T he Z-transform method enables us to determine the transient response of sampled-data control systems only at the sampling instants. To obtain the values of the response between sampling instants, modified Z transforms are useful. They are also helpful in analyzing sampled-data control systems containing transportation lag (i.e., dead time).

12.1. Definition and Evaluation of Modified Z Transforms[1, 2, 3]

Suppose that the transfer function of a process with dead time is represented by the following expression:

$$G_p(s) = G(S)e^{-\theta_d s} \qquad (12.1)$$

where $G(s)$ contains no dead time and θ_d = dead time. If we substitute for θ_d the quantity $NT + \theta$, where N is the largest integer number of sampling intervals in θ_d and T is the sampling period, Equation 12.1 becomes

$$G_p(s) = G(s)\, e^{-(NT+\theta)s} \tag{12.2}$$

For example, if $\theta_d = 0.5$ and $T = 0.11$, then, $N = 4$ and $\theta = 0.50 - (4)(0.11) = 0.06$. It can be easily verified that for a given θ_d and T, θ lies between 0 and T.

Now, let us take the Z transform of Equation (12.2).

$$G_p(Z) = Z\left\{ G(s)\, e^{-(NT+\theta)s} \right\} \tag{12.3}$$

$$= Z^{-N} Z\left\{ G(s)\, e^{-\theta s} \right\}$$

The quantity $Z\left\{ G(s)\, e^{-\theta s} \right\}$ is defined as the modified Z-transform of $G(s)$ and is denoted by $Z_m\{G(s)\}$ or $G(Z, m)$. Thus

$$G(Z, m) = Z_m\left[G(s) \right] = Z\left\{ G(s)\, e^{-\theta s} \right\}$$

Let us evaluate the modified Z transform of some simple functions.

1. Unit Step Function

$$f(t) = \begin{cases} u(t) & t \geq 0 \\ 0 & t < 0 \end{cases}$$

$$Z_m\{F(s)\} = Z_m\left\{ \frac{1}{s} \right\} = Z\left\{ \frac{e^{-\theta s}}{s} \right\}$$

$$= Z\{u(t - \theta)\}$$

$$= \sum_{n=0}^{\infty} u(nT - \theta)\, Z^{-n}$$

$$= 0 + u(T - \theta)\, Z^{-1} + u(2T - \theta)\, Z^{-2} + u(3T - \theta)\, Z^{-3}$$

$$= Z^{-1}\{ 1 + Z^{-1} + Z^{-2} \} + \cdots$$

Therefore

$$Z_m\left[u(t) \right] = \frac{Z^{-1}}{1 - Z^{-1}} = \frac{1}{Z - 1}$$

2. Evaluate $z_m \{e^{-at}\}$

$$Z_m\{e^{-at}\} = Z_m\left\{\frac{1}{s+a}\right\} = Z\left\{\frac{e^{-\theta s}}{s+a}\right\}$$

$$= Z\{u(t-\theta)\, e^{-a(t-\theta)}\}$$

$$= \sum_{n=0}^{\infty} u(nT-\theta)\, e^{-a(nT-\theta)}\, Z^{-n}$$

$$= 0 + e^{-a(T-\theta)}\, Z^{-1} + e^{-a(2T-\theta)}\, Z^{-2} + e^{a(3T-\theta)}\, Z^{-3} + \cdots$$

Let

$$mT = T - \theta \text{ or } m = 1 - \frac{\theta}{T}.$$

Then,

$$Z_m\{e^{-at}\} = e^{-amT}\, Z^{-1} + e^{-amT}\, e^{-aT}\, Z^{-2} + e^{-amT}\, e^{-2aT}\, Z^{-3}$$

$$Z^{-1}\, e^{-amT}\left\{1 + Z^{-1}\, e^{-aT} + e^{-2aT}\, Z^{-2} + \cdots\right\}$$

Therefore,

$$Z_m\{e^{-at}\} = \frac{Z^{-1}\, e^{-amT}}{1 - Z^{-1}\, e^{-aT}}$$

This procedure can be applied to functions whose modified Z transforms are desired. A table of modified Z transforms is included in Appendix A.

12.2 Application of Modified Z Transforms to Systems with Dead Time

The use of modified Z transforms simplifies the analysis of systems containing a dead-time element. The procedure is best illustrated by an example.

Example 1. Determine the response of the system shown in Figure 12.1 to a unit step change in set point. Assume $T = 0.5$ and $D(Z)$ is a PI control algorithm with $K_c = 0.43$, $\tau_I = 1.57$. The closed-loop pulse transfer function of this system is

$$\frac{C(Z)}{R(Z)} = \frac{D(Z)\, G_{h_0} G_p(Z)}{1 + D(Z)\, G_{h_0} G_p(Z)} \qquad (12.4)$$

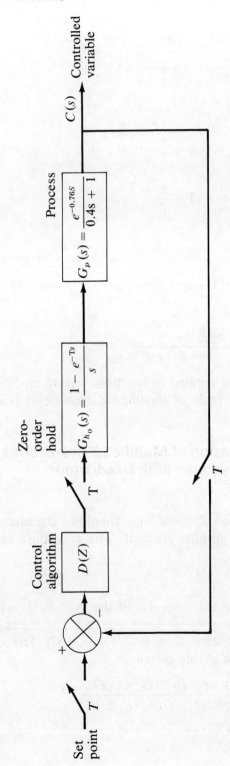

Figure 12.1
Sampled-Data Control System for a Process with Dead Time

where

$$G_{h_0} G_p(Z) = Z\{G_{h_0}(s) G_p(s)\}$$

Let us evaluate $G_{h_0} G_p(Z)$ by modified Z transforms.

$$G_{h_0} G_p(Z) = Z\left[\frac{1 - e^{-sT}}{s} \cdot \frac{e^{-0.76s}}{0.4s + 1}\right]$$

$$= Z\left[\frac{e^{-0.76s}}{s(0.4s + 1)}\right] - Z\left[\frac{e^{-0.76s} e^{-Ts}}{s(0.4s + 1)}\right]$$

(12.5)

Recall the theorem on translation of the function, which states that if the Z transform of $f(s)$ is $F(Z)$, then the Z transform of $e^{-Ts} f(s)$ is $Z^{-1} F(Z)$. Therefore,

$$G_{h_0} G_p(Z) = Z\left[\frac{e^{-0.76s}}{s(0.4s + 1)}\right] - Z^{-1} Z\left[\frac{e^{-0.76s}}{s(0.4s + 1)}\right]$$

(12.6)

$$= (1 - Z^{-1}) Z\left[\frac{e^{-0.76s}}{s(0.4s + 1)}\right]$$

Now, let us introduce modified Z transforms to simplify the calculation of $G_{h_0} G_p(Z)$.

In this example $\theta_d = 0.76$ and $T = 0.5$.

Therefore, N, the integral number of sampling intervals in θ_d, equals 1 and

$$\theta = \theta_d - NT = 0.76 - (1)(0.5) = 0.26$$

(12.7)

Thus

$$G_{h_0} G_p(Z) = (1 - Z^{-1}) Z\left[\frac{e^{-\theta_d s}}{s(0.4s + 1)}\right]$$

(12.8)

$$= (1 - Z^{-1}) Z\left[\frac{e^{-\theta s} e^{-NTs}}{s(0.4s + 1)}\right]$$

$$= (1 - Z^{-1}) Z^{-N} Z\left[\frac{e^{-\theta s}}{s(0.4s + 1)}\right]$$

$$= (1 - Z^{-1}) Z^{-1} Z\left[\frac{e^{-0.26s}}{s(0.4s + 1)}\right]$$

$$= (1 - Z^{-1}) Z^{-1} Z_m\left[\frac{1}{s(0.4s + 1)}\right]$$

From the table of modified Z transforms,

$$Z_m \left\{ \frac{a}{s(s+a)} \right\} = Z^{-1} \left(\frac{1}{1-Z^{-1}} - \frac{e^{-amT}}{1-e^{-aT}Z^{-1}} \right) \qquad (12.9)$$

where

$$a = \frac{1}{0.4} = 2.5$$

$$m = 1 - \frac{\theta}{T} = 1 - \frac{0.26}{0.5} = 0.48$$

Substitution of these values in Equation (12.9) gives

$$Z_m \left\{ \frac{1}{s(0.4s+1)} \right\} = Z_m \left\{ \frac{2.5}{s(s+2.5)} \right\} \qquad (12.10)$$

$$= Z^{-1} \left(\frac{1}{1-Z^{-1}} - \frac{0.5488}{1-0.2865\,Z^{-1}} \right)$$

$$= \frac{Z^{-1} \left[1 - 0.2865\,Z^{-1} - 0.5488 + 0.5488\,Z^{-1} \right]}{(1-Z^{-1})(1-0.2865\,Z^{-1})}$$

$$= \frac{Z^{-1}(0.4512 + 0.2623\,Z^{-1})}{(1-Z^{-1})(1-0.2865\,Z^{-1})}$$

The expression for $G_{h_0} G_p(Z)$ thus becomes

$$G_{h_0} G_p(Z) = \frac{(1-Z^{-1})Z^{-1}Z^{-1}(0.4512 + 0.2623\,Z^{-1})}{(1-Z^{-1})(1-0.2865\,Z^{-1})} \qquad (12.11)$$

$$G_{h_0} G_p(Z) = \frac{0.4512\,Z^{-2} + 0.2623\,Z^{-3}}{1-0.2865\,Z^{-1}}$$

Now let us return to Equation (12.4) and evaluate $D(Z)$. For a *PI* controller the output is related to the error by the equation

$$v_n = v_{n-1} + K_c (e_n - e_{n-1}) + \frac{K_c}{\tau_I} T e_n \qquad (12.12)$$

Taking the Z transform of Equation 12.12 gives

$$V(Z) = Z^{-1} V(Z) + K_c E(Z) - Z^{-1} K_c E(Z) + \frac{K_c}{\tau_I} T E(Z) \qquad (12.13)$$

Therefore,

$$D(Z) = \frac{V(Z)}{E(Z)} = \frac{\left(K_c + \frac{K_c}{\tau_I} T\right) - K_c Z^{-1}}{1 - Z^{-1}} \tag{12.14}$$

Substituting the given values for the controller constants

$$K_c = 0.43$$
$$\tag{12.15}$$
$$\tau_I = 1.57$$

into Equation (12.14), we get

$$D(Z) = \frac{0.57 - 0.43 Z^{-1}}{1 - Z^{-1}} \tag{12.16}$$

Now, for a unit step change in set point

$$R(t) = u(t)$$

and $\tag{12.17}$

$$R(Z) = \frac{1}{1 - Z^{-1}}$$

Thus Equation (12.4) becomes

$$C(Z) = \frac{1}{1 - Z^{-1}} \times$$

$$\left[\frac{\left(\dfrac{0.57 - 0.43Z^{-1}}{1 - Z^{-1}}\right)\left(\dfrac{0.4512Z^{-2} + 0.2623Z^{-3}}{1 - 0.2865 \, Z^{-1}}\right)}{1 + \left(\dfrac{0.57 - 0.43Z^{-1}}{1 - Z^{-1}}\right)\left(\dfrac{0.4512Z^{-2} + 0.2623Z^{-3}}{1 - 0.2865 \, Z^{-1}}\right)}\right] \tag{12.18}$$

The right side can be simplified and then expressed in power series of Z^{-1} by long division to give

$$C(Z) = 0 + 0Z^{-1} + 0.2565 \, Z^{-2} + 0.542 \, Z^{-3} + 0.658 \, Z^{-4}$$
$$+ \, 0.664 \, Z^{-5} + 0.649 \, Z^{-6} + \cdots \tag{12.19}$$

A plot of $C^*(t)$ versus time is shown in Figure 12.2

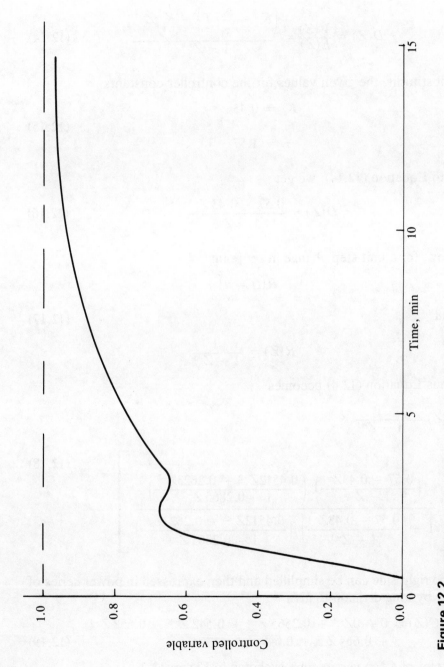

Figure 12.2
Response of System to Unit Step Change in Set Point

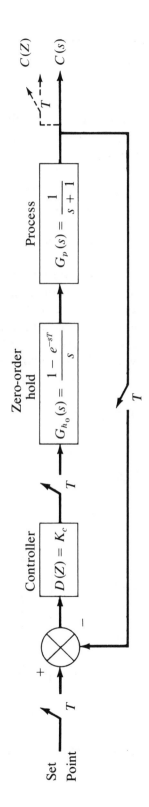

Figure 12.3
Proportional Sampled-Data Control System for a First-Order Process

12.3 Application of Modified Z Transforms to Determine Output Between Sampling Instants

Modified Z transforms offer an extremely valuable method for evaluating the response of sampled-data control systems between sampling instants. This is important, because a stable system and an unstable system can yield identical responses at the sampling instants. Since the Z-transform technique provides information only at the sampling instants, it would be impossible to tell which of the two systems is unstable. The procedure describing the use of modified Z transforms for determining the output between sampling instants is outlined in the following paragraphs.

Consider the block diagram of a sampled-data control system shown in Figure 12.3. Suppose the continuous response of the controlled variable to a step change in set point is as shown in Figure 12.4a. The Z-transform technique will give us the response of $C^*(t)$ as shown in Figure 12.4b. We would like to generate additional information about $C^*(t)$ between sampling instants so that the complete time response of C may be developed. To begin with, let us define a quantity m as

$$m = 1 - \frac{\theta}{T} \tag{12.20}$$

or

$$T = mT + \theta$$

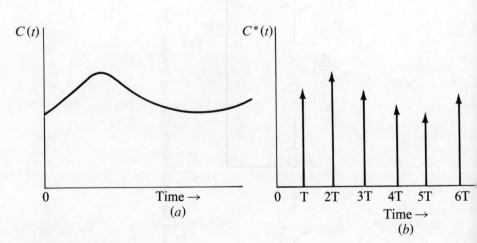

C(t)

0   Time →

(a)

C*(t)

0   T   2T   3T   4T   5T   6T

Time →

(b)

Figure 12.4
Response of System Shown in Figure 12.3

Equation (12.20) shows that when θ equals 0, m will equal 1, and when $\theta = T$, m will equal 0.

Now, let us add a fictitious dead-time element to the block diagram of Figure 12.3, as shown in Figure 12.5. Since the delay element is outside the block diagram, the response of $C(t)$ is unaffected. The fictitious delay element will enable us to delay the controlled variable so that we may obtain the inter-sampling information. For example, if the signal $C(t)$ at $t = 0$ is delayed θ time units and then sampled, we will obtain a different value of C^* as shown in Figure 12.6. Advancing the output of the sampler some more gives another value of C^*. In fact, as θ is varied between O and T (i.e., m between 0 and 1 as per Equation 12.20), the entire signal between these sampling instants can be reproduced. The repeated application of this procedure then will enable us to obtain the complete time response of the controlled variable. Let us illustrate the technique by applying it to an example problem.

Example 2[4]. Determine the response $C(t)$ of the system shown in Figure 12.3 for all values of time t, for a unit step change in set point. The closed-loop pulse transfer function of this system is

$$\frac{C(Z, m)}{R(Z)} = \frac{D(Z)\, G_2(Z)}{1 + D(Z)\, G_{h_0} G_p(Z)} \tag{12.21}$$

where

$$C(Z, m) = Z\{e^{-\theta s}\, C(s)\}$$

$$G_2(Z) = Z\{G_{h_0}(s)\, G_p(s)\, e^{-\theta s}\}$$

$$= Z_m\{G_{h_0}(s)\, G_p(s)\}$$

$$= G_{h_0} G_p(Z, m)$$

and

$$G_{h_0} G_p(Z) = Z\{G_{h_0}(s)\, G_p(s)\}$$

Thus Equation (12.21) can be written as

$$\frac{C(Z, m)}{R(Z)} = \frac{D(Z)\, G_{h_0} G_p\, (Z, m)}{1 + D(Z)\, G_{h_0} G_p(Z)}$$

Let us evaluate the Z transforms: For this illustration assume that

$$D(Z) = 1.5 \tag{12.22}$$

$$G_2(Z) = G_{h_0}\, G_p\, (Z, m) = Z_m\left[\frac{1 - e^{-sT}}{s} \cdot \frac{1}{s + 1}\right] \tag{12.23}$$

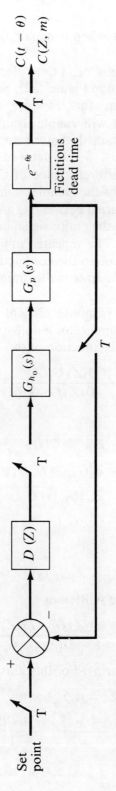

Figure 12.5
Sampled-Data System of Figure 12.3 with a Fictitious Dead-Time

$$= Z\left[\frac{(1 - e^{-sT})}{s(s + 1)} e^{-\theta s}\right]$$

$$= \frac{Z^{-1}\left[(1 - e^{-mT}) + (e^{-mT} - e^{-T})Z^{-1}\right]}{1 - e^{-T} Z^{-1}}$$

and
$$G_{h_0}G_p(Z) = Z\{G_{h_0}(s) G_p(s)\}$$

$$= Z\left\{\frac{1 - e^{-sT}}{s(s + 1)}\right\} \qquad (12.24)$$

$$= \frac{(1 - e^{-T}) Z^{-1}}{1 - e^{-T} Z^{-1}}$$

For a step change in $R(t)$

$$R(Z) = \frac{1}{1 - Z^{-1}}$$

Substituting in Equation 12.21 for $R(Z)$ gives

$$C(Z, m) = R(Z)\frac{D(Z) G_{h_0}G_p (Z, m)}{1 + D(Z) G_{h_0}G_p (Z)} \qquad (12.25)$$

$$= \frac{1}{1 - Z^{-1}} \cdot \frac{1.5Z^{-1}\left[(1 - e^{-mT}) + (e^{-mT} - e^{-T})Z^{-1}\right]}{\left[1 - e^{-T} Z^{-1} + 1.5(1 - e^{-T})Z^{-1}\right]}$$

For $T = 1$, this equation gives, after some simplification,

$$C(Z, m) = \frac{1.5Z^{-1}\left[(1 - e^{-mT}) + (e^{-mT} - e^{-T})Z^{-1}\right]}{1 - 0.42Z^{-1} - 0.58Z^{-2}} \qquad (12.26)$$

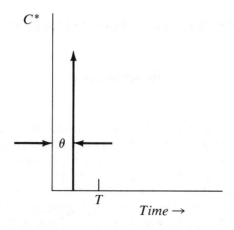

Figure 12.6 Response Delayed by θ Time Units and then Sampled

Equation (12.26) can be expressed as a power series by long division. Thus

$$C(Z, m) = 1.5\,(1 - e^{-mT})Z^{-1} + 1.5(0.58\,e^{-mT} + 0.05)Z^{-2}$$
$$+ 1.5\,(0.602 - 0.341\,e^{-mT})Z^{-3} + \cdots \qquad (12.27)$$

The coefficient in front of Z^{-n} gives the response C^* between $(n - 1)$th and nth sampling instant. By varying m between 0 and 1, the complete response between any two sampling instants can be evaluated. Thus between 0 and T the response is given by

$$C^* = 1.5\,(1 - e^{-mT}) \qquad (12.28)$$

By varying m between 0 and 1, the complete response between O and T can be evaluated. The complete response thus obtained for the problem is sketched in Figure 12.7.

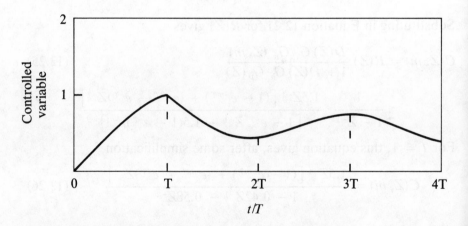

Figure 12.7 Complete Response of System of Figure 12.3

References

1. Barker, R. H., The Pulse Transfer Function and Its Application to Sampling Servo Systems, *Proc. IEE, London*, **99**, Part 4, 1952, 302–317.
2. Jury, E. I., Syntheses and Critical Study of Sampled-data Control Systems, *Trans. A.I.E.E.*, **75**, Part 2, 1956, 141–151.
3. Jury, E. I., Additions to the Modified *Z*-Transform Method, *IRE WESCON Conventional Record*, Part 4, August 21, 1957. pp. 136–156.
4. Smith, C. L., *Digital Computer Process Control*, Intext Publishers, Scranton, Pa., 1972.

PART II

Design and Application of Advanced Control Concepts

CHAPTER 13

Process Modeling and Identification

From our study of Chapter 10 it should be clear that the design of Z transform-based computer-control algorithms requires that the process transfer function be known. The knowledge of the process transfer function is also required for the implementation of advanced control strategies. A process transfer function may be developed from a theoretical analysis of the process or from experimental tests in the plant. In the former approach, we write appropriate unsteady-state balances (e.g., those involving conservation of mass, energy, and momentum), linearize the resulting differential equations if the differential equations turn out to be nonlinear, express the variables in deviation form, and finally take the Laplace transform so as to obtain the process transfer function. In the experimental approach we introduce a suitable change in the input and record the output response. These data are subsequently analyzed to determine an approximate process transfer function.

The development of dynamic mathematical models for specific

process systems (e.g., distillation columns, heat exchangers, chemical reactors) is beyond the scope of this text. For the interested reader we have included at the end of this chapter a bibliography on mathematical modeling of chemical processes.

The dynamic mathematical models are extremely useful in the design and analysis of process-control systems. A common application of the models is in the design phase of the project. In this instance, the process-control system based on the model may be simulated on a digital, analog, or hybrid computer so as to assess the relative benefits of different control strategies. Of course, this approach suffers somewhat, because some factors that can affect the operation of the real process may not be included in the model.

In dealing with existing plants that are in operation, a more direct approach would be to employ experimental tests to develop a process model or a process transfer function. This is the subject of this chapter.

Chemical engineering processes may be described either as being open-loop stable or open-loop unstable. Most processes that we as control engineers must work with are open-loop stable and are therefore self-regulating. This means that in manual control if a step change in the process input is made, the process output will ultimately reach a new steady-state level. On the other hand, some commercially important processes exhibit open-loop unstable behavior and are termed nonself-regulating processes. Examples of this class of processes include some level control systems and exothermic chemical reactors. In an exothermic chemical reactor a change in the coolant flow causes a temperature upset, which affects reaction rate and in turn temperature and so on; thus, in theory at least, the temperature can rise without a bound. In some level control systems a step change in the manipulated variable causes the level to change indefinitely. The experimental identification of process parameters of open-loop unstable systems is generally difficult. Deshpande[1] has presented a technique for obtaining process models of open-loop unstable systems from closed-loop tests, and the reader interested in the modeling of such systems is referred to that article for details. In this chapter we describe three experimental techniques, in order of increasing complexity, for identifying the dynamics of open-loop stable processes. The first is based on step response and gives the parameters of a first- or second-order model having a dead-time element. The second, called pulse testing, is a frequency-domain method, which yields the frequency response diagram of the open-loop process. The third technique is a time-domain method which yields the gain and time constants of the process model.

13.1 Process Models from Step-Test Data

This technique consists of subjecting the process, while it is operating under steady-state conditions with the feedback controller in manual, to a step change in input and recording the resulting transient response. The notion of a transfer function is involved in the development of the model parameters, and therefore, some care should be exercised in selecting the size of the step input. The step size should be sufficiently large so that the output data is distinguishable from the process noise but not so large as to drive the system out of the linear range. One should also make certain that the process is free of load upsets throughout the duration of the test. Generally, at least a couple of tests should be conducted, one involving a positive step change and another involving a negative step change, to ascertain that the process loads are absent and that the range of linearity is not exceeded.

The plot of step response of an open-loop process versus time is called a process reaction curve. Numerous methods are available for the determination of model parameters from the process reaction curve. Ziegler and Nichols[2] and Miller[3] have described a simple procedure for finding the parameters of a first-order plus dead-time model. Oldenbourg and Sartorius[4] were probably the first investigators to propose a procedure for determining the parameters of a second-order model, which was later extended by Sten[5] to include the estimation of dead time. Other approaches to model parameter evaluation have been proposed by Smith[6], Cox[7], Smith and Murrill[8], Meyer et al.[9], Csaki and Kis[10], Naslin and Miossec,[11] and Sundaresan et al.[12] Most of the available methods for parameter estimation of second-order models depend on the accurate location of the inflection point, which is prone to error. Sundaresan et al.[12] appear to have overcome this difficulty. We describe their procedure, with the permission of the publisher, in the following paragraphs. The interested reader may wish to consult the earlier references to learn about the previous methods.

The transfer function of an overdamped second-order plus dead-time model is given by

$$G_p(s) = \frac{e^{-\theta_d s}}{(\tau_1 s + 1)(\tau_2 s + 1)} \tag{13.1}$$

Whereas the transfer function of an underdamped second-order plus dead-time model is given by

$$G_p(s) = \frac{e^{-\theta_d s}}{\dfrac{1}{\omega^2_n} s^2 + \dfrac{2\zeta}{\omega_n} s + 1} \qquad (13.2)$$

The objectives are to determine the three parameters θ_d, τ_1, and τ_2 or θ_d, ω_n, ζ of the model from the process reaction curve. The steady-state gain K_p is easily determined from the plot as the ultimate change in the controlled variable per unit change in the process input and therefore is not one of the unknown parameters in this analysis. The plot of a typical process response to which we wish to fit the second-order plus dead-time model is shown in Figure 13.1. The method of Sundaresan, which is based on the work of Lees[13] and Gibilaro and Waldram,[14] recognizes the fact that the process reaction curve resembles certain statistical distribution functions and therefore can be described by its first moment, which in turn could be utilized for parameter evaluation. First, let us consider the evaluation of the overdamped system parameters K_p, τ_1, and τ_2.

As pointed out by Sundaresan et al., the first moment of the response function $c(t)$ is

$$m_1 = \int_0^\infty (1 - c(t)) dt \qquad (13.3)$$

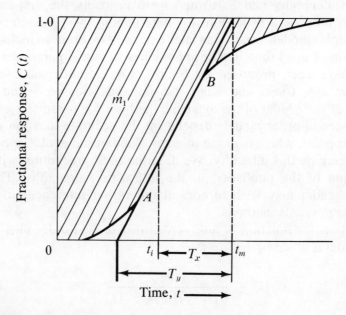

Figure 13.1
Evaluation of Second-Order Dead-Time Parameters from an Overdamped Curve

Thus m_1 is simply the shaded area in Figure 13.1.

The authors also point out that the Laplace transform is related to the moment-generating function. The relationship between the first moment m_1 and the transfer function $G_p(s)$ is

$$m_1 = - \left. \frac{d\,G_p(s)}{d\,s} \right|_{s=0} \tag{13.4}$$

$$= \theta_d + \tau_1 + \tau_2$$

Now, consider the time-domain solution of Equation (13.1) for a step change in input, which is given by the equation

$$C(t) = \left[1 - \frac{\tau_1}{(\tau_1 - \tau_2)} e^{- \frac{t - \theta_d}{\tau_1}} \right. $$

$$\left. + \frac{\tau_2}{(\tau_1 - \tau_2)} e^{- \frac{t - \theta_d}{\tau_2}} \right] u(t - \theta_d) \tag{13.5}$$

By differentiating Equation (13.5) twice and setting the resulting second derivative equal to zero, we may find the expression for point of inflection

$$t_i = \theta_d + \alpha \ln \eta \tag{13.6}$$

where

$$\eta = \frac{\tau_1}{\tau_2} \tag{13.7}$$

and

$$\alpha = \frac{\tau_1 \tau_2}{\tau_1 - \tau_2} \tag{13.8}$$

The slope M_i at the point of inflection is

$$M_i = \frac{(\eta)^{1/(1-\eta)}}{(\eta - 1)\alpha} \tag{13.9}$$

This slope is that of a tangent that passes through t_i and intersects the final value of $c(t)$ at time t_m, as shown in Figure 13.1. This value of time is given by

$$t_m = \theta_d + \alpha \left[\ln \eta + \frac{\eta^2 - 1}{\eta} \right] \tag{13.10}$$

Equations (13.4), (13.9), and (13.10) may be combined so as to get

$$(t_m - m_1)M_i = \frac{\eta^{1/(1-\eta)}}{(\eta - 1)} \ln \eta \tag{13.11}$$

Note that if in this last equation η is changed to $1/\eta$, the right side remains unchanged. Therefore, in the solution of Equation (13.11) we need to consider only the range of 0 to 1 for η. Let us now write Equation (13.11) in the form

$$\lambda = \chi e^{-\chi} \tag{13.12}$$

where

$$\lambda = (t_m - m_1)M_i \tag{13.13}$$

and

$$\chi = \ln \eta / (\eta - 1) \tag{13.14}$$

It is clear from Equation (13.12) that the maximum value of λ is e^{-1} which occurs when the system is critically damped, that is, $\eta = 1$ or $\chi = 1$. For the overdamped case, η is less than 1, and therefore, χ lies between 0 and e^{-1}. At the extreme, where η equals zero and, therefore, $\lambda = 0$, the model reduces to a first-order system. Thus for approximating the true response by a second-order model with or without dead time the values of λ lie between 0 and e^{-1}. A plot of Equation (13.12) is shown in Figure 13.2. The procedure for parameter evaluation is as follows:

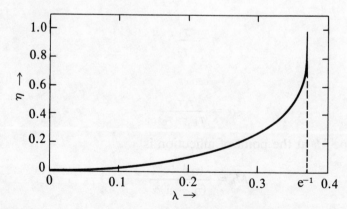

Figure 13.2
Plot of λ for Overdamped Second-Order Approximations

1. Determine the shaded area of Figure 13.1 by numerical integration.

2. Draw a tangent that passes through the straight line portion of the response curve.

3. Denote the slope of this line as M_i and its intersection with the final value of $C(t)$ as t_m.

4. Compute λ as per Equation (13.13).

5. From Figure 13.2 determine η.

6. Solve Equation (13.9) for α.

7. Solve Equations (13.4), (13.7), and (13.8) for θ_d, τ_1, and τ_2.

The equations resulting from the application of this procedure are

$$\tau_1 = \frac{\eta^{\eta/(1-\eta)}}{M_i}$$

$$\tau_2 = \frac{\eta^{1/(1-\eta)}}{M_i} \tag{13.15}$$

$$\theta_d = m_1 - \frac{\eta^{1/(1-\eta)}}{M_i}\left(\frac{\eta+1}{\eta}\right)$$

For critically damped systems $\eta = 1$ and these equations reduce to

$$\tau_1 = \tau_2 = \frac{1}{M_i e} \tag{13.16}$$

$$\theta_d = m_1 - \frac{2}{M_i e}$$

For underdamped second-order approximations, the following procedure has been recommended. Note that in this case λ is greater than e^{-1}. Figure 13.3 shows a typical response for an underdamped system. The time domain solution of the system of Equation (13.2) to a step change in input is

$$c(t) = u(t - \theta_d)\left[1 - e^{-(t-\theta_d)\omega_n \xi}\left\{\frac{\xi}{\sqrt{1-\xi^2}}\sin\sqrt{1-\xi^2}\,\omega_n\right.\right.$$

$$\left.\left. \times\,(t-\theta_d)) + \cos\left(\sqrt{1-\xi^2}\,\omega_n(t-\theta_d)\right)\right\}\right] \tag{13.17}$$

Using this equation and the application of a procedure similar to that employed in the overdamped case, the authors have derived the following equations for the system parameters

$$\lambda = (t_m - m_1)M_i = \frac{\cos^{-1}\xi}{\sqrt{1-\xi^2}}\exp\left(\frac{-\xi}{\sqrt{1-\xi^2}}\cos^{-1}\xi\right) \tag{13.18}$$

$$\omega_n = \frac{\cos^{-1}\xi}{\sqrt{1 - \xi^2}} \frac{1}{(t_m - m_1)} \tag{13.19}$$

$$\theta_d = m_1 - \frac{2\xi}{\omega_n} \tag{13.20}$$

It may be noted that in the computation of m_1, the areas below the ordinate of 1 are taken as positive and those above 1 are taken as negative. Also, although Equation (13.18) provides for an infinite number of roots for ζ for a given value of λ, in the region of our interest $(0 \leqslant \zeta \leqslant 1)$ λ is a monotonic function of ζ, decreasing from $\pi/2$ at $\zeta = 0$ to e^{-1} at $\zeta = 1$. A plot of ζ versus λ is shown in Figure 13.4. The other system parameters are then determined from Equations (13.19) and (13.20).

The method described in this chapter is simple and easy to use, but its application would require a digital computer program for numerical integration of Equation (13.3). Let us apply the method to a couple of example problems.

Example 1. The process transfer function of a third-order system is given by

$$G(s) = \frac{C(s)}{X(s)} = \frac{1}{(s + 1)(0.5s + 1)(2s + 1)} \tag{13.21}$$

where the time constants are expressed in minutes. The true response

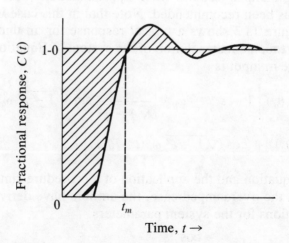

Figure 13.3
Response of Second-Order Underdamped System with Dead Time

of $C(t)$ to a step change in input $X(t)$ is shown in Figure 13.5. Approximate this process by a second-order plus dead-time model.

From Figure 13.5 the shaded area m_1 is computed to be 3.5 min. The tangent line drawn along the straight-line portion of the plot has a

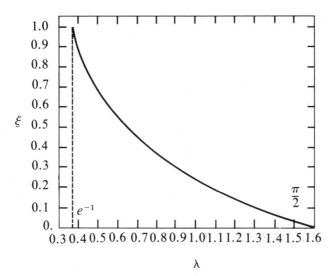

Figure 13.4
Plot of λ versus ξ for Second-Order Underdamped Approximations

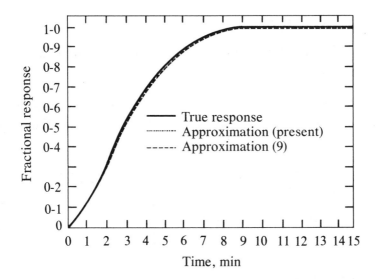

Figure 13.5
Comparison of True Response with Approximate Response of an Overdamped Second-Order System

slope $M_i = 0.23$ min^{-1}, and it intersects $C(t) = 1$ line at $t_m = 5.1$ min. For $\lambda = (t_m - m_1)M_i = 0.368$, Figure 13.2 gives $\eta = 0.9$. Substituting these values into Equation (13.15) gives $\theta_d = 0.33$, $\tau_1 = 1.5$, and $\tau_2 = 1.67$. Thus the approximating transfer function is

$$G_p(s) = \frac{e^{-0.33s}}{(1.67s + 1)(1.5s + 1)} \tag{13.22}$$

Example 2. Consider the block diagram of an open-loop cascade control system having an inner closed-loop shown in Figure 13.6. The response of $C(t)$ to a step change in $R(t)$ is shown in Figure 13.7. From this figure $M_i = 0.417$ min^{-1}, $t_m = 3.4$ min, and $m_1 = 1.99$ min, which gives a value of λ equal to 0.588. From Figure 13.4 at this value of λ the damping parameter ζ equals 0.588. These values together with Equations (13.19) and (13.20) give $\theta_d = 0.566$ min and $\omega_n = 0.826$ min^{-1} (or $\tau = 1/0.826 = 1.21$ min). Thus the system of Figure 13.6 may be approximated as

$$R(s) \rightarrow \boxed{\frac{e^{-0.566s}}{1.465s^2 + 1.42s + 1}} \rightarrow C(s)$$

The step response of the model is compared with the true response of the process in Figure 13.7. The results show good agreement.

13.2. Pulse Testing for Process Identification

Unlike the reaction curve method, this method is a frequency-domain method yielding a frequency-response diagram of the open-

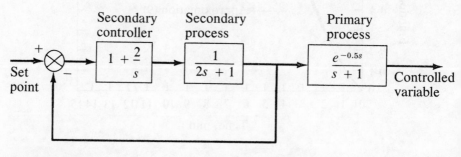

Figure 13.6
Open-loop Cascade System with a Closed Inner Loop

loop process. As in the case of the reaction curve method, one must ascertain that load upsets are absent during the test. Also, since the notion of Laplace transforms is involved, the range process linearity should not be exceeded throughout the duration of the test.

In the pulse-testing method, a pulse of arbitrary shape is applied at the input of the process $[X(s)$ in Figure 13.8$]$, while it is operating under steady-state conditions with the controller in manual, and the transient response of the process is recorded. These records (see Figure 13.9) assist us in the development of a frequency-response diagram, as described in the following paragraphs. For a historical discussion and examples of early applications of pulse testing, the interested reader is referred to Hougen[15a] and Clements and Schnelle.[15b] The Laplace transform of the input variable $X(s)$ is related to the Laplace transform of the output $Y(s)$ via the transfer function $G(s)$, which is the product of $G_u(s)$, $G_p(s)$, and $G_m(s)$. Thus

$$G(s) = \frac{Y(s)}{X(s)} \tag{13.23}$$

If the definition of Laplace transforms is substituted in Equation (13.23), it can be written as

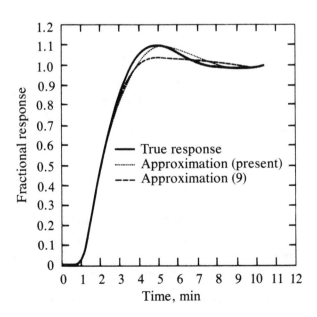

Figure 13.7
Comparison of Time Response with Approximate Response for an Underdamped Second-Order Process

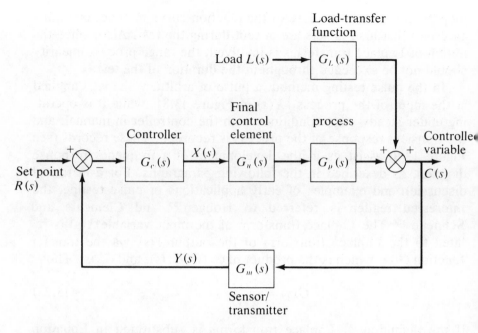

Figure 13.8
Typical Closed-Loop Control System

$$G(s) = \frac{\displaystyle\int_0^\infty Y(t)\, e^{-st}\, dt}{\displaystyle\int_0^\infty X(t)\, e^{-st}\, dt} \tag{13.24}$$

From Figure 13.9 it is clear that the upper limits on the two integrals can be replaced by t_y and t_x, the time periods during which changes in Y and X occur. If, in addition, $j\omega$ is substituted for s, Equation (13.24) becomes

$$G(j\omega) = \frac{\displaystyle\int_0^{t_y} Y(t)\, e^{-j\omega t}\, dt}{\displaystyle\int_0^{t_x} X(t)\, e^{-j\omega t}\, dt} \tag{13.25}$$

The numerator in Equation (13.25) is the Fourier transformation of the time function $Y(t)$ and the denominator is the Fourier transformation of the time function $X(t)$. Since $e^{-j\omega t}$ can be replaced by $\cos \omega t - j \sin \omega t$, Equation (13.25) can be written as

$$G(j\omega) = \frac{\displaystyle\int_0^{t_y} Y(t)\cos(\omega t)\, dt - j\int_0^{t_y} Y(t)\sin(\omega t)\, dt}{\displaystyle\int_0^{t_x} X(t)\cos(\omega t)\, dt - j\int_0^{t_x} X(t)\sin(\omega t)\, dt} \tag{13.26}$$

$$= \frac{A - jB}{C - jD} \tag{13.27}$$

where A, B, C, and D represent the four integrals in Equation (13.26). If the right-hand side is multiplied and divided by the complex conjugate of the denominator, Equation (13.27) becomes

$$G(j\omega) = \frac{A - jB}{C - jD} \cdot \frac{C + jD}{C + jD}$$

$$= \frac{AC + BD + j(AD - BC)}{C^2 + D^2} \tag{13.28}$$

$$= \frac{(AC + BD)}{C^2 + D^2} + j \frac{(AD - BC)}{C^2 + D^2}$$

$$= \text{Re } G(j\omega) + j \text{ Im } G(j\omega)$$

From this equation the amplitude ratio AR and the phase angle ϕ are determined by taking the magnitude and the angle of the complex number, respectively.

Thus

$$AR = |G(j\omega)| = \sqrt{\text{Re } G(j\omega)^2 + \text{Im } G(j\omega)^2}$$

$$= \sqrt{\left(\frac{AC + BD}{C^2 + D^2}\right)^2 + \left(\frac{AD - BC}{C^2 + D^2}\right)^2} \tag{13.29}$$

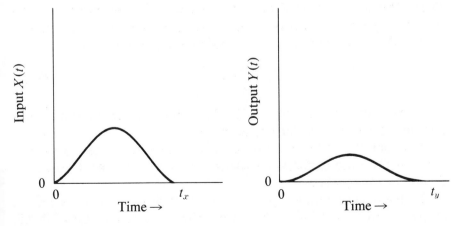

Figure 13.9
Typical Input-Output Plots from Pulse Tests

and

$$\phi = \arctan \left(\frac{AD - BC}{AC + BD} \right) \tag{13.30}$$

To use these equations, a specific value of frequency ω is chosen. The integrations are performed, yielding a single value of A, B, C, and D. Substituting these values in Equations (13.29) and (13.30) then gives a point on the frequency-response diagram, corresponding to the chosen frequency. The procedure is repeated for numerous values of ω, and the complete frequency response diagram is prepared. Since the time functions $Y(t)$ and $X(t)$ would be available in graphical form, the integrals in Equation (13.26) are most conveniently evaluated using numerical integration techniques on the digital computer. The listing of a digital computer program[16] written in Fortran IV for analyzing pulse test data is included in Appendix D. The program deck and the accompanying user's manual are available for purchase at ISA headquarters.

The pulse-testing technique gives us a frequency-response diagram of the open-loop process. This is as far as we need go if our purpose is to determine the tuning constants of conventional controllers. If, on the other hand, we require a process transfer function for the purpose of implementing computer control or advanced control applications, it may be fitted to the frequency response diagram as described in the following examples.

Example 3. The results from a pulse test on an industrial open-loop cascade control system with a closed inner loop are shown in Figure 13.10. The use of these data in conjunction with the pulse analysis program results in frequency-response data which are plotted in Figure 13.11. We wish to fit a transfer function of the form

$$G_p(s) = \frac{K_p \, e^{-\theta_d s}}{\tau_p s + 1} \tag{13.31}$$

to the data shown in Figure 13.11. The following step-by-step procedure may be employed for this purpose:

1. From the pulse analysis program the steady-state gain K_p, which is the ratio of the areas under the input and output curves in Figure 13.10, equals 1.

2. The corner frequency ω_c for the first-order process occurs at an AR equal to $0.707 \, K_p$. Thus, from Figure 13.11a at $AR = 0.707$, the corner frequency ω_c equals 2.5 radians/min. Therefore, the time constant of the first-order process is

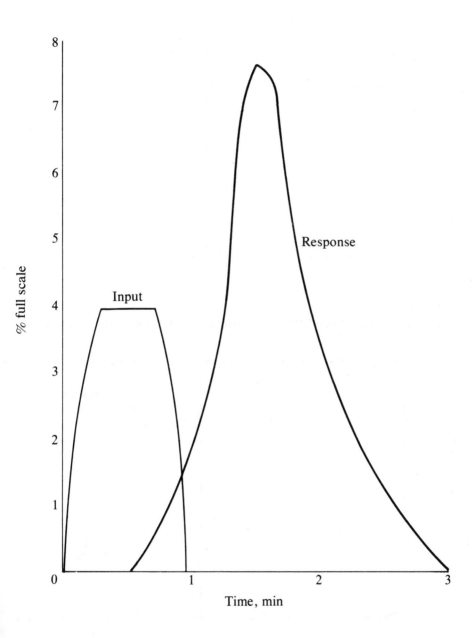

Figure 13.10
Input Pulse and Transient Response

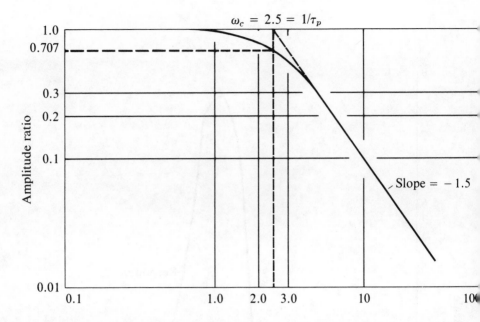

Figure 13.11*a*
Frequency-Response Diagram—Amplitude Ratio versus Frequency

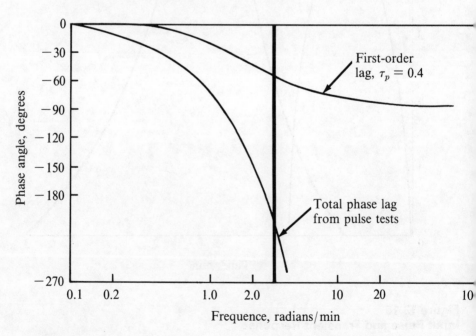

Figure 13.11*b*
Frequency-Response Diagram — Phase Angle versus Frequency

$$\tau_p = \frac{1}{\omega_c} = \frac{1}{2.5} = 0.4$$

3. Now we plot the phase angles for this first-order lag with $\tau_p = 0.4$, as shown in Figure 13.11b.

4. For a first-order process the phase angle at the corner frequency equals 45°. Therefore, at this frequency the total phase angle is the sum of the phase angles of first-order lag and that of dead time. Thus, from Figure 13.11b

$$\phi_{total} = -154 = \left.\begin{array}{c} \text{phase angle due} \\ \text{to dead time} \end{array}\right| \begin{array}{c} -45° \\ \\ \omega_c \end{array}$$

or

$$-154 = -\omega_c \, \theta_d \left(\frac{180}{\pi}\right) - 45$$

$$= -(2.5) \left(\frac{180}{\pi}\right) \theta_d - 45$$

Therefore,

$$\theta_d = \frac{(109)\,(\pi)}{(180)\,(2.5)} = 0.76 \text{ min}$$

and the approximate model is

$$G_p = \frac{1e^{-0.76s}}{(0.4s + 1)}$$

In the previous example we considered the fitting of frequency-response curves to a first-order lag with dead time. To fit second-order models we note the following characteristics of the second-order systems:

1. As in first-order systems the steady-state gain is the ratio of areas under the input and output pulse curves. The zero-frequency asymptote on the amplitude-ratio portion of the Bode plot also gives the steady-state gain of the process.

2. For overdamped second-order approximations, the two time constants are found by noting the corner frequencies (also called breakpoint frequencies) on the amplitude curve of the Bode plot. To do this, we fit the amplitude data as far out in frequency as possible to a first-order curve by trial and error. We then note the corner frequency of the first-order curve, which is the reciprocal of the time constant τ_1. Then we graphically subtract the first-order

curve from the amplitude data which gives the amplitude data for another first-order lag to which a first-order lag is fitted so as to obtain τ_2.

3. For underdamped second-order systems the time constant is found from the breakpoint frequency, and the damping factor is found by comparing the resonant peak with the amplitude data of known second-order systems. The Bode plot of an underdamped second-order system for several values of the damping factor ζ is shown in Figure 13.12.

4. Once the amplitude data are fitted to a known transfer function, its phase angles are plotted and compared with the experimen-

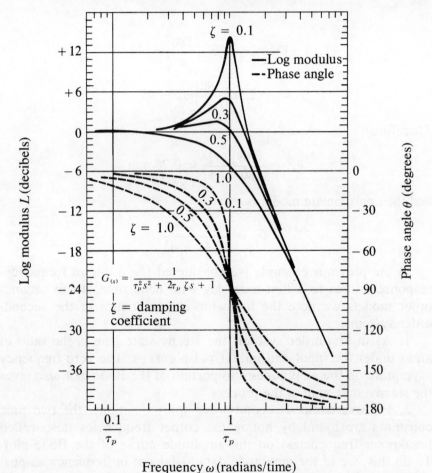

Figure 13.12
Second-Order System Bode Plots [From Ref. 17 with Permission]

tal phase angle data. The difference, if any, is the contribution of dead time. Since for a dead-time element the phase angle equals $\omega\theta_d$, θ_d can be easily evaluated.

In Figures 13.13*a* and 13.13*b*, taken from Luyben,[17] we show two examples of fitting higher-order transfer functions. The details of the fitting procedure have been left out as an exercise for the reader.

The test input during a pulse test begins and ends at the same value. Generally, the output will also return to the initial value. However, under certain circumstances the output will not return to the initial steady state. This occurs when the open-loop process contains a pure integrator.[17] For example, if the inflow to a tank is

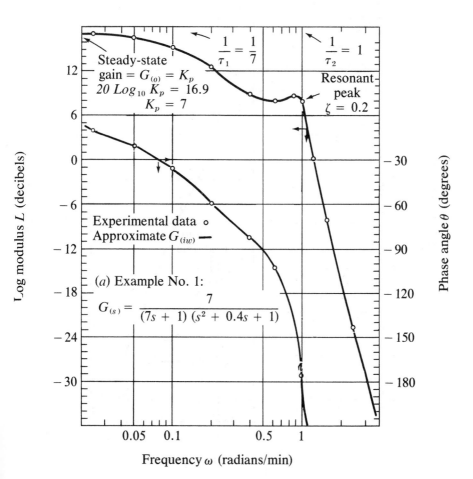

Figure 13.13*a*
Fitting Approximate Transfer Function to Experimental Frequency-Response Data (From Ref. 17 with Permission)

pulsed while the flow out of the tank is held constant, the tank level will rise to a new level and will not return to its initial value on conclusion of the test. The computer program listing of Appendix D is capable of handling nonclosing pulses that arise when working with processes containing an integrating element.

Let us now discuss the criteria for pulse selection and for the selection of the step size in the numerical integration of Equation (13.26).

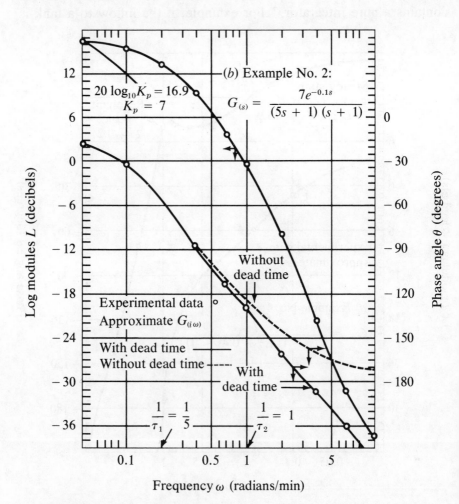

Figure 13.13b
Fitting Approximate Transfer Function to
Experimental Frequency-Response Data (From Ref. 17 with Permission)

Shape of the Input Pulse. Consider the definition of $G(j\omega)$

$$G(j\omega) = \frac{\int_0^{t_y} Y(t)\, e^{-j\omega t}\, dt}{\int_0^{t_x} X(t)\, e^{-j\omega t}\, dt} \tag{13.31}$$

The shape of the pulse should be such that its frequency content (i.e., amplitude of FIT, the abbreviation for Fourier integral transforms) should be finite over the frequency range of interest. For a rectangular pulse, FIT goes to zero at $\omega = 2\pi/T$, $4\pi/T$, and so on. For a displaced cosine* pulse FIT goes to zero at $4\pi/T$, $6\pi/T$, and so on. Therefore, the latter may be a more suitable input. However, a near rectangular pulse is so much easier to generate and therefore is used in most applications.

Magnitude of the Pulse. Recall that we are after a linear dynamic model in the form of $G(j\omega)$. It must be a linear model, because the notion of a transfer function applies only to a linear system. In general, the process is nonlinear, and we are obtaining a model that is linearized around the steady-state operating level. If the height of the pulse is too high, we may drive the process out of the linear range. Therefore, pulses of various heights should be tried. ISA[16] recommends 10% of span as a starting point for magnitude of the input pulse. It is a good idea to test with positive and negative pulses. The computed $G(j\omega)$ should be identical if the region of linearity is not exceeded.

Duration of the Input Pulse. The FIT of a rectangular pulse of width D is:

$$\text{FIT} = h\, \frac{\sin \omega D}{\omega} - i\, \frac{h}{\omega}(1 - \cos \omega D) \tag{13.32}$$

when frequency $\omega = 2\pi/D$, FIT goes to zero, and the calculation of the transfer function is meaningless. Therefore, the smaller D can be made, the higher is the frequency to which $G(j\omega)$ can be found.

According to Luyben,[17] a good rule of thumb is to keep the width of the pulse less than about half the smallest time constant of interest.

---

*For a Displaced cosine function

$$f(t) = 1 - \cos \frac{2\pi}{T} t \qquad 0 \leq t \leq T$$

If the dynamics of the process are completely unknown, it may take a few trials to establish a reasonable pulse width.

If D is too small for a given pulse height the system is disturbed very little, and it becomes difficult to separate the real output signal from noise and experimental error.

Size of ΔT in Numerical Integration

According to Messa et al.,[18] the criterion for picking the number of data points is that enough data points must be used so that the approximating function matches the real function over all intervals. Based on the data from several investigators, they suggest the size of Δt should be such that $\omega_0 (\Delta t) \leqslant \pi/2$ where ω_0 is the upper limit on the desired frequency and Δt is the time interval between points. The upper limit of frequency is determined by the frequency content of the input signal. Thus, for a rectangular pulse of width D, we have

$$\text{FIT} = h \, \frac{\sin \omega D}{\omega} - i \, \frac{h}{\omega} \, (1 - \cos \omega D) \qquad (13.33)$$

FIT goes to zero when $\omega = 2\pi/D$. Therefore, calculation of FIT and the transfer function near this frequency is meaningless. Also, Murill et al.[19] have shown that FIT amplitude beyond $\omega = 2\pi/D$ is very small. Therefore, the upper limit of frequency for a rectangular pulse is $\omega_0 = 2\pi/D$. As an example, if D = 5 min, the upper limit of frequency equals (2) (3.14)/5 = 1.26 rad/min. Now if we select Δt of 1 min, $\omega_0 \Delta t$ for this case will be (1.26) rad/min $\times$ (1.0) min = 1.26 rad. This is well below the suggested value of $\pi/2$ (equals 1.56 rad). Therefore, 1 min for the size of Δt is entirely satisfactory.

13.3 Time Domain Process Identification

The need for time domain identification arises because dynamic parameters of processes can change owing to changes in raw materials and production levels, equipment fouling, and so on. If the controller parameters are not adjusted in response to these changes in dynamic parameters, a degradation in control quality can occur. As a first step in this adaptive control scheme, it is necessary to be able to identify the dynamic parameters on line, if not instantaneously, then at least within a reasonable time period so that corrective actions can be taken. In this section we describe two simple procedures for time domain process identification. We will consider adaptive control concepts in more detail in the next chapter.

Random Search Procedure [20,21]

We illustrate the method by means of an overdamped second-order plus dead time process model although the method is not at all restricted to models of this type.

The transfer function of overdamped second-order processes containing dead time is of the form

$$G(s) = \frac{Y(s)}{X(s)} = \frac{K_p e^{-\theta_d s}}{(\tau_1 s + 1)(\tau_2 s + 1)} \tag{13.34}$$

The objective of the identification procedure is to determine K, θ_d, τ_1 and τ_2. The method utilizes the time domain equivalent of Equation (13.34). The time domain equation relating Y to X may be derived in one of two ways. One is to begin with the differential equation representing this class of processes, i.e., the inverse of Equation (13.34),

$$\tau_1 \tau_2 \frac{d^2 Y}{dt^2} + (\tau_1 + \tau_2) \frac{dY}{dt} + Y(t) = K_p X(t - \theta_d) \tag{13.35}$$

and use the numerical solution as the expression relating Y to X. Alternately, we may relate Y and X in the Z-domain by the relationship

$$\frac{Y(Z)}{X(Z)} = Z\{G_{h_0}(s)\, G(s)\}$$

$$= Z\left\{ \frac{1 - e^{-sT}}{s} \cdot \frac{K_p e^{-\theta_d s}}{(\tau_1 s + 1)(\tau_2 s + 1)} \right\} \tag{13.36}$$

$$= \frac{b_1 Z^{-(N+1)} + b_2 Z^{-(N+2)} + b_3 Z^{-(N+3)}}{1 - a_1 Z^{-1} + a_2 Z^{-2}}$$

and invert it to give

$$Y_i = a_1 Y_{i-1} - a_2 Y_{i-2} + b_1 X_{i-N-1} + b_2 X_{i-N-2} + b_3 X_{i-N-3} \tag{13.37}$$

where

$$a_1 = e^{-T/\tau_1} + e^{-T/\tau_2}$$
$$a_2 = e^{-T/\tau_1} \cdot e^{-T/\tau_2}$$

$$b_1 = K_p\left(1 + \frac{\tau_2 e^{-mT/\tau_2} - \tau_1 e^{-mT/\tau_1}}{\tau_1 - \tau_2}\right)$$

$$b_2 = K_p\left[-a_1 + \frac{\tau_1 e^{-mT/\tau_1}(1 + e^{-T/\tau_2}) - \tau_2 e^{-mT/\tau_2}(1 + e^{T/\tau_1})}{\tau_1 - \tau_2}\right]$$

and

$$b_3 = K_p\left[a_2 - \frac{\tau_1 e^{-mT/\tau_1} \cdot e^{-T/\tau_2} - \tau_2 e^{-mT/\tau_2} \cdot e^{-T/\tau_1}}{\tau_1 - \tau_2}\right]$$

The material in this section is based on Equation (13.37). In either case we will obtain an equation relating the process output at the ith sampling instant to the past outputs and inputs. The constants a_1, a_2, b_1, b_2, b_3 are functions of the model parameters K_p, θ_d, τ_1, and τ_2, which are to be determined. The procedure consists of the following steps:

1. A suitable disturbance (e.g., step or pulse) is applied to the input of the process while the process is operating at steady state in manual control.

2. Input and output data are recorded.

3. Trial values of K_p, θ_d, τ_1, τ_2 are assumed.

4. Using these trial values and the input data of step 1, the output data are predicted via Equation (13.37).

5. The difference between the experimental output and the predicted output, an indication of error in the trial parameters, is calculated by the expression

$$\text{ERROR} = \sum_{i=0}^{N} (Y_{ia} - Y_{ip})^2 \tag{13.38}$$

6. The parameters are updated and steps 4 and 5 repeated until the ERROR is minimized to an acceptable level.

From these six steps it should be clear that this is a classical optimization problem having Equation (13.38) as the objective function. A considerable body of literature exists that deals with the solution of nonlinear optimization problems (e.g., see References 22, 23, 24). Among the available methods are gradient searching techniques, methods of steepest descent, and random searching methods. The method selected in this section was proposed by Luus and Jaakola,[21] and will be referred to as the LJ optimization procedure. The method is simple, relatively fast, and is insensitive to starting guesses of the parameters. Their program listing in FORTRAN is available in the literature, so only the method will be described. The LJ optimization procedure employs the following eight steps:

1. Specify initial values of the parameters X_1, X_2, X_3, X_4 to be optimized and denote them as X_1^0, X_2^0, X_3^0, X_4^0 and specify an initial range for each r_1^0, r_2^0, r_3^0, r_4^0. Set iteration index $J = 1$.

2. Read in a sufficient number of random numbers between -0.5 and $+0.5$ (2000 numbers are suggested by Luus and Jaakola.[21]

3. Take $4\,P$ random numbers from step 2 (Luus and Jaakola suggest $P = 100$) and assign them to X_1, X_2, X_3, X_4 so that there are P sets of values, which are calculated as

$$X_i^J = X_i^{J-1} + Y(K,i)\, r_i^{J-1}$$

where

$i = 1,\ldots 4$

$K = 1,\ldots P.$

4. Test constraint equations and calculate a value of ERROR for each set. In this instance the constraints are that K_p, $\theta_d \geq 0$. An examination of the definition of the constants a_1, a_2, b_1, b_2, b_3 indicates that to avoid numerical problems on the computer it is necessary to specify that

$$\tau_1 - \tau_2 < \epsilon_1$$
$$\tau_1 > \epsilon_2$$
$$\tau_2 > \epsilon_2$$

In the example that follows $\epsilon_1 = 10^{-6}$ and $\epsilon_2 = 0.025$.

5. From the P sets find that set which minimizes ERROR. Write this value of ERROR and the corresponding X_i^J. Increment J by one to $J + 1$.

6. If the number of iterations has reached the maximum (Luus and Jaakola suggest 200 iterations), then end the problem. If not, go to the next step.

7. Reduce the range by an amount ϵ such that

$$r_i^J = (1 - \epsilon)\, r_i^{J-1}$$

Usually $\epsilon = 0.05$.

8. Go to step 3 and continue.

A flow chart of the program for the implementation of the LJ procedure is shown in Figure 13.14. Program capability has been assessed by numerous examples, one of which is presented for illustration.

Example 4. The process has a fourth-order transfer function

$$G(S) = \frac{1}{(0.5s + 1)(s + 1)^2(2s + 1)} \tag{13.39}$$

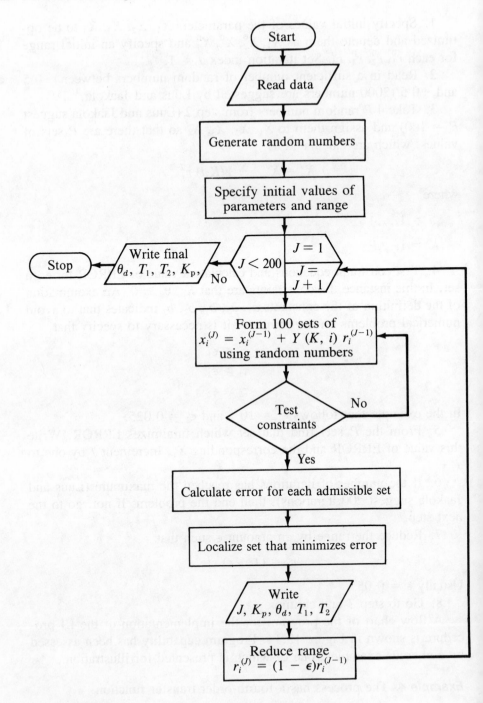

Figure 13.14
Flow Chart for *LJ* Optimization Procedure

It is desired to develop an overdamped second-order plus dead time process model that best fits this transfer function. To solve this problem, the step response of the process whose transfer function is given by Equation (13.39) is determined. The step input and the resulting output data are read into the program as inputs. The optimum parameters determined by the program are shown in Table 13.1. The results shown indicate that the method is not sensitive to starting guesses. For comparison purposes, Table 13.1 also shows the results obtained by the method of Brantley, et al.,[25] which, as was mentioned in their article, is sensitive to initial guesses. The transient response of the process is shown in Figure 13.15 alongside the response of the reduced model, indicating a good fit.

Recently, several enhancements of the LJ procedure have appeared in the literature.[26] One involves the use of different values of $\epsilon > 0.05$ to reduce execution times. The reader is encouraged to refer to these papers for details.

Although the foregoing optimization procedure was employed to find optimum process parameters, the same procedure can be readily adapted to find optimum tuning constants of digital PID-type controllers, as will be explained. In this instance the objective is to determine the optimal

Table 13.1
Process Identification Results

Initial Guess	Brantley et al. Final Value (212 cycles)	LJ Method Final Value (200 Iterations)
$K_p = 1.0$	1.008	1.008
$\theta_d = 1.0$	1.814	0.8370
$\tau_1 = 6.0$	3.947	4.035
$\tau_2 = 2.0$	3.423	3.310
	Error = 0.00109	Error = 0.00111
	CPU = 14 seconds	CPU = 1 minute, 25 seconds

Initial Guess	Final Value (37 cycles)	Final Value (200 Iterations)
$K_p = 1.0$	1.045	1.011
$\theta_d = 1.0$	3.443	0.8971
$\tau_1 = 2.0$	6.544	4.520
$\tau_2 = 0.5$	0.02	2.825
	Error = 0.011	Error = 0.00137
	CPU = 4 seconds	CPU = 1 minute, 25 seconds

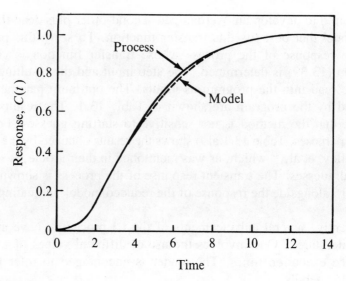

Figure 13.15
Transient Step Response of Process and Model

controller constants of a P, PI, or PID control algorithm that minimize one of the following user-specified objective functions:

$$IAE = \int_0^\infty |E|\, dt \qquad (13.40a)$$

where E = error, set point-measured value of the process output

$$ITAE = \int_0^\infty t\, |E|\, dt \qquad (13.40b)$$

or

$$ISE = \int_0^\infty E^2\, dt \qquad (13.40c)$$

It is assumed that the process can be described by an overdamped second-order or a first-order model containing dead time and that the model parameters are known, perhaps from a previous process identification study. The previously described LJ optimization technique is used in this application. The procedure consists of the following steps:

1. Enter the model parameters and the sampling period.

2. Assume trial values of each tuning constant, say X_i^0 and a range of search for each, say r_i^0. Set index $J = 1$.

3. Read in 2000 random numbers between -0.5 and 0.5.

4. Take $3P$ random numbers from step 2 ($P = 100$) and assign them to X_1, X_2, X_3 ($X_1 = K_c$, $X_2 = \tau_i$, $X_3 = \tau_D$ for a PID controller chosen for illustration of the technique) so that there are P sets of values that are calculated by the equation

$$X_i^J = X_i^{J-1} + Y(K,i) \, r_i^{J-1}$$

where

$$i = 1,\ldots 3 \text{ (for a PID controller)}$$

$$K = 1,\ldots P$$

5. For any of the P sets, if any $X_i < 0$, then discard that set and return to step 4 and calculate a new set.

6. Obtain the closed-loop response of the system to, say, a unit step change in set point by the Z-transform method or more simply by the impulse response method.

7. Compute the value of the objective function by numerical integration for each set. From the P sets find that set which minimizes the objective function. Write this value of the objective function and the associated X_i^J. Increment J by 1 to $J + 1$.

8. If the number of iterations J has reached 200, end the problem. If not, go to the next step.

9. Reduce the range according to

$$r_i^J = (1 - \epsilon) \, r_i^{J-1}$$

where

$$i = 1,\ldots 3$$

$$\epsilon = 0.05$$

10. Go to step 4 and continue.

A flowchart of this program is shown in Figure 13.16. A FORTRAN program was used to test the LJ optimization procedure. One of the test examples is based on a transfer function

$$G(s) = \frac{0.312e^{-0.4s}}{(0.96s + 1)(2.781s + 1)}$$

and the sampling period is 0.1 time units. In this example it was desired to determine the tuning constant of a PID controller that minimized the IAE criterion.

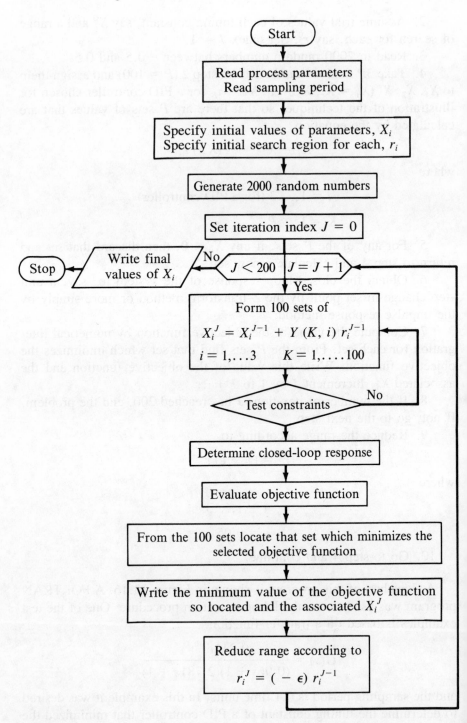

Figure 13.16
Flow Chart of *LJ* Program to Determine Tuning Constants

The results are summarized in Table 13.2. Note that two different initial trial values were used to ascertain that different starting values did not influence the results. Also shown for comparison are the optimal constants determined by the method of Fertik[27] and Gallier and Otto.[28] The resulting closed-loop responses are shown in Figure 13.17. Here, too, the execution times can be reduced by incorporating the enhancements proposed by Gaddy and coworkers.[26]

Least Squares Method[29, 30, 31, 32]

In the previous section we described a random search procedure for process identification and optimal tuning. The LJ method has been successfully applied to numerous problems. However, we have found the execution times to be quite large in some applications. In this section a least squares algorithm that gives smaller execution times is presented. The method will be applied to find the optimal tuning constants of a PID controller, although the same method can be applied to find optimal process constants as well.

The algorithm seeks to minimize the objective function

$$I(x) = \sum_{i=1}^{N} E_i^2 \qquad (13.41)$$

where E_i represents the difference, at time instant i, between the set point and the predicted output based on the assumed tuning constants $x(x_1 = K_c, x_2 = \tau_I, x_3 = \tau_D)$. Let G denote the column vector of N functions, E_i, so that

$$I = G^T G \qquad 13.41(a)$$

The objective function I is minimized when its gradient is zero. In this approach the gradient becomes zero at a point $x + \Delta x$ where Δx is computed according to

Table 13.2
Optimal Tuning Constants

Method	K_c	τ_I	τ_D	*IAE*
Gallier and Otto	13.1	3.7	0.7	1.0419
Fertik	17.5	2.7	0.3	2.1769
This work				
Initial guess				
(10.0, 10.0, 10.0)	13.6274	4.0892	0.7630	1.0619
(5.0, 5.0, 5.0)	13.6165	4.0538	0.7689	1.0540

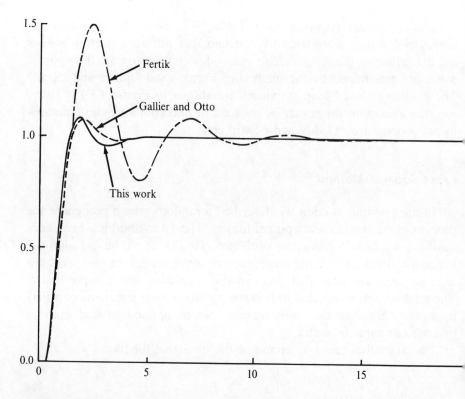

Figure 13.17
**Closed-loop Response of the Second-order Plus Dead-time Process
Model where $K_p = 0.312$, $\tau_1 = 0.960$, $\tau_2 = 2.781$, $\theta_d = 0.400$, and a
Sampling Rate = 0.100 to a Unit Step Change in Set Point**

$$\Delta x = -(J^T J + \lambda I) J^T G \qquad (13.42)$$

where x is a positive real number and J is an $N \times 3$ Jacobian matrix,
whose elements are computed according to

$$\frac{\partial E_i}{\partial x_1} = \frac{E_i(x_1 + \Delta x_1, x_2, x_3) - E_i(x_1, x_2, x_3)}{\Delta x_1}$$
$$i = 1, N \qquad (13.43)$$

The partial derivatives of E_i with respect to x_2 and x_3 are computed in
a similar manner. The computation of J, initially, is based on an arbi-
trarily selected Δx. For subsequent iterations, the values of Δx are available from Equation (13.42).

The iterations are stopped when the gradient of I becomes sufficiently
small. This is synonymous to stopping the computer program when Δx
falls below some user-specified tolerance level (say, 0.0001).

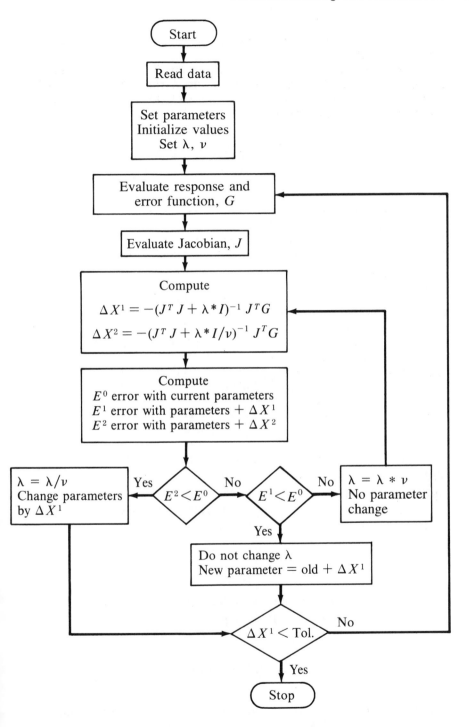

Figure 13.18
Least-squares Optimization Algorithm Flow Chart

Table 13.3
Optimum Controller Tuning Parameters for a Process with the Transfer Function.
(Reprinted with permission from "Controller Tuning by a Least Squares Method," Patwardhan, A.A., and Karim, M.N. from *AIChE Journal*, vol. 33, no. 10, pp. 1735–1739, Oct. 1987, © American Institute of Chemical Engineers.)

$$G(s) = \frac{3.5e^{-0.5s}}{(5s + 1)(2.5s + 1)}$$

Method[1]	IAE	K_c	τ_I	τ_D	CPU time (secs)	# of iterations
(a) Initial guesses: 1.0, 1.0, 1.0						
LSQ	6.2386	1.6823	7.3561	1.7460	2.490	18
LJ	6.2611	1.6980	7.4559	1.7454	12.615	10
(b) Initial guesses: 5.0, 5.0, 5.0						
LSQ	6.2511	1.7992	7.3558	1.8078	3.456	29
LJ	6.2669	1.8480	7.4483	1.7414	12.434	10
(c) Initial guesses: 10.0, 10.0, 10.0						
LSQ	6.2438	1.7723	7.3115	1.7792	5.417	42
LJ	6.2437	1.6736	7.3346	1.7382	12.349	10
(d) Initial guesses: 0.1, 20.0, 0.1						
LSQ	6.2388	1.6861	7.3668	1.7445	2.213	14
LJ	6.2680	1.6885	7.4760	1.7712	12.636	10

[1]LSQ: Least-squares method; LJ: Luus-Jaakola method.

A flow chart of the least squares strategy is shown in Figure 13.18. The flow chart shows the initial value of $\lambda = 0.01$ and that of a multiplier, $\nu = 5$ selected according to Marquardt.[30] Patwardhan et al.[29] have presented an example consisting of an overdamped second-order process with dead time having the transfer function

$$G_p(s) = \frac{3.5e^{-0.5s}}{(5s + 1)(2.5s + 1)} \tag{13.44}$$

The tuning constants and the execution times are shown in Table 13.3. Improvements in execution times are evident.

References

1. Deshpande, P. B., Process Identification of Open-loop Unstable Systems, *A. I. Ch. E. J.*, **26**, 2, 1980.

2. Ziegler, J. G., Nichols, N. B., Optimum Settings for Automatic Controllers, *Trans. ASME*, **64**, 11, November 1942, 759.
3. Miller, J. A., et al., A Comparison of Controller Tuning Techniques, Control Engineering, **14**, 2, December 1967, 72.
4. Oldenbourg, R. C., Sartorius, H., *The Dynamics of Automatic Controls*, American Society of Mechanical Engineers, New York, 1948, p. 276.
5. Sten, J. W., Evaluating Second-Order Parameters, *Instrumentation Tech.* **17**, 9, September 1970, 39–41.
6. Smith, O. J. M., A Controller to Overcome Dead-time, *ISA J.*, **6**, 2, February 1959, 28–33.
7. Cox, J. B., et al., A Practical Spectrum of DDC Chemical-Process Control Algorithms, *ISA J.*, **13**, 10, October 1966, 65–72.
8. Smith, C. L., Murrill, P. W., *ISA J.*, **13**, 9, 1966, 48.
9. Meyer, J. R., et al., Simplifying Process Response Approximations, *Instruments and Control Systems*, **40**, 12, December 1967, 76–79.
10. Csaki, F., Kis, P., *Period. Polytech.*, **13**, 1969, 73.
11. Naslin, P., Miossec, C., Automatisme, **16**, 1971, 513.
12. Sundaresan, K. R., et al., Evaluating Parameters from Process Transients, *Ind. Eng. Chem. Process Des. Dev.*, **17**, 3, 1978, 237–241.
13. Lees, F. P., *Chemical Engineering Science*, **26**, 1971, 1179.
14. Gibilaro, L. G., Waldram, S. P., *Chemical Engineering Science*, **4**, 1972, 197.
15a. Hougen, J. O., *Experiences and Experiments with Process Dynamics*, CEP Monograph Series, Vol. 60, No. 4, 1964.
15b. Clements, W. C., Schnelle, K. B., Ind. Eng. Chem. Process Des. Dev. 2, 1963, 94.
16. Dynamic Response Testing of Process Control Instrumentation, ISA-S26 Standard, Instrument Society of America, 400 Stanwix Street, Pittsburgh, Pa., October 1968.
17. Luyben, W. L., *Process Modeling, Simulation, and Control for Chemical Engineers*, McGraw-Hill, New York, 1973.
18. Messa, C. J., et al., Criteria for Determining the Computational Error in Numerically Calculated Fourier Integral Transforms, *I & EC Fundamentals*, **8**, 4, 1969.
19. Murrill, P. W., Pulse Testing Methods, *Chemical Engineering*, February 24, 1969.
20. Stark, P. A., D. L. Ralston, and P. B. Deshpande, "Comparative Assessment of Two Recent On-Line Process Identification Techniques," American Control Conference, Arlington, VA, June 1982.
21. Luus, R., Jaakola, T. H. I., *AIChE J.*, **19**, 760 (1973).
22. Himmelblau, D. M., *Applied Nonlinear Programming*, McGraw-Hill, New York, 1972.
23. Asghar Husain and Kota Gangiah, *Optimization Techniques for Chemical Engineers*, Macmillan, India, New Delhi, 1976.

24. Reklaitis, G. V., A. Ravindran, K. M. Ragsdell, *Engineering Optimization*, Wiley-Interscience Publishers, New York, 1983.
25. Brantley, R. O., R. A. Schaefer, and P. B. Deshpande, "On-Line Process Identification," *Ind. Eng. Chem. Proc. Des. Dev.*, **21**:297–301, 1982.
26. Martin, D. L. and J. L. Gaddy, "Process Optimization with the Randomly Directed Search," *Chem. Eng. Prog. Symp. Ser.* **24**, AIChE 1982.
27. Fertik, H. A., "Tuning Controllers for Noisy Processes," *ISA Trans.*, **14**:292–304, 1975.
28. Gallier, P. W. and R. F. Otto, "Self-Tuning Computer Adapts DDC Algorithms," *InTech*, pp. 65–70, February 1968.
29. Patwardhan, A., A. and M. N. Karim, "Controller Tuning by a Least Squares Method," to appear in *A.I.Ch.E. J.*, 1988.
30. Marquardt, D. W., *J. Soc. Indust. Appl. Math.*, **11**, 1963, p. 431.
31. Beightler, C. S., D. T. Phillips, and J. Wilde, *Foundation of Optimization*, Prentice-Hall, Englewood Cliffs, NJ, 1979.
32. Kumar, S., "A Comparative Study of Two Optimization Techniques for Simplified Model Predictive Control," M.S. Thesis, University of Louisville, 1987.

SHORT BIBLIOGRAPHY ON MODELING OF CHEMICAL ENGINEERING SYSTEMS

1. Luyben, W. L., *Modeling, Simulation, and Control for Chemical Engineers*, McGraw-Hill, New York, 1973.
2. Franks, R. G. E., *Modeling and Simulation for Chemical Engineers*, Wiley, New York, 1972.

CHAPTER **14**

Adaptive Control and Self-Tuning

In the last chapter we described a few simple process identification techniques and alluded to how they might be used to keep a loop well tuned. In this chapter we take a somewhat more detailed look at such procedures and offer an introductory treatment of self-tuning control.

An adaptive control system is one that automatically adjusts controller settings to accommodate changing process characteristics such that the controlled variable follows its desired path. A few examples may be cited to illustrate the need for adaptive control. One has to do with pH control in a reactor. Consider the titration curve for the system shown in Figure 14.1. Notice that small additions of acid near the desired set point result in large changes in pH, but large additions of acid farther away from the set point cause only small changes in pH, thus rendering control difficult. Another example consists of catalyst decay in a reactor, and a third is about fouling in heat exchange devices. In both of these examples, parameter variations are the net result, causing difficulties for a controller.

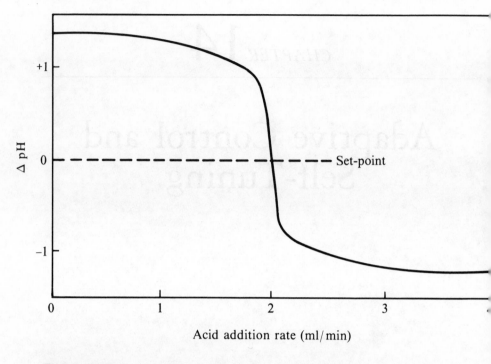

Figure 14.1
Titration Curve for pH Control

Literature on adaptive control and self-tuning is abundant. There are several hundred technical papers, including several excellent reviews, and several books. The reviews[1,2,3,4] contain not only source material but also a compilation of a large body of references.

It may be mentioned that successful applications of adaptive control and self-tuning have been reported in industry, and dedicated devices to implement them are available in the marketplace.

The three basic representations of adaptive control and self-tuning are gain scheduling, model reference adaptive control, and self-tuning regulators. These schemes are described in the following paragraphs.

14.1 Gain Scheduling

Consider the block diagram of a single-input single-output control system as shown in Figure 14.2(a). The objective is to adjust controller parameters in response to changing process dynamics or disturbance dynamics. In gain scheduling, an auxiliary variable is used to find the best values of controller parameters, as shown in Figure 14.2(b). The procedure requires an *a priori* knowledge of the relationship between the auxiliary variable and the dynamic parameters. The method is called gain scheduling because the scheme was originally used to accommodate changes in the process gain only. Examples of gain scheduling include the pH control problem cited earlier.

14.2 Model-Reference Adaptive Control (MRAC)

A model-reference adaptive control system is shown in Figure 14.3. In this instance the "reference model" specifies how the MRAC system

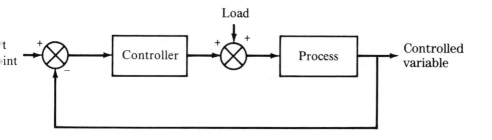

Figure 14.2*a*
Typical SISO Control System

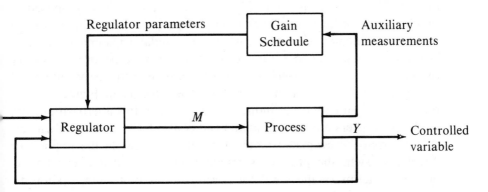

Figure 14.2*b*
The Gain Scheduling Concept

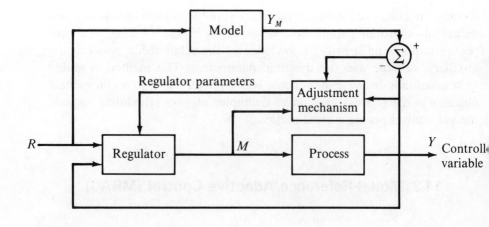

Figure 14.3
Model-Reference Adaptive Control System

output is to respond to a change in set point. The objective of the control effort is to adjust the controller parameters such that the deviation between the actual process output and the model output is minimized.

14.3 Self-Tuning Regulators (STR)[1,2,3,4]

STRs have been gaining popularity in recent years. A block diagram of an STR system is shown in Figure 14.4. The system is composed of two loops: an ordinary feedback loop plus an outer loop that contains a parameter estimator and a block where the regulator parameters are calculated. To obtain good parameter estimates it may become necessary to introduce external perturbation signals, although this is not indicated in Figure 14.4 to maintain simplicity. The scheme in Figure 14.4 is referred to as an explicit or indirect STR, since in this approach process model parameters are first estimated and controller settings are then calculated from the estimated parameters. By contrast, in the implicit or direct approach, the process is reparameterized so that it can be expressed in terms of the regulator parameters. This results in simplification since the design calculations are eliminated. For parameter estimation, many different schemes have been used. They include least squares, generalized and extended least squares, instrumental variables,

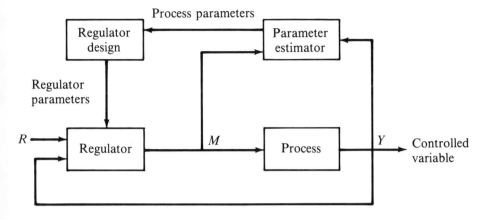

Figure 14.4
Block diagram of a self-tuning regulator (STR).

extended Kalman filtering, and the maximum likelihood method. In the design calculations a variety of approaches have been employed. They include gain and phase margins, pole placement, and minimum variance approaches.

In this chapter we have attempted to introduce the reader to adaptive control and self-tuning. A quantitative treatment is beyond the scope of this text, but the interested reader may consult the references for more information.

References

1. Astrom, K. J. "Theory and Applications of Adaptive Control—A Survey," *Automatica*, **19, 5**, 1983, pp 471–486.
2. Seborg, D. E., "Adaptive Control Strategies for Process Control: A Survey," *A.I.Ch.E. J.*, **32, 6**, 1986, pp 881–913.
3. Astrom, K. J., "Adaptive Feedback Control," *Proc. of the IEEE*, **75, 2**, 1987, pp. 185–217.
4. Seborg, D. E., "The Prospects of Advanced Process Control," *Proc. of the IFAC Conference*, Munich, July 26–31, 1987.

Feedforward Control

15.1. Introduction and Design Fundamentals

Feedforward control is probably one of the more widely used advanced control techniques in the process industries. It is implementable with analog hardware, although somewhat more specialized equipment is needed for its implementation. Its purpose is to protect the control system against the detrimental effects of changing process loads. When properly designed and tuned, feedforward control can produce amazing results. In this chapter we develop the design equations for feedforward control and present two experimental applications that show the benefits of this control technique. As an introduction to feedforward control, consider a simple process consisting of heating a continuous stream of water in a tank, as shown in Figure 15.1.

In this case the process loads are the flow rate and temperature *303* of the supply water. In the usual industrial situation the incoming stream to a process comes from an upstream portion of the plant and is not subject to manipulation or control. If the temperature or flow rate of the incoming water changes, it upsets the controlled variable, the temperature of water in the tank. The system remains disturbed until the feedback system brings the controlled variable back to the set point. Feedforward control can be used to improve the response of the system under these circumstances. The basic principle of this technique is to measure the disturbances as they occur and make adjustments in the manipulated variable so as to prevent

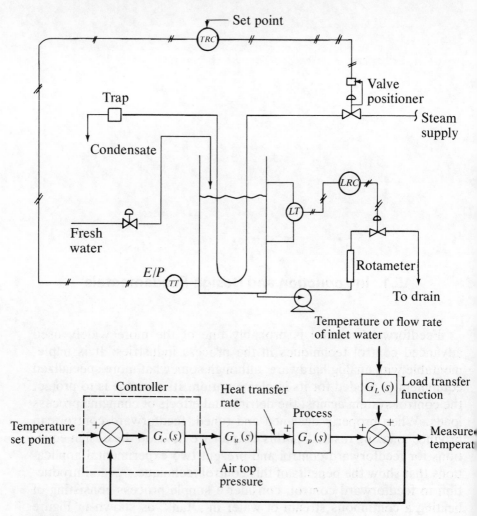

Figure 15.1
Schematic and Block Diagram of Temperature Control System

them from upsetting the controlled variable. The design equations for the feedforward controller are derived as follows:

The response of the feedback control system shown in Figure 15.1 to load changes is

$$C(s) = \frac{G_L(s)}{1 + G_c(s)G_u(s)G_p(s)} L(s) \tag{15.1}$$

Ideally, C should be zero for any load L (note that C is a deviation variable); C is reduced by a large G_c (i.e., high controller gains), but the magnitude of G_c is constrained by physical limitations and system stability.

In feedback control we feed the error signal back to the controller, which in turn updates the manipulated variable so as to reduce the error. But suppose we could measure the signal which has the potential of upsetting the process if no action is taken. Then we would measure and transmit this signal to a controller. This controller would act on this signal and compute the new value of the manipulated variable and forward the output to the final control element. If we did everything correctly, the controlled variable would not be affected if a process load should occur. The block diagram of this control system is shown in Figure 15.2. Note that in this arrangement errors in the controlled variable are not fed back, but the changes in process loads are fed forward. This arrangement is re-

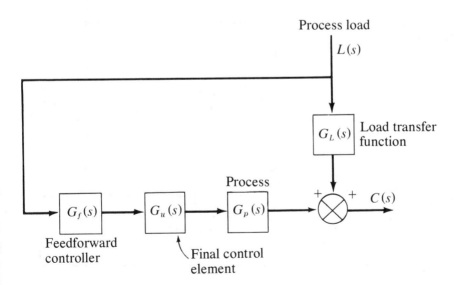

Figure 15.2
Feedforward Control Loop

ferred to as a feedforward control system. From the block diagram of this feedforward control system we can write

$$C(s) = L(s)\{G_L(s) + G_p(s)G_u(s)G_f(s)\} \tag{15.2}$$

where $G_f(s)$ is the transfer function of the feedforward controller. If we set

$$G_f(s) = -\frac{G_L(s)}{G_u(s)G_p(s)} \tag{15.3}$$

Then C will be zero for all L. This equation gives us the basis for feedforward controller design. Equation (15.3) also shows that accurate models of the elements G_L and $G_u G_p$ are required. If the models are not accurate, the terms in the brackets in Equation (15.2) will not vanish, and the controlled variable will show a deviation from set point. Therefore, feedforward control is seldom used alone, but rather in combination with feedback control, as shown in Figure 15.3. Note that this arrangement implies that the sensor dynamics are included in the transfer functions $G_p(s)$, $G_L(s)$, and $G_f(s)$. The transfer function of the feedback/feedforward combination to load changes is

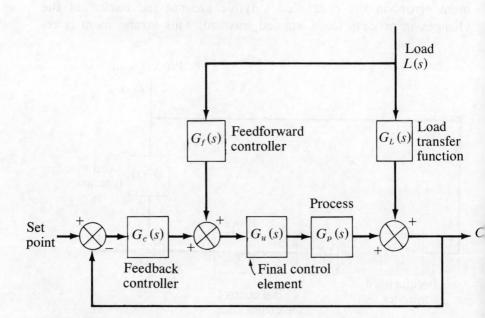

Figure 15.3
Combined Feedback/Feedforward Control System

$$C(s) = \frac{\{G_L(s) + G_p(s)G_u(s)G_f(s)\}}{1 + G_c(s)G_u(s)G_p(s)} L(s) \qquad (15.4)$$

The feedforward controller transfer function is still given by Equation (15.3). The feedforward controller eliminates the effects of process loads, whereas the feedback mechanism eliminates the effects of inaccuracies in the feedforward controller and other unmeasured disturbances.

Equation 15.3 shows that to determine the transfer function of the feedforward controller we require the transfer functions $G_p(s)$, $G_u(s)$, and $G_L(s)$. These, of course, are the open-loop transfer functions as shown in Figure 15.4. These transfer functions may be determined from a mathematical model or from experimental pulse or step tests. To determine $[G_u(s)G_p(s)]$, for example, by pulse testing, we would employ the following step-by-step procedure:

1. Place the feedback controller in manual and disconnect the feedforward controller as shown in Figure 15.4.

2. Start up the process and adjust the feedback controller output until the desired steady state operation is achieved.

3. Introduce a suitable pulse in M and record the transient response of C.

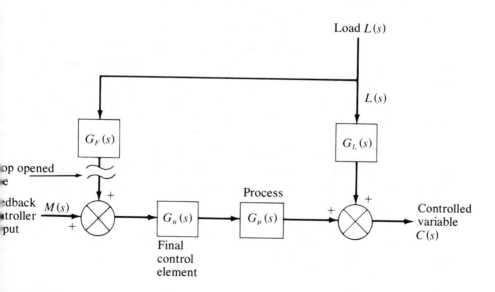

Figure 15.4
Open Loop Transfer Function

4. Analyze the pulse test data by Fourier transforms and generate a frequency-response diagram of the open-loop system.

5. Fit an approximate transfer function to the Bode' plot. This transfer function represents the product of $G_u(s)$ and $G_p(s)$.

6. With the process still operating at steady state and with the controller in manual, introduce a suitable pulse into L and record C.

7. The analysis of the pulse test data as before gives us $G_L(s)$.

Consider the design of a feedforward controller for a system in which the process transfer function as well as the load transfer function can be adequately described as first-order models with dead time, that is, suppose

$$G_u(s) \, G_p(s) = \frac{K_p \, e^{-\theta_d s}}{\tau_p s + 1}$$

and

$$G_L(s) = \frac{K_L \, e^{-\theta_L s}}{\tau_L s + 1}$$

(15.5)

Then the feedforward controller will be represented by the equation

$$G_f(s) = - \frac{G_L(s)}{G_u(s) \, G_p(s)} = - \frac{K_L}{K_p} \left[\frac{\tau_p s + 1}{\tau_L s + 1} \right] e^{(\theta_d - \theta_L)s}$$

(15.6)

This representation of the feedforward controller shows that this controller has three essential parts:

1. The feedforward controller gain is $-K_L/K_p$.

2. $(\tau_p s + 1)/(\tau_L s + 1)$ represents a lead-lag element. If this element is to be physically realizable, τ_L must be greater than zero. If the load dynamics are very fast, $\tau_L \simeq 0$. Then the lead-lag network reduces to a pure differentiator element, which is physically unrealizable. In this case we would select $\tau_L \ll \tau_p$ but still greater than zero.

3. If the process dead time θ_d exceeds the load deadtime θ_L, a pure predictor will be needed. This, of course, is also physically unrealizable. In this case we would set $\theta_d = \theta_L$ and eliminate the predictor term.

If we drop the dynamic terms from Equation 15.6, we obtain the design equation of what is referred to as the steady-state feedforward controller, G_{fss}. Thus

$$G_{fss} = - \frac{K_L}{K_p}$$

(15.7)

$$= - \frac{\Delta C}{\Delta L}\bigg|_{M = \text{constant}} \bigg/ \frac{\Delta C}{\Delta M}\bigg|_{L = \text{constant}}$$

A steady-state mathematical model or two simple step tests are all that are needed to implement the steady-state feedforward controller. Let us now illustrate the application of feed-forward control to a couple of experimental systems.

15.2. Example 1

Shinskey[1] presented an application of feedforward control to a heat exchanger. A schematic of the system is shown in Figure 15.5.

In this example we are interested in controlling the fluid outlet temperature T_o by manipulating steam flow F. Let us assume that the fluid inlet temperature T_I is constant and that the primary load disturbance is W, the flow rate of the process fluid.

The design equation for the steady-state feedforward controller can be derived from a steady-state mathematical model of the process. Thus if we apply the basic steady-state energy balance, we will get

$$W C_p (T_o - T_I) = F\lambda \tag{15.8}$$

where

W = process fluid flow rate, lb/hr

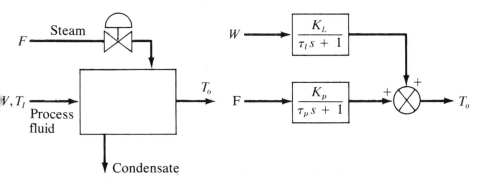

Figure 15.5
Process Schematic of the Heat Exchanger

C_p = specific heat of fluid Btu/1b°F

T_I = fluid inlet temperature

T_o = fluid outlet temperature

F = steam flow, 1b/hr

λ = latent heat of vaporization (i.e., heat released by steam), Btu/lb.

The solution of Equation (15.8) for T_o gives

$$T_o = \frac{F}{W}\,\frac{\lambda}{c_p} + T_I \tag{15.9}$$

The steady-state gains can be computed as the following partial derivatives at the steady-state operating point F_o and W_o. Thus,

$$K_p = \left.\frac{\partial T_o}{\partial F}\right|_{W_o,\,F_o} = \frac{\lambda}{W_o\,c_p} \tag{15.10}$$

$$K_L = \left.\frac{\partial T_o}{\partial W}\right|_{W_o,\,F_o} = -\frac{F_o}{W_o^2}\,\frac{\lambda}{c_p}$$

Thus the steady-state feedforward controller will be given by

$$G_{fss} = -\frac{K_L}{K_p} = \frac{F_o}{W_o} \tag{15.11}$$

Note that at steady state, the energy balance, Equation (15.8) becomes

$$W_o\,c_p\,(T_o - T_I) = F_o\lambda \tag{15.12}$$

or

$$\frac{F_o}{W_o} = \frac{c_p}{\lambda}\,(T_o - T_I) \tag{15.13}$$

Thus

$$G_{fss} = \frac{c_p}{\lambda}\,(T_{sp} - T_I) \tag{15.14}$$

where T_{sp} is the desired value of T_o or set point.

In this investigation Shinskey compared the conventional three-mode feedback control of T_o to feedforward control alone [Equation (15.14)] as well to the combined feedback/feedforward system. Both the steady-state feedforward and the dynamic feedforward controllers were implemented. The dynamic feedforward controller had the transfer function

$$G_f(s) = \frac{F(s)}{W(s)} = \left[\frac{C_p}{\lambda}(T_{sp} - T_I)\right]\frac{\tau_p s + 1}{\tau_L s + 1} \qquad (15.15)$$

The time constants τ_p and τ_L were selected by trial and error.

Figure 15.6 shows the results of the investigation for a 40% step change in W. In Figure 15.6a the three-mode controller was tuned for optimal recovery at 80% load. The nonlinearity of the process is evident in the overdamped recovery at 40% load. Figure 15.6b shows the results of steady-state feedforward control alone, and Figure 15.6c shows the results when the dynamics are added. The benefits of feedforward control are quite evident from these plots. Figure 15.6d

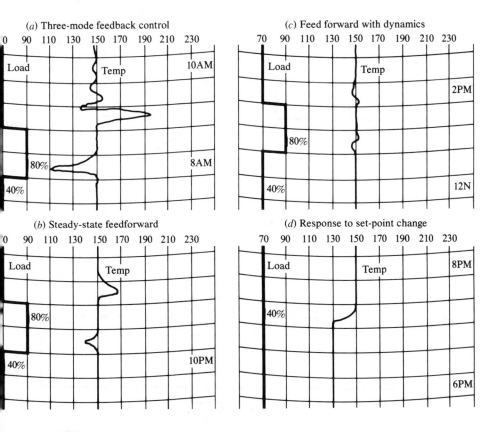

Figure 15.6
Comparison of Feedback and Feedforward Control System for the Heat Exchanger (Reproduced with Permission of Ref. 1)

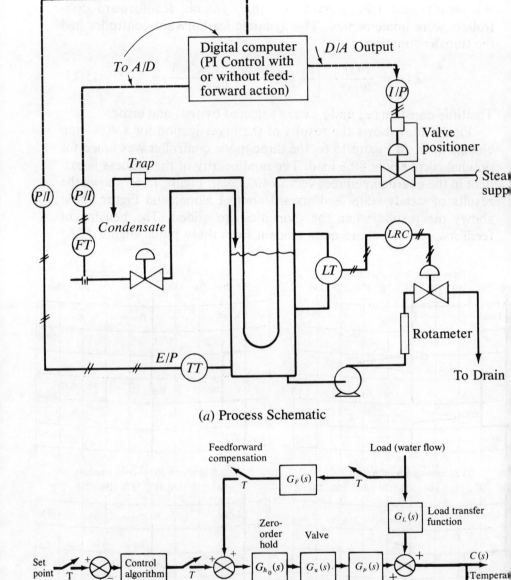

(a) Process Schematic

(b) Block diagram

Figure 15.7
The Combined Feedback Feedforward Control System

shows the response of the system having a feedforward-only controller when the set point is changed from 130°F to 150°F. However, for the reasons we mentioned earlier, we would normally use a combined feedback/feedforward system for good control of processes.

15.3. Example 2

This example shows a laboratory application of computer feedback/feedforward control of temperature. The schematic of the process for this computer control application as well as the resulting block diagram are shown in Figure 15.7. The simpler steady-state feedforward control algorithm has been selected to illustrate the technique. In this application the steady-state feedforward control algorithm is given by

$$G_{fss} = -\frac{K_L}{K_p} = -\left.\frac{\Delta T}{\Delta W}\right|_{F\,=\,\text{constant}} \bigg/ \left.\frac{\Delta T}{\Delta F}\right|_{W\,=\,\text{constant}} \tag{15.16}$$

where

T = temperature transmitter output, psig

W = flow transmitter output, psig

F = air-top pressure to steam control valve, psig

The open-loop step tests suggest that G_{fss} be set equal to 1 psig/psig. That is, for every psig change in the flow transmitter output, the feedforward controller must increment the air-top pressure on the steam control valve by 1 psig. The steady-state operating conditions are shown in Table 15.1.

Table 15.1
Operating Conditions

Tank level set point	18.4 in.
Steady-state water flow	2.54 gal/min
Step changed to	3.3 gal/min
Feedback controller settings:	
Proportional band	2%
Reset, τ_I	0.83 min
Sampling period	5 sec

The records of closed-loop response of the control system, both in the feedback control mode as well as in the combined feedback/feedforward control mode are shown in Figure 15.8. These results, too, show the excellent performance of the feedforward control system.

Reference

1. Shinskey, F. G., Feedforward Control Applied, *ISA* J., November 1963.

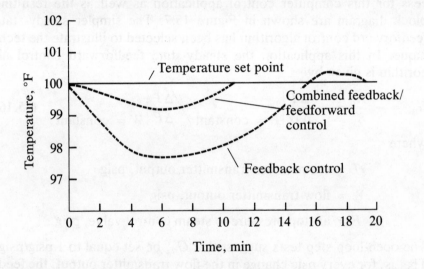

Figure 15.8
Response of Combined Feedback/Feedforward System to a Positive Step Change in Water Flow Rate

CHAPTER 16

Cascade Control

I n process-control applications variations in the manipulated variable frequently cause deterioration of the performance of feedback control loops. An example of such a control loop is shown in Figure 16.1. In this application the temperature of the water in the tank is maintained constant by adjusting the flow rate of steam. If a disturbance in the steam-supply pressure occurs, the flow rate of steam changes, which in turn upsets the controlled variable, temperature. Of course, once the temperature measuring device senses the upset, it feeds the information back to the controller, which takes corrective action so as to bring the temperature back to the set point. Meanwhile, the disturbance has entered the process and has upset the controlled variable. One can visualize that if the variations in the steam-supply pressure are frequent, the controlled variable may not remain at the set point for very long.

To correct this problem, a second control loop can be added, as shown in Figure 16.2. In the presence of steam-supply pressure

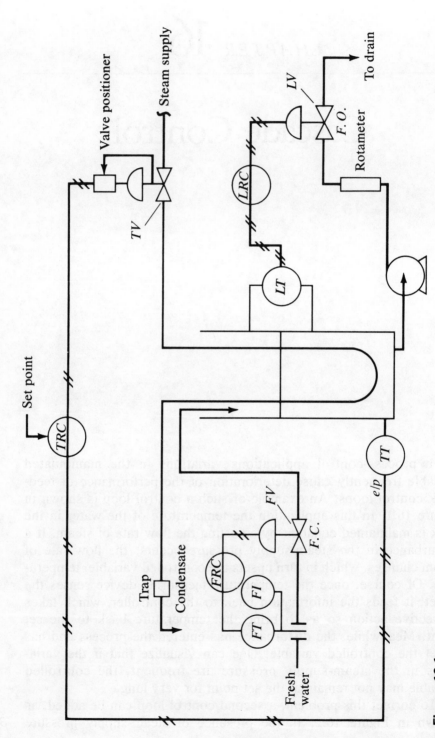

Figure 16.1
Typical Feedback Control System

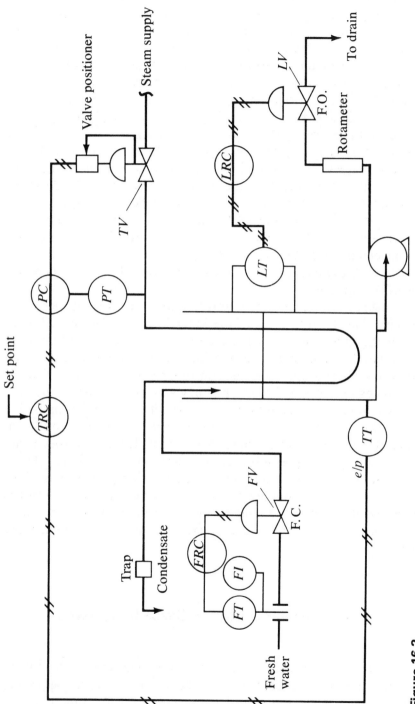

Figure 16.2
Cascade Control of Temperature

fluctuations, the pressure sensor senses the changes, and the pressure controller manipulates the steam valve so as to hold the downstream pressure constant. This way, the effect of supply pressure fluctuations on the steam flow rate can be eliminated. The performance of the feedback control loop is thus greatly improved.

The control situations, such as the one discussed here, in which the output of one controller manipulates the set point of another controller are called *cascade control systems*. The terminology commonly used in describing the cascade control systems is shown on the block diagram of the system in Figure 16.3. It may be noted from this figure that each controller is served by its own measurement device, but only one controller—the primary or master controller—has an independent set point, and only one controller—the secondary or slave controller—has an output to the process. The secondary controller, the manipulated variable, and its measurement device constitute the elements of the *inner* or *secondary loop*. The outer loop consists of all the elements of the cascade control system, including those of the inner loop.

To ensure that the cascade control system functions properly it is necessary that the dynamics of the inner loop be at least as fast as those of the outer loop, and preferably faster. This should be intuitively clear. If the dynamics of the inner loop are much faster than those of the outer loop, the inner controller will correct the effect of disturbances in that loop before they have a chance to upset the controlled variable. If this condition is not met, it is generally impossible to tune the master controller satisfactorily. The commonly encountered process control loops, in order of decreasing speed, are flow, liquid-level and pressure, temperature, and composition. However, these are general observations, and specific process situations must be analyzed thoroughly to assess whether cascade control is needed and if so, which should be the outer loop and which should be the inner loop. Several industrial examples of cascade control are shown in Figure 16.4.

16.1. Controller Design of Cascade Systems

There are two approaches available for determining the tuning constants of cascaded controllers. One is an analytical approach which can be used if the open-loop process transfer functions are

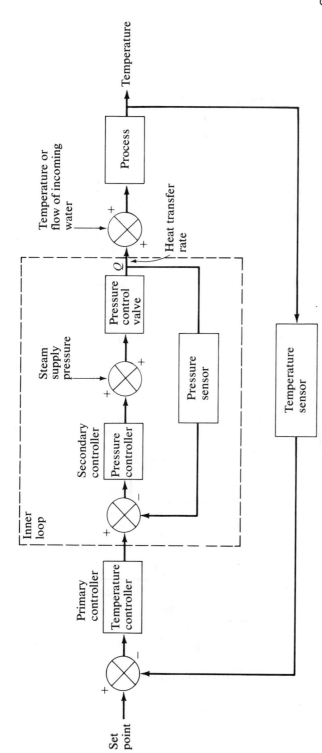

Figure 16.3
Block Diagram of Cascade Control System for Process of Figure 16.2

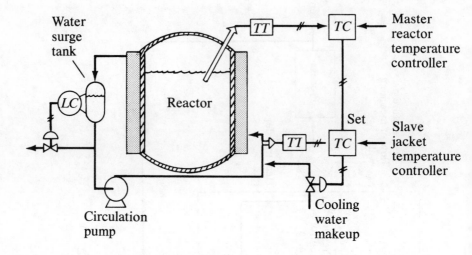

(a) Exothermic batch reactor

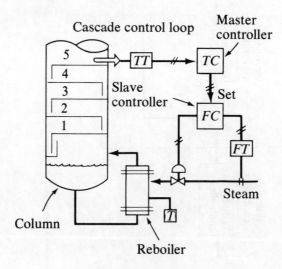

(b) Distillation column

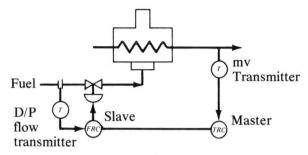

The *TRC-FRC* cascade control loop.

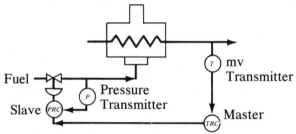

The *TRC-PRC* cascade control loop.

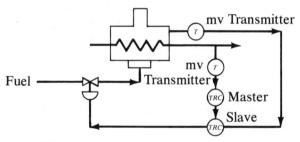

The *TRC-TRC* cascade control loop.

(c) Fuel-fired heater

Figure 16.4
Some Industrial Examples of Cascade Control Systems

available. These transfer functions can be developed from a dynamic mathematical model of the process. The second approach is an experimental one in which we would conduct open-loop step or pulse tests on the process, analyze the input-output data, and plot the frequency-response diagram, from which we would determine the tuning constants.

The analytical determination of controller settings for a cascade control system is straightforward once it is recognized that the elements of the inner loop can be reduced to a single block by the block-diagram-reduction techniques. The procedure is as follows:

1. Prepare the frequency-response diagram of all the elements of the inner loop, excluding the secondary controller.

2. Add the curves generated in step 1, graphically, to determine the composite Bode plot of these elements in series.

3. Design the secondary controller according to Cohen and Coon or Ziegler–Nichols criteria or by using Fertik's controller parameter charts (see Chapter 10). Integral action on the secondary controller is often unnecessary, since the inner-loop gain is often large and the effect is eventually corrected by the integral action of the primary controller. If inner-loop gain is low (a frequent occurrence in flow control), integral action is incorporated.

4. Using block-diagram-reduction techniques 1, reduce the closed inner loop to a single block.

5. Prepare the Bode plot of all the elements of the outer loop, excluding the primary controller, but including the block found in step 4.

6. Add the curves found in step 5 to develop the Bode plot of all the elements (except the primary controller) of the outer loop in series.

7. Design the primary controller using the Ziegler–Nichols or some other method.

Let us now illustrate this design method by an example.

Example 1 (with Permission of Ref. 2). Determine the controller settings of the primary controller with and without the inner loop. Measurement lags are negligible.

Solution Part A Cascade Control System.

Step 1. Prepare Bode plot of each element of $G_3(s)$ and $H_2(s)$. In this case the elements are $1/(s + 1)^2$ and $1/(10s + 1)$. The results are shown in Figure 16.6.

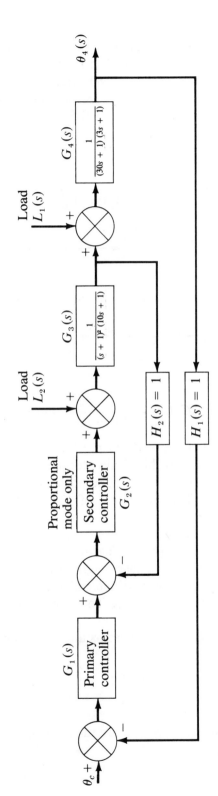

Figure 16.5
Cascade Control System

Step 2. Add the curves generated in step 1 to obtain the Bode plot of $1/((s + 1)^2 (10s + 1))$ as shown in Figure 16.6.

Step 3. From the figure, at $\phi = -180°$, $AR = 0.0416$. Therefore Ultimate gain, $Ku = 1/0.0416 = 24$. Therefore the Ziegler–Nichols gain for the secondary controller is $K_c = (0.5)(24) = 12$.

Step 4. Consider the inner loop

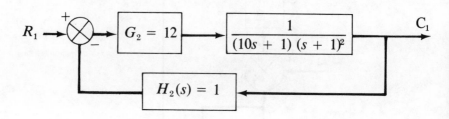

This block diagram can be replaced by its equivalent down below

$$R_1 \longrightarrow \boxed{\frac{G}{1 + G}} \longrightarrow C_1$$

where

$$G(s) = G_2 G_3(s) = \frac{12}{(10s + 1)(s + 1)^2}$$

From step 2 we already have a Bode plot of $1/(10s + 1)(s + 1)^2$. Incorporation of $K_c = 12$ into this Bode plot shifts the entire amplitude ratio portion of the plot upward. The phase angle portion of the plot remains unaffected. The Bode plot of $12/(s + 1)^2 (10s + 1)$ is also shown in Figure 16.6.

Now recall that given a $G(s)$, $G(s)/1 + G(s)$ can be determined from Nichols chart, Figure 16.7.[3] Therefore, read off numerous values of amplitude ratio and phase angles of $G(s) = 12/(s + 1)^2 (10s + 1)$ and go to Nichols chart to determine the amplitude ratios and phase angles of $G(s)/(1 + G(s))$. These results are shown in Table 16.1.

Step 5. Determine the amplitude ratios and phase angles of $G_4(s)$ from the relationships

$$AR = 1/(\sqrt{1 + \omega^2 (30)^2})(\sqrt{1 + \omega^2 3^2})$$

and

$$\phi = \arctan(-3\omega) + \arctan(-30\omega)$$

and enter in Table 16.1.

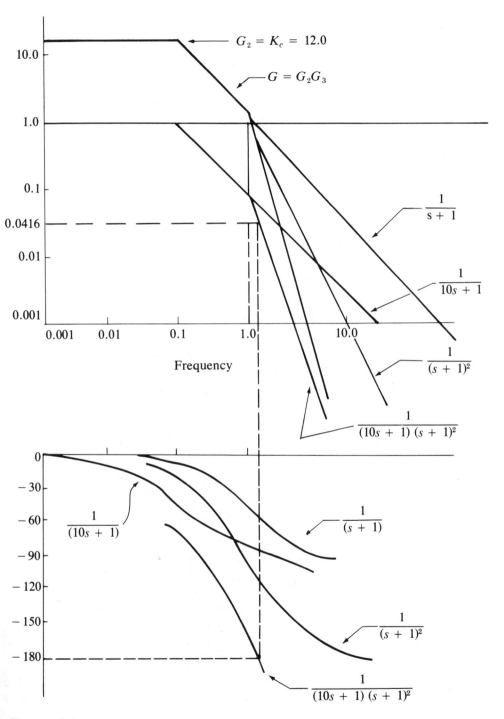

Figure 16.6
Bode Plot of Inner Loop Elements

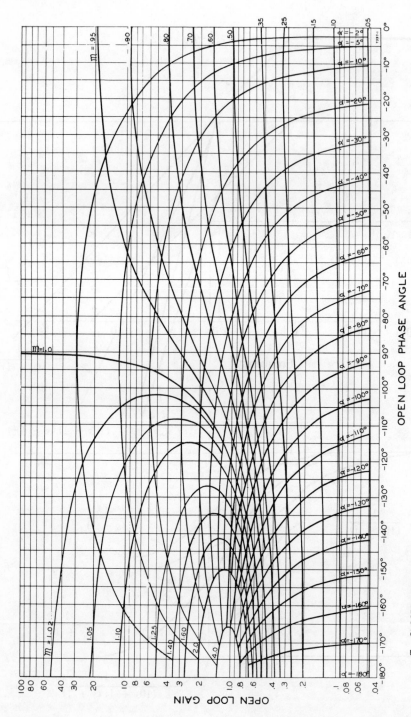

m - CLOSED LOOP GAIN
α - CLOSED LOOP PHASE ANGLE

**Figure 16.7
Nichols Chart**

Table 16.1
Frequency-Response Data For Cascade System

Frequency, ω Radians/min	$G(s) = G_2(s)G_3(s)$		$G(s)/1 + G(s)$		$G_4(s)$		$G_4(s)\left[\dfrac{G(s)}{1 + G(s)}\right]$	
	AR	$\phi°$	AR	$\phi°$	AR	$\phi°$	AR	$\phi°$
0.1	8.4	−55	0.93	−5	0.29	−88		
0.15	6.4	−73	0.945	−8	0.21	−101		
0.2	4.9	−85	0.97	−12	0.14	−112		
0.4	2.4	−120	1.15	−25	0.053	−135	0.061	−160
0.5	1.92	−133	1.33	−32	0.037	−143	0.049	−175
0.6	1.44	−143	1.7	−50	0.027	−148	0.046	−198
1.0	1.20	−170	4.5	−60				
1.2	0.55	−180	1.4	−180				
3.0	0.03	−210	0.04	−220				

Step 6. Combine the amplitude ratios and phase angles of $G_4(s)$ with those of $G(s)/[1 + G(s)]$ to obtain the magnitude ratios and phase angles of $G_4\, G(s)/[1 + G(s)]$, as shown in Table 16.1.

Step 7. At $\phi = -180°$ the amplitude ratio of the combined elements is equal to 0.048. Therefore, the ultimate gain is

$$K_u = \frac{1}{0.048} = 21$$

The Ziegler–Nichols settings for the primary controller are, therefore,

$$K_c = (0.45)\,(21) = 9.5$$

$$\tau_I = \frac{1}{1.2}\,\frac{2\pi}{\omega\big|_{\phi = -180°}} = \frac{1}{1.2}\,\frac{2\pi}{(0.53)} = 9.9 \text{ min}$$

$$\text{reset} = \frac{1}{\tau_I} = 0.101$$

Solution Part B: Feedback Control System. For simple feedback control, $G_2(s)$ would not exist. The elements of $G_3(s)$ and $G_4(s)$ are combined to obtain the composite Bode plot of $G_3(s)G_4(s)$, as shown in Table 16.2. From the table

$$\omega\big|_{\phi = -180°} = 0.16 \text{ and } AR = 0.093$$

Therefore, the Ziegler–Nichols controller settings are

$$K_c = 0.45\,Ku = \frac{0.45}{0.093} = 5$$

Table 16.2
Frequency-Response Data For Feedback System

Frequency, ω Radians/min	$G_3(s)$		$G_4(s)$		$G_3(s)\,G_4(s)$	
	AR	$\phi°$	AR	$\phi°$	AR	$\phi°$
0.1	0.70	-55	0.29	-88	0.2	-143
0.15	0.53	-73	0.21	-101	0.11	-174
0.2	0.41	-85	0.14	-112	0.057	-197
0.4	0.20	-120	0.053	-135		
0.5	0.16	-133	0.037	-143		
0.6	0.12	-143	0.027	-148		
1.0	0.10	-170				
1.2	0.045	-180				
3.0	0.0025	-210				

$$\tau_I = \frac{1}{1.2} \; \frac{2\pi}{\omega|_{\phi=-180°}} = \frac{2\pi}{(1.2)\,(0.16)} = 32.6$$

$$\text{reset} = \frac{1}{32.6} = 0.0306$$

A comparison of the results of the two systems shows that the cascade control system allows much higher gain and reset as compared to the feedback system.

To assess whether integral action is desirable on either controller of the cascade system, let us evaluate the response of the cascade control loop to load changes. Refer to block diagram, Figure 16.5.

a. For a unit step change in L_2 find the offset in θ_4.

$$\frac{\theta_4(s)}{L_2(s)} = \frac{G_3 G_4}{1 + H_2 G_2 G_3 + G_1 G_2 G_3 G_4 H_1}$$

Let

$$G_2 = 12 \text{ and } G_1 = 10$$

$$\lim_{t \to \infty} \theta_4(t) = \lim_{s \to 0} s\theta_4(s) = \frac{(1)\,(1)}{1 + (1)\,(12)\,(1) + (10)\,(12)\,(1)\,(1)\,(1)}$$

$$= \frac{1}{1 + 12 + 120} = 0.008$$

$$\text{offset} = 0.008$$

Therefore, integral actional on secondary controller may not be necessary.

b. For a unit change in L_1 find the offset in θ_4

$$\frac{\theta_4(s)}{L_1(s)} = \frac{G_4(1 + H_2G_2G_3)}{1 + H_2G_2G_3 + G_4G_3G_2G_1H_1}$$

$$\text{offset} = \lim_{t \to \infty} \theta_4(t) = \lim_{s \to 0} s\theta_4(s) = \frac{1(1 + 12)}{1 + 12 + 120} = .0978$$

Therefore, integral action on primary controller may be desirable.

Krishnaswamy and Rangaiah[6] have shown that it is possible to identify primary and secondary process dynamics with a single step or pulse test. The procedure consists of introducing a step or pulse change in the set point of the primary controller, R_1, with both loops in automatic but in the absence of disturbances, and recording the response of primary and secondary process outputs, C_1 and C_2, respectively. The frequency response of C_1/R_1 and C_2/R_1 are then prepared as has been explained. Division of C_1/R_1 by C_2/R_1 gives the frequency response diagram of the primary process transfer function. The secondary process transfer function can then be easily determined. The mathematical details of this procedure have been left out as an exercise for the reader.

Now we describe the experimental approach to determining the tuning constants of cascaded controllers.

Example 2 (by Permission from Ref. 4). The industrial application described here involves a double-cascade control system. The innermost controller is analog, and the two outer control algorithms are executed on a digital computer. The purpose of the double-cascade system is to maintain the controlled variable C at set point despite changes in loads L_1 and L_2. A block diagram of this system is shown in Figure 16.8.

The purpose is to identify the dynamics of the open-loop process and that of each successive level of the closed-loop system. The resulting dynamic data are used to design the innermost analog controller and the two outer digital control algorithms. Then the controller settings are applied to the process operating under closed-loop control, in the presence of load disturbances, to evaluate the adequacy of their design.

The pulse-testing technique has been used in this application. To prepare for the plant tests, the input X_1 and the output Y_1 (see Figure

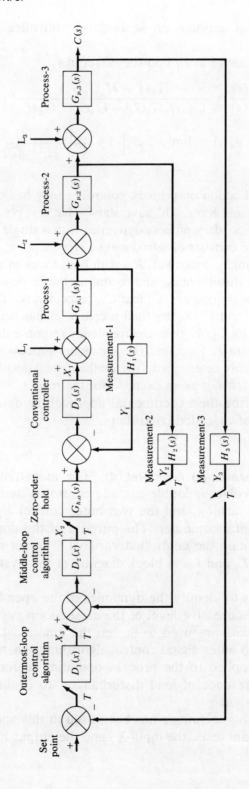

**Figure 16.8
The Test System Used in Double-Cascade Control Application**

16.8) were connected to separate pens of a two-pen strip-chart recorder. When the desired steady state was achieved and the process was free from load disturbances, all control loops were switched to manual. A pulse of desired magnitude and duration was introduced into the innermost loop by rotating the manual knob on the innermost controller. By observing the time record of the input pulse, it was possible to return the input to the initial steady-state operation. The test was repeated for different pulse widths and heights (above and below the initial steady-state values). For each run numerous values of the input and output were read from their respective time records into an off-line computer program (Appendix D), which solved the appropriate equations so as to determine frequency response. From these data the tuning constants for the innermost PI controller were determined by the Ziegler–Nichols method. These settings were implemented on the innermost controller and that loop was switched to automatic. The transients were allowed to disappear, and when the steady state was reached, the second pulse test was conducted.

The innermost controller is a computer-set control station. The controller is equipped with a stepping motor which accepts a pulse train from the process-control computer at a rate of up to 30 pulses per second. The stepping motor moves the setpoint needle of the controller. Each pulse into the stepping motor moves the setpoint needles one thousandth of full scale. The stepping motor responds to discontinuities in voltage between "closure" and "common" terminals. Thus, by disconnecting the computer from the control station and by connecting two leads to the "closure" and "common" terminals, the set point can be pulsed any desired amount by repeatedly contacting the two leads.

In this manner, several pulses of desired magnitude and duration were introduced into the middle loop at X_2, and the transient response Y_2 was recorded. As before, time records for X_2 and Y_2 were processed on the off-line computer as to determine the Ziegler–Nichols settings for the middle-loop PI algorithm. These settings were divided by two (the reason for this is explained in a subsequent paragraph) and then read into the on-line control computer as tuning parameters for the middle-loop algorithm. The middle loop was then switched to automatic by reconnecting the pulse contacts from the computer to the control station.

With the two inner loops in automatic, the outermost loop was switched to manual. A pulse of desired magnitude and duration was generated on the on-line control computer and introduced into the input to the outermost loop at X_3, and the transient response Y_3 was

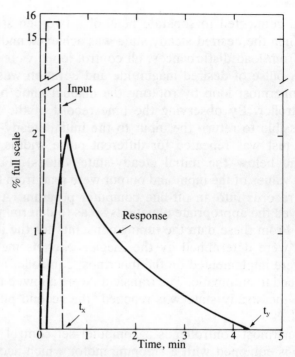

Figure 16.9
The Input and Response for the Inner Loop

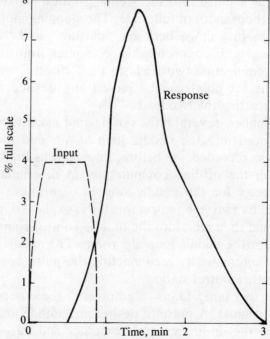

Figure 16.10
The Input and Response for the Middle Loop

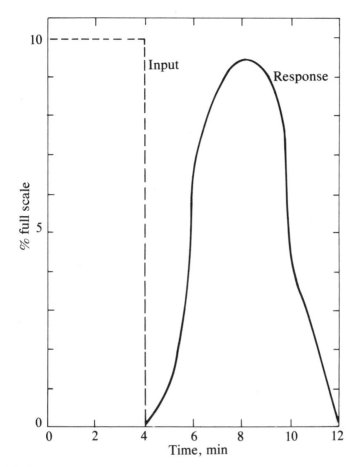

Figure 16.11
The Input and Response for the Outer Loop

recorded. The time records for Y_3 and X_3 assisted in the determination of the PI control algorithm constants of the outermost loop. The test was repeated using a pulse of different magnitude.

These settings were halved and then read into the control-computer program as the tuning parameters of the outermost algorithm. All three loops were switched to automatic, and the transient response of the process, in the presence of load disturbances, was obtained and analyzed.

Sample time records of input pulse and the associated transient response for each loop are shown in Figures 16.9, 16.10, and 16.11 for one of the tests. The computer-generated Bode plots from the analysis of these figures are shown in Figures 16.12, 16.13, and 16.14, respectively. The tuning parameters determined from these Bode plots are shown in Table 16.3.

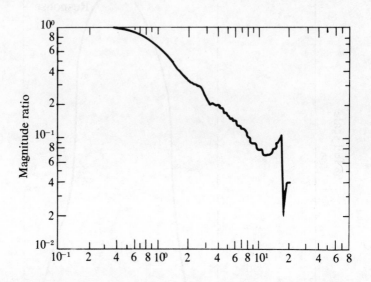

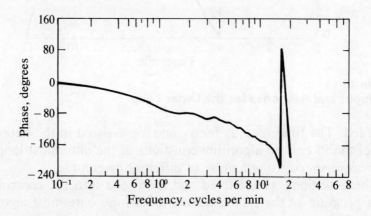

Figure 16.12
Frequency Response for the Inner Loop

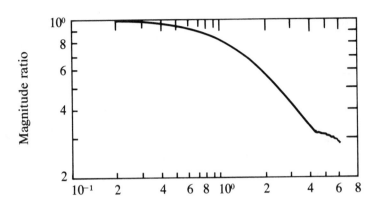

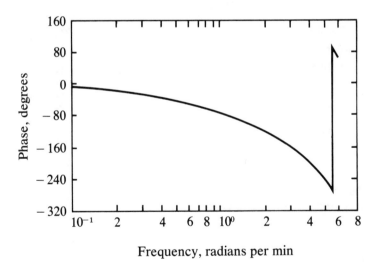

Frequency, radians per min

Figure 16.13
Frequency Response for the Middle Loop

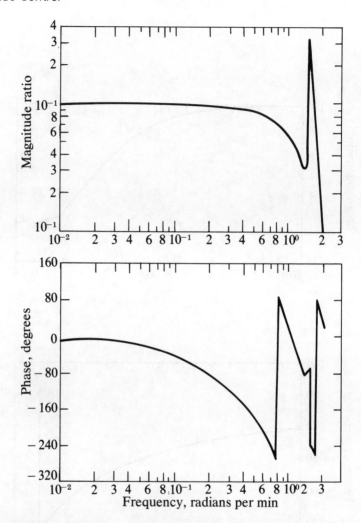

Figure 16.14
Frequency Response for the Outer Loop

Table 16.3
Tuning Parameters for Pulse Data

Loop	Inner	Middle	Outer
Natural period, minutes	0.50	1.88	15.4
Maximum controller, gain	14.00	1.92	8.60
Ziegler–Nichols gain	6.30	0.86	3.87
Actual gain implemented	6.30	0.43	1.94
Reset time, min	0.42	1.57	12.83

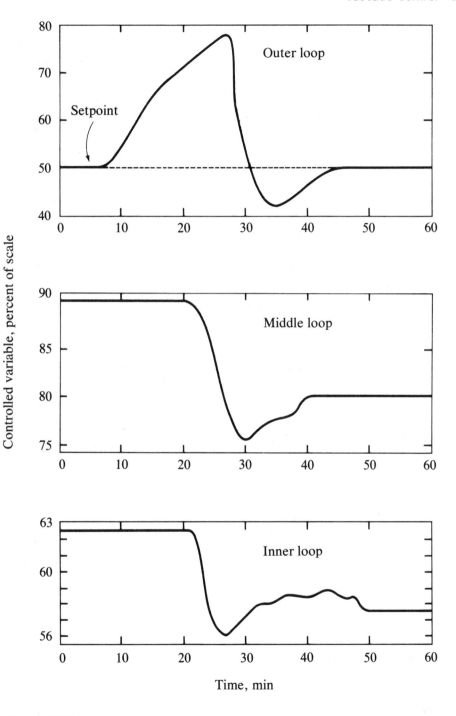

Figure 16.15
The System Response to a Load Change

If all three loops of the double-cascade system were analog, these settings would have been satisfactory. However, since the two outer loops are under computer control, their performance depends not only on the tuning constants but also on the sampling period. Indeed, it can be shown by the Z-transform techniques, that a second-order system, which is stable for all values of the gain for a conventional system, can become unstable for some values of the sampling period in a sampled-data loop. The performance of the two outer loops with the Ziegler–Nichols settings, found from pulse testing, was found to exhibit unstable behavior. Since it was not possible to alter the sampling period due to other constraints, the Ziegler–Nichols settings had to be reduced. Digital computer simulations using Z transforms showed that good transient response could be obtained if Ziegler–Nichols settings were halved.

Finally, the closed-loop response of the system, with all loops in automatic, is shown in Figure 16.15. Subsequent observations have shown these settings to be quite adequate.

16.2. An Industrial Application of Cascade-Control Technique

The block diagram of an industrial polymerization control system is shown in Figure 16.16. In this application the temperature in the

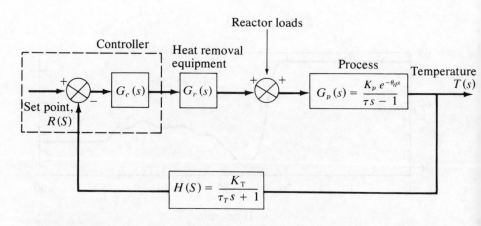

Figure 16.16
Block Diagram of Polymerization Control System (Published by Permission of E. I. Du Pont de Nemours and Company, Louisville, Kentucky)

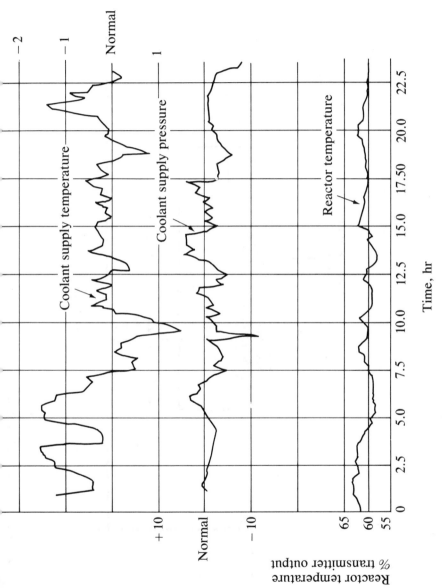

Figure 16.17
Response of Feedback Control System in the Presence of Disturbances

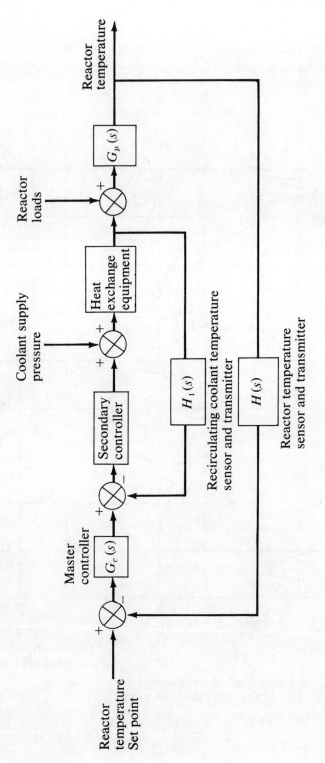

Figure 16.18
Block Diagram of Cascade Control System

jacketed reactor is controlled by manipulating the flow of coolant in the jacket. The temperature response shown in Figure 16.17 has oscillations caused primarily by the variations in supply temperature and pressure of the cooling medium. Thus this system is an ideal candidate for implementing cascade control.

The block diagram of a temperature-on-temperature cascade-control system is shown in Figure 16.18. When this cascade system was implemented, the temperature response of the system, shown in Figure 16.19, was considerably improved.

From Figure 16.17 note that the supply pressure and temperature disturbances of the coolant system go right through the system and upset the reactor temperature. The feedback system is not able to cope with these upsets. In the cascade-control results shown in Figure 16.19 the secondary loop detects the changes in coolant temperature resulting from these upsets and constantly manipulates the flow of coolant so that the temperature of the reactor is not affected.

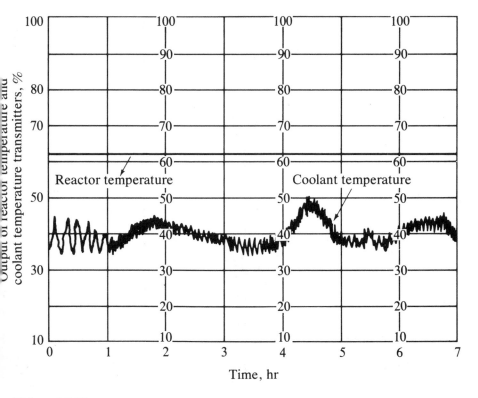

Figure 16.19
Response of Cascade Control System

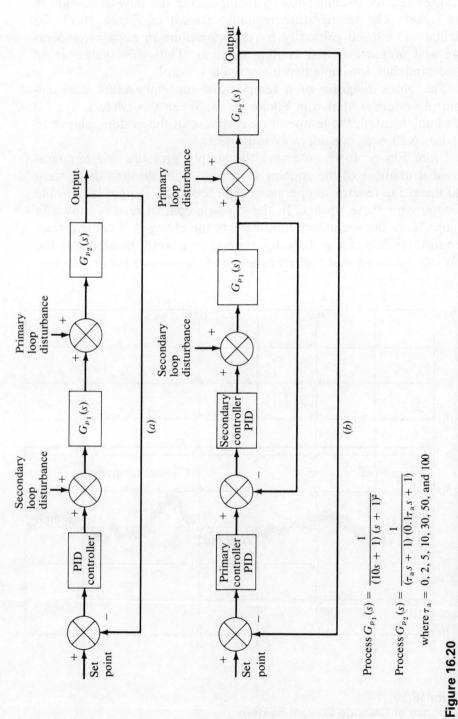

Process $G_{p_1}(s) = \dfrac{1}{(10s + 1)(s + 1)^2}$

Process $G_{p_2}(s) = \dfrac{1}{(\tau_a s + 1)(0.1\tau_a s + 1)}$

where $\tau_a = 0, 2, 5, 10, 30, 50,$ and 100

Figure 16.20
Block Diagrams and System Time Constants Used in the Simulation Study (a) Feedback System (b) Cascade System

16.3. When to Use Cascade Control

Franks and Worley[5] conducted an analog computer simulation study to assess the benefits of cascade control. Figure 16.20 shows the block diagrams and the system time constants. The results of their study are summarized in Figure 16.21. This figure shows that the improvement for set-point changes and for master-loop disturbances is much less than for disturbances introduced into the slave

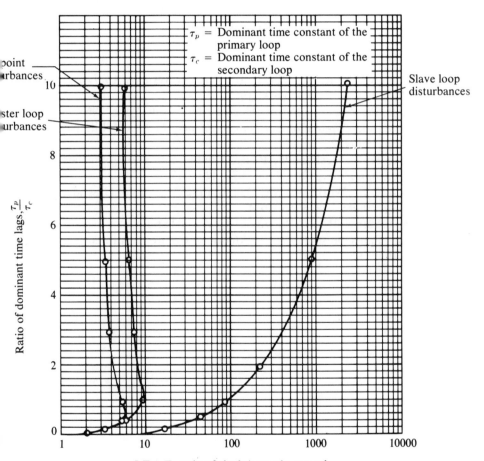

Figure 16.21
Relative Improvement of Cascade Control over Single Loop Control.
(Franks, R.G., and C.W. Worley; "Quantitative Analysis of Cascade
Control", *Ind. Eng. Chem.***, 48:1074, 1956 (Reproduced with Permission)**

loop. Maximum benefit from cascade control for set-point changes and for master-loop disturbances occurs when the dominant time constants of the two loops are roughly equal. The improvement in the response of the control system to slave-loop disturbances increases rapidly as the ratio of the dominant time constants increases.

The applicability of Figure 16.21 is somewhat limited, because it does not show the effect of dead time on the control performance of cascade systems. However, it does show quantitatively, the benefits of cascade control for the types of systems that were studied.

References

1. Coughanowr, D. R., Koppel, L. B., *Process Systems Analysis and Control*, McGraw-Hill, New York, 1965.
2. Harriott, P., *Process Control*, McGraw-Hill, New York, 1965.
3. James, et al., *Theory of Servomechanisms*, McGraw-Hill, New York, 1947, pp. 180–182.
4. Schork, F. J., Deshpande, P. B., Double Cascade Controller Tested, *Hydrocarbon Processing*, June 1978, 113–117.
5. Franks, R. G., Worley, C. W., Quantitative Analysis of Cascade Control, *Ind. Eng. Chem.*, **48**, 1956, 1074.
6. Krishnaswamy, P. R., Rangaiah, G. P., "Tuning of Cascade Control Loops," Paper presented at the American Control Conference, St. Paul, MN, June 1987.

CHAPTER 17

Multivariable Control Systems

A multivariable system is one in which an input not only affects its own output but also one or more other outputs in the plant. In a 2 × 2 system, for instance, *coupling* is said to exist if one input affects both outputs but the second input influences only one output. If both inputs affect both outputs, then, *interaction* is said to exist.

A vast majority of control loops in processing plants can be designed by the SISO methodologies that we have already considered in this text. However, a class of commercially important processes are multivariable in nature. Figure 17.1 shows several examples. Quality control considerations, safety, energy utilization, and overall plant economics dictate that multivariable processes must be operated under good control strategies for optimum return on investment. This, combined with the availability of powerful, low cost digital computers, has spurred considerable research and development activities in recent years into the development of multivariable control methodologies.

Numerous multivariable control methods are available. Edgar[1] sur- 345 veyed methods available up to 1976, but several powerful, computer-based methods have appeared in the literature since then. The methods available to date may be grouped in three categories. In the first category are the Laplace domain or frequency domain methods, which employ process gain matrices, process transfer functions, and frequency response data. Examples of the methods in this category are relative gain analysis/decoupling, singular value decomposition, and direct/inverse Nyquist arrays. In the second category are methods that employ state-space models. The state-space methods are generally grouped under the heading of optimal control or modern control theory. Examples of the techniques in this category are linear quadratic design, Lyapunov function, and incomplete state feedback. Finally, in the third category are methods that utilize impulse response models of the multivariable process as a basis for control systems design. These are powerful methods that take full advantage of the computational power of modern control computers. They require only process step responses for implementation, and they directly take process interaction into account. They are ideally suited to handle constraints. Examples in this category are model algorithmic control[2] and dynamic matrix control.[3] Garcia and Morari[4,5,6] studied the theoretical properties of these algorithms and have shown that, although independently developed, they are two manifestations of the same technique they referred to as "internal model control." These methods have been applied successfully in industrial applications.

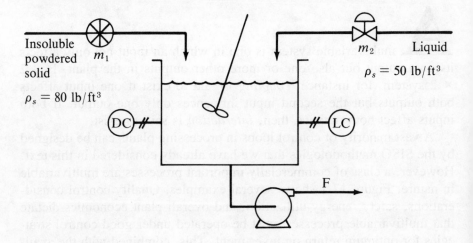

Figure 17.1a
Slurry Blending Tank
(Reprinted by permission from *Interaction Analysis: Principles and Applications*, Thomas J. McAvoy, © 1983, Instrument Society of America.)

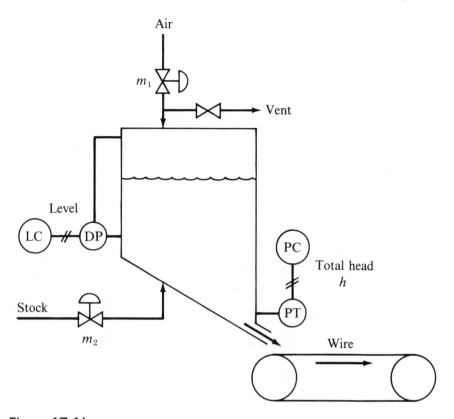

Figure 17.1*b*
Head Box System
(Reprinted by permission from *Process Control Systems*, 2nd Ed.,
F.G. Shinskey, © 1979, Instrument Society of America.)

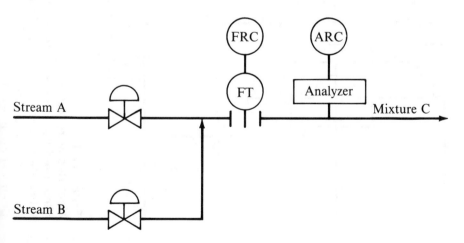

Figure 17.1*c*
Schematic of Blending System

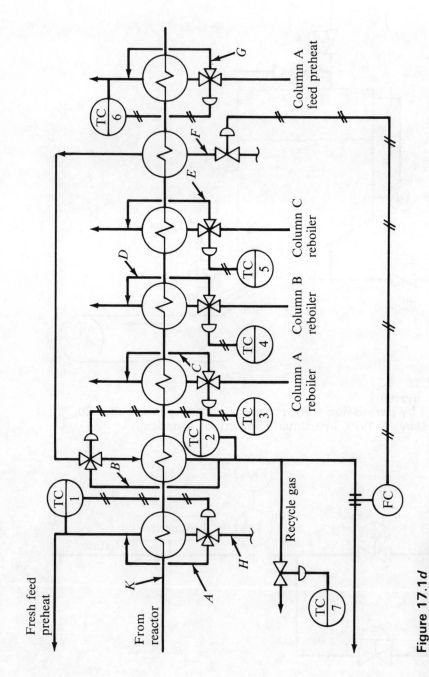

Figure 17.1d
Heat Recovery Network
(Reprinted by permission from "Applying Control Computers to an Integrated Plant," A. Eli Nisenfeld, from *Chemical Engineering Progress*, vol. 69, no. 9, p. 45, © 1973, American Institute of Chemical Engineers.)

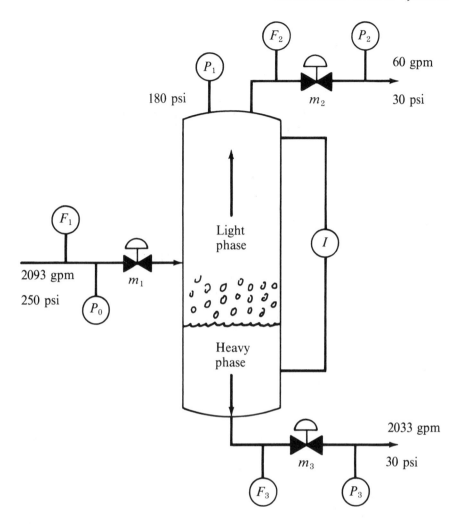

Figure 17.1e
Decanter System
(Reprinted by permission; © 1980, Instrument Society of America.)

Although numerous methods are available for multivariable control systems design, as is evident from the foregoing discussion, not all of them can be applied with equal ease in chemical process plants. Their use in chemical engineering applications is complicated for the following reasons:

• Many variations in design and operation exist in process units. Thus, a strategy that is perfectly adequate for one distillation column may not work at all for another. This is in sharp contrast to electrical

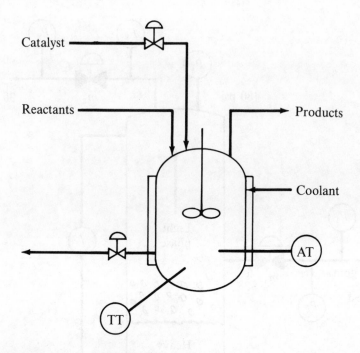

Figure 17.1f
Exthermic Chemical Reactor

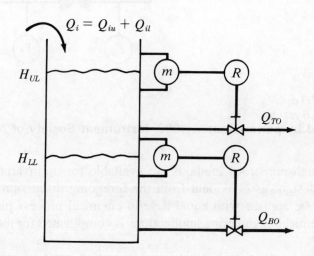

Figure 17.1g
Decanter System

and aeronautical/aerospace applications where the same strategy may be applied in every unit made.

● Process plants exhibit significant nonlinearities. As a result, special attention must be given to the problem of varying process gains as the operating level changes in response to market forces. Furthermore, many process disturbances, which affect product quality unpredictably, may be unidentifiable and/or unmeasurable.

● Process units invariably exhibit large apparent dead time characteristics, and process noise is often a problem.

● In chemical engineering systems not all the state variables can be readily measured. This presents certain difficulties for state-space methods requiring the reconstruction or estimation of the missing variables. The use of state-space models in deriving control laws can lead to problems since such representation is not unique and is very sensitive to the order of the system.[7]

● The extension of the single-loop design procedure for PID-type controllers to multivariable chemical engineering processes is not straightforward due to interaction unless the latter is transformed into a decoupled system. The task of developing a satisfactory design procedure for multivariable PID-type process controllers remains a difficult problem.

● Multivariable processes often must be operated near operating constraints, and the number of inputs and outputs is not always the same.

In view of the foregoing complications, the designer must exercise caution in selecting the proper method for a specific application.

The scope of this work does not permit the inclusion of all the multivariable control methods currently available. However, a text specifically devoted to multivariable control methods is available.[8]

The first task in designing a multivariable control strategy is to determine the extent of interaction present. The objective in this instance is to determine the best set of manipulated and controlled variables from among several competing sets. Let us take an example. A distillation control system can often be represented as a 2 × 2 problem having three combinations of controlled and manipulated variables, which are shown in Table 17.1. The question is, which of the three sets should be selected? The tool for arriving at an answer to this problem is the so-called interaction analysis. Once a proper selection of manipulated and controlled variables has been made, the control systems engineer may find that, even in the finally selected set, sufficient interaction exists to warrant further work. The next step is either to use a decoupling approach, which aims to eliminate interaction, or to use a multivariable control strategy, which inherently compensates for interactions. Factors, such

Table 17.1
Possible Pairings in Distillation Control

	Control distillate composition with	*Control bottoms product composition with*
Set I	R	V
Set II	D	V
Set III	R	B

Note: R = reflux
       B = bottoms product flow
       V = vapor boilup

as presence of moving constraints or inequality of inputs and outputs, may well dictate that the latter be employed. If the selected final pairing exhibits only modest interaction, it may be sufficient to employ SISO PID-type controllers. A relatively simple procedure for tuning these controllers in a multiloop environment is available.[9]

The approach we follow in this chapter is as follows. We first introduce the reader to interaction analysis and discuss how it is used and applied. We then describe how SISO controllers in a multiloop environment may be designed. This is followed by design of decouplers. Finally, we provide a broad coverage of three approaches to multivariable control, viz., internal model control, dynamic matrix control, and simplified model predictive control.

17.1 Interaction Analysis

The purpose of interaction analysis is to determine the extent of interaction present in a multivariable system. Thus, interaction analysis can be used to determine the best set of manipulated and controlled variables from among competing sets. If a set having only a small amount of interaction can be identified, then SISO or multivariable techniques may be used to design the controllers. If, on the other hand, the best set exhibits significant interaction, then the system must either be decoupled and SISO controllers applied or multivariable techniques that inherently compensate for interaction must be employed. As has been mentioned, the presence of constraints and complexities in process dynamics, as evidenced by the presence of inverse response, long and unequal time delays in the transfer function matrix elements, etc., may well dictate which strategy is the best for a given application.

A number of methods for interaction analysis are available. These include the relative gain array[10] (RGA), singular value analysis,[11,12] IMC interaction measure,[13] the interaction quotient,[14] the relative dynamic gain array,[15,16,17] the average dynamic gain array,[18] the relative transient response functions,[19] direct and Nyquist arrays,[20,21,22,8] and the Jacobi eigenvalue criterion.[23] Several of these methods have been surveyed and a comparative study made by Jensen et al.,[24] Yu and Luyben,[25] and Mijares, et al.[23] In this section we describe two of these methods and show how they are used. It appears that no one simple method presently available is satisfactory for all problems. This suggests that a control engineer may well have to analyze the problem by the numerous approaches available. Should the result by any one of them indicate a potential problem in terms of interaction, further analysis must be carried out.

Bristol's Relative Gain Analysis

A measure of the extent of interaction in multivariable control systems is obtained by Bristol's method.[10] It is based on the steady-state input-output relationships for the process. The method seeks to determine the best single-input/single-output connections (i.e., best pairing of manipulated and controlled variables). It yields a measure of steady-state gain between a given input-output pairing and thus, by using the most sensitive input-output connections, control interaction can be minimized. Consider a system, shown in Figure 17.2, with two controlled variables C_1 and C_2 and two manipulated variables M_1 and M_2 where each C is affected by both Ms. Around some steady-state operating point we can express the steady-state relationships between the Cs and Ms as follows:

$$\Delta C_1 = \left.\frac{\partial C_1}{\partial M_1}\right|_{M_2} \Delta M_1 + \left.\frac{\partial C_1}{\partial M_2}\right|_{M_1} \Delta M_2 = K_{11}\Delta M_1 + K_{12}\Delta M_2 \quad (17.1)$$

$$\Delta C_2 = \left.\frac{\partial C_2}{\partial M_1}\right|_{M_2} \Delta M_1 + \left.\frac{\partial C_2}{\partial M_2}\right|_{M_1} \Delta M_2 = K_{21}\Delta M_1 + K_{22}\Delta M_2 \quad (17.2)$$

The Ks (partial derivatives evaluated at the steady-state operating point) are called the *open-loop steady-state gains* of the process. They quan-

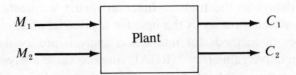

Figure 17.1
A 2 × 2 Multivariable System

titatively describe how the Ms affect the Cs. They can be determined from a mathematical model or by experiment tests on the process. To evaluate K_{11}, for example, we would make a small change in M_1 while the process was operating at steady state and while holding M_2 constant. After C_1 and C_2 reached their new steady-state values, we would evaluate

$$K_{11} = \frac{\Delta C_1}{\Delta M_1}\bigg|_{\substack{M_2 \text{ constant} \\ (\Delta M_2 = 0)}} \quad ; K_{21} = \frac{\Delta C_2}{\Delta M_1}\bigg|_{\substack{M_2 \text{ constant} \\ (\Delta M_2 = 0)}} \quad (17.3)$$

The gain K_{11}, then, determines the change in C_1 due to a change in M_1 while M_2 is held constant. Suppose, instead of holding M_2 constant while we make a small change in M_1, we simultaneously manipulate M_2 to bring C_2 back to the original value it had before the change in M_1 was made. We can then define another "gain" between C_1 and M_1:

$$a_{11} = \frac{\Delta C_1}{\Delta M_1}\bigg|_{C_2 \text{ constant}} \quad (17.4)$$

The a_{11} is a measure of how much M_1 would affect C_1 if all other controlled variables were under closed-loop control (i.e., held constant). The ratio of K_{11} to a_{11} is called the *relative gain, λ_{11}*. Thus,

$$\lambda_{11} = \frac{K_{11}}{a_{11}} = \frac{\dfrac{\Delta C_1}{\Delta M_1}\bigg|_{M_2 \text{ constant}}}{\dfrac{\Delta C_1}{\Delta M_1}\bigg|_{C_2 \text{ constant}}} \quad (17.5)$$

By comparing the relative gains for each manipulated variable, we can get a quick quantitative assessment of which M has the most effect on a given controlled variable C, and thus how we should pair the Ms and Cs.

The a_{ij} gains can be computed from the K_{ij} gains as follows. Rewriting the open-loop relationship between Ms and Cs from Equations (17.1) and (17.2), we have

$$\Delta C_1 = K_{11} \, \Delta M_1 + K_{12} \, \Delta M_2 \qquad (17.6)$$

$$\Delta C_2 = K_{21} \, \Delta M_1 + K_{22} \, \Delta M_2 \qquad (17.7)$$

The a_{11}, by definition, is

$$\frac{\Delta C_1}{\Delta M_1} \text{ with } \Delta C_2 = 0.$$

Thus, from Equation (17.7)

$$\Delta C_2 = 0 = K_{21} \, \Delta M_1 + K_{22} \, \Delta M_2$$

solving for ΔM_2:

$$\Delta M_2 = - \frac{K_{21}}{K_{22}} \Delta M_1 \qquad (17.8)$$

Substituting Equation (17.8) into (17.6) we get

$$\Delta C_1 = K_{11} \, \Delta M_1 - \frac{K_{12} \, K_{21}}{K_{22}} \Delta M_1 = \left(K_{11} - \frac{K_{12} \, K_{21}}{K_{22}} \right) \Delta M_1 \quad (17.9)$$

Thus,

$$a_{11} = \frac{\Delta C_1}{\Delta M_1} \bigg|_{\Delta C_2 = 0} = K_{11} - \frac{K_{12} \, K_{21}}{K_{22}} = \frac{K_{11} K_{22} - K_{12} K_{21}}{K_{22}} \qquad (17.9a)$$

The relative gain λ_{11} is then

$$\lambda_{11} = \frac{K_{11}}{a_{11}} = \frac{K_{11} \, K_{22}}{K_{11} K_{22} - K_{12} K_{21}} \qquad (17.10)$$

The other three gains a_{ij} can be computed similarly; they are:

$$a_{12} = \left.\frac{\Delta C_1}{\Delta M_2}\right|_{\Delta C_2 = 0} = \frac{K_{12}K_{21} - K_{11}K_{22}}{K_{21}}$$

$$\Rightarrow \lambda_{12} = \frac{K_{12}K_{21}}{K_{12}K_{21} - K_{11}K_{22}}$$

$$a_{21} = \left.\frac{\Delta C_2}{\Delta M_1}\right|_{\Delta C_1 = 0} = \frac{K_{12}K_{21} - K_{11}K_{22}}{K_{12}} \qquad (17.11)$$

$$\Rightarrow \lambda_{21} = \frac{K_{12}K_{21}}{K_{12}K_{21} - K_{11}K_{22}}$$

$$a_{22} = \left.\frac{\Delta C_2}{\Delta M_2}\right|_{\Delta C_1 = 0} = \frac{K_{11}K_{22} - K_{12}K_{21}}{K_{11}}$$

$$\Rightarrow \lambda_{22} = \frac{K_{11}K_{22}}{K_{11}K_{22} - K_{12}K_{21}}$$

To facilitate the pairing of manipulated and controlled variables, it is convenient to arrange the relative gains as follows:

	M_1	M_2	$\cdots$	M_n
C_1	λ_{11}	λ_{12}	$\cdots$	λ_{1n}
C_2	λ_{21}	λ_{22}	$\cdots$	λ_{2n}
$\vdots$	$\vdots$	$\vdots$	$\vdots$	$\vdots$
C_n	λ_{n1}	λ_{n2}	$\cdots$	λ_{nn}

$$(17.12)$$

For each controlled variable C_i the manipulated variable selected is the one having the largest positive relative gain closest to 1. For example, the manipulated variable for C_1 would be the one corresponding to the largest relative gain λ_{ij} in the first row of the relative gain matrix above. The relative gain matrix for the two-variable system is as follows:

	M_1	M_2
C_1	$\dfrac{K_{11}K_{22}}{K_{11}K_{22} - K_{12}K_{21}}$	$\dfrac{K_{12}K_{21}}{K_{12}K_{21} - K_{11}K_{22}}$
C_2	$\dfrac{K_{12}K_{21}}{K_{12}K_{21} - K_{11}K_{22}}$	$\dfrac{K_{11}K_{22}}{K_{11}K_{22} - K_{12}K_{21}}$

$$(17.13)$$

A useful property of the relative gain matrix is that each row and each column sums to 1. Thus, in a 2×2 system, only one of the four relative gains needs to be explicitly computed.

It is possible for relative gains to be negative. For a 2×2 system, if an m and a C with a negative relative gain are paired, the system will be uncontrollable and unstable—each variable will be driven to its limit. Poor control may also result if $\lambda_{ij} >> 1$.

For systems with more than two Cs, the a_{ij}s can be conveniently computed from the K_{ij}s as follows. First, arrange the Ks in a matrix as

$$
\mathbf{K} = \begin{bmatrix}
K_{11} & K_{12} & \cdots & K_{1n} \\
K_{21} & K_{22} & \cdots & K_{2n} \\
\vdots & & & \\
K_{n1} & K_{n2} & \cdots & K_{nn}
\end{bmatrix}
\tag{17.14}
$$

Then, compute a *complimentary matrix* $\mathbf{C}$, by first inverting, then transposing the matrix $\mathbf{K}$:

$$
\mathbf{C} = (\mathbf{K}^{-1})^T = \begin{bmatrix}
C_{11} & C_{12} & \cdots & C_{1n} \\
C_{21} & C_{22} & \cdots & C_{2n} \\
C_{n1} & C_{n2} & \cdots & C_{nn}
\end{bmatrix}
\tag{17.15}
$$

The element on the ith row and jth column of $\mathbf{C}$ is the *reciprocal of a_{ij}*; that is,

$$
C_{ij} = \frac{1}{a_{ij}}
\tag{17.16}
$$

Thus each relative gain term λ_{ij} is found by multiplying each element in matrix $\mathbf{K}$ by its corresponding term in matrix $\mathbf{C}$:

$$
\lambda_{ij} = \frac{K_{ij}}{a_{ij}} = K_{ij} C_{ij}
\tag{17.17}
$$

Example 1 (By Permission from Ref. 26)

As shown in Figure 17.3, two liquids are mixed in line to produce a mixture of desired composition X^{set}. The total flow rate is also to be controlled. The problem is to find the proper pairing of the controlled and manipulated variables if the composition X is to be controlled at 0.3 mass fraction A.

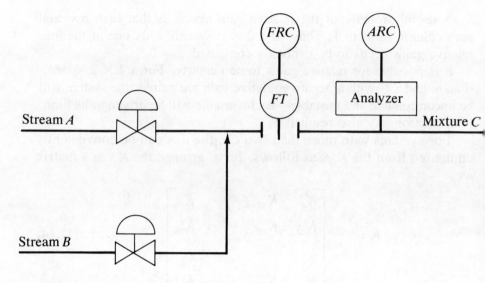

Figure 17.3
Schematic of Blending System

The equations relating controlled variables to the manipulated variables are:

$$C = A + B$$

and

$$X = \frac{A}{A + B}$$

The steady-state gains are

$$K_{11} = \frac{\partial C}{\partial A} = \frac{\partial (A + B)}{\partial A} = 1; \quad K_{12} = \frac{\partial C}{\partial B} = \frac{\partial (A + B)}{\partial B} = 1$$

$$K_{21} = \frac{\partial X}{\partial A} = \frac{\partial \left[A/(A + B) \right]}{\partial A} = \frac{(A + B) - A}{(A + B)^2} = \frac{B}{C^2} = \frac{1 - X}{C}$$

$$K_{22} = \frac{\partial X}{\partial B} = \frac{\partial \left[A/(A + B) \right]}{\partial B} = \frac{O - A}{(A + B)^2} = -\frac{A}{C^2} = -\frac{X}{C}$$

Therefore, with reference to Equation (17.13), λ_{11} is computed as

$$\lambda_{11} = \frac{K_{11}K_{22}}{K_{11}K_{22} - K_{12}K_{21}} = \frac{(1)\left(-\dfrac{X}{C}\right)}{(1)\left(-\dfrac{X}{C}\right) - (1)\left(\dfrac{1-X}{C}\right)} = X$$

As has been pointed out earlier, only one element of the relative gain matrix need be explicitly computed, since the rows and columns of the matrix sum to 1. Thus,

$$\lambda = \begin{array}{c|cc} & A & B \\ \hline C & X & 1-X \\ X & 1-X & X \end{array}$$

If $X^{\text{set}} = 0.3$, then

$$\lambda = \begin{array}{c|cc} & A & B \\ \hline C & 0.3 & 0.7 \\ X & 0.7 & 0.3 \end{array}$$

Since the elements with the largest positive relative gain are to be selected, this matrix suggests that we control the total flow by manipulating B and composition by manipulating A.

Example 2[27]

We now present a laboratory application of Bristol's approach to multivariable pairing. We operate the closed-loop process with correct pairing and with incorrect pairing of the controlled and manipulated variables to assess the benefits of proper pairing. The hardware for this experiment is shown in Figure 17.4.

The process objective is to control the level (in effect, total flow, m_t) and temperature of water in the tank, T. There are two inputs to the process namely, the flow of cold water, m_c and the flow of hot water, m_h into the tank. So, the controlled variables are temperature and total flow, and the manipulated variables are the cold water and hot water flow rates. The question is, should the temperature be controlled by manipulating hot water flow and level (i.e., total flow) by cold water flow or vice versa? Bristol's method provides the answer.

The functional steady-state relationship between temperature, total flow and the flow stream is

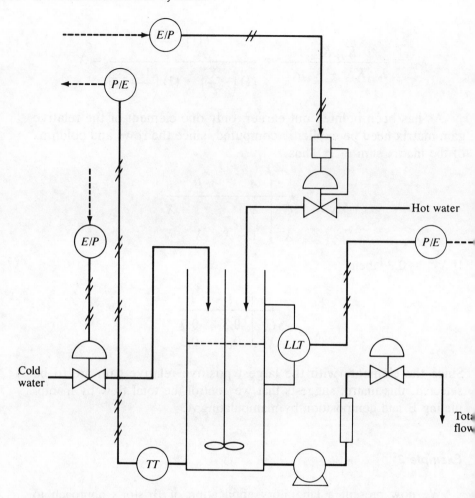

Figure 17.4
Schematic of Multivariable System (Dashed Lines Indicate Signal Transmission Between Process and Computer

$$T = f(m_c, m_h) = (m_c T_c + m_h T_h)/m_t \qquad (17.18)$$

$$m_t = f(m_c, m_h) = m_c + m_h$$

Around some steady-state operating point these relationships can be expressed as

$$\Delta T = \left. \frac{\partial T}{\partial m_c} \right|_{m_h = \text{constant}} \Delta m_c + \left. \frac{\partial T}{\partial m_h} \right|_{m_c = \text{constant}} \Delta m_h$$

$$= K_{11} \Delta m_c + K_{12} \Delta m_h$$

and $\qquad\qquad\qquad\qquad\qquad\qquad\qquad\qquad\qquad\qquad\qquad$ (17.19)

$$\Delta m_t = \frac{\partial m_t}{\partial m_c}\bigg|_{m_h = \text{constant}} \Delta m_c + \frac{\partial m_t}{\partial m_h}\bigg|_{m_c = \text{constant}} \Delta m_h$$

$$= K_{21}\Delta m_c + K_{22}\Delta m_h$$

Recall that the relative gains matrix for a 2 × 2 system is

$$\lambda = \begin{array}{c|cc} & m_c & m_h \\ \hline T & \lambda_{11} & \lambda_{12} \\ m_T & \lambda_{21} & \lambda_{22} \end{array} \qquad (17.20)$$

where

$$\lambda_{11} = \frac{K_{11}K_{22}}{K_{11}K_{22} - K_{12}K_{21}}$$

Since each row and column sums to one, only one λ need be computed in this application. Thus, taking the partial derivatives of the terms in Equation (17.18) as indicated in Equation (17.19), we obtain the following relative gains matrix.

$$\lambda = \begin{array}{c|cc} & m_c & m_h \\ \hline T & \dfrac{m_h}{m_t} & \dfrac{m_c}{m_t} \\[2ex] m_T & \dfrac{m_c}{m_t} & \dfrac{m_h}{m_t} \end{array} = \begin{array}{c|cc} & m_c & m_h \\ \hline T & 0.172 & 0.828 \\ m_t & 0.828 & 0.172 \end{array} \qquad (17.21)$$

This equation shows that T should be controlled by manipulating m_h and m_t by manipulating m_c. In this application both loops use a PI control algorithm on the digital computer as the control element. The algorithm was tuned by trial and error. The steady-state operating conditions were: level set point, 50% (which corresponded to total outlet flow of 11.6 lit/min); temperature set point, 24.4°C; cold-water flow, 9.61 lit/min; hot water flow, 1.99 lit/min. The process was operated with correct pairing as well as with incorrect pairing. The benefits of proper pairing are clearly evident in the set point responses shown in Figures 17.5 and 17.6. These results show that Bristol's approach is a simple and powerful tool in the control systems design of multivariable processes.

For higher than 2 × 2 systems, a modification of the RGA rule has been suggested.[28,29] The modified procedure suggests that one would pair on the positive relative relative gains closest to 1.0 as before, but then

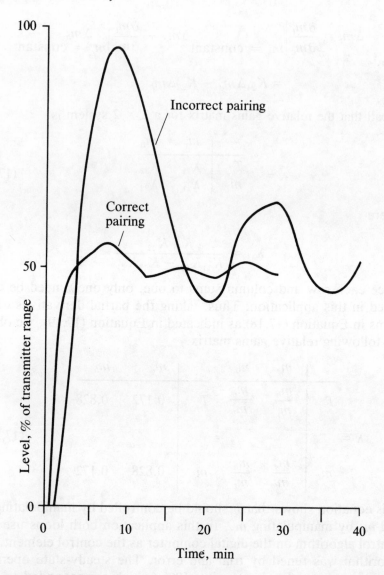

Figure 17.5
Transient Response of Level

the resulting pairings should be checked for stability using Niederlinski's theorem; if the pairings turn out to be unstable, other pairings having positive relative gains closest to 1 should be selected. Pairing on negative relative gain elements should be avoided. Niederlinski's stability theorem states that a closed-loop system containing the pairing M_1-C_1, M_2-C_2,..., M_n-C_n will be unstable if

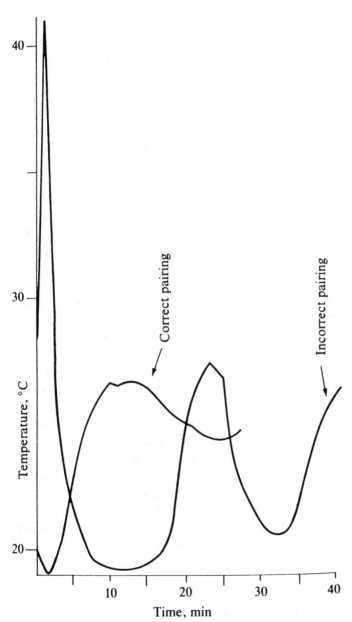

Figure 17.6
Transient Response of Temperature

$$\frac{|\mathbf{K}|}{\displaystyle\prod_{i=1}^{n} K_{ii}} < 0 \tag{17.22}$$

where $|\mathbf{K}|$ is the determinant of the process gain matrix.

Niederlinski's theorem can be used to show that, to achieve stability, one may have to pair on a negative RGA element, directly in contrast to the original pairing rule. However, in this instance, if the loop that is paired on the negative RGA element is opened, the lower-order closed-loop system becomes unstable.[29] Furthermore, in such a case the order in which the loops are switched from manual to automatic becomes important. Due to these reasons, pairing on negative RGA elements should be avoided whenever possible. Recent research appears to indicate that Niederlinski's theorem gives the necessary condition for stability but is not a sufficient criterion for stability.[23] This means that, although unstable pairings can definitely be ruled out by the theorem, the pairings that satisfy the theorem conditions may or may not be stable. Mijares et al.[23] illustrate these problems with several examples.

The RGA method is based on steady-state information; thus, the conclusions may or may not be valid under upset (transient) conditions. McAvoy[29] has developed the concept of dynamic relative gain to accommodate the effect of dynamics. The procedure is similar to the RGA method, but in this instance process transfer functions instead of process gains are used. If the Laplace transform operator s in the transfer function elements is replaced by $j\,\omega$, where ω is frequency, and the relative gain matrix is determined by the usual procedure, then the elements of the λ matrix will be functions of frequency. Now a plot of $|\lambda|$ versus ω is prepared and examined. The results of the steady-state relative gain analysis with regard to a particular suggested pairing may be considered as valid only if a small change in frequency around the ultimate frequency of the loop does not produce a large change in the magnitude of the dynamic relative gain, $|\lambda|$. If this were to occur, the degree of interaction would vary and poor control may result.

Singular Value Decomposition[11,12,30,31]

In this section we describe how singular value decomposition (SVD) may be used to find proper pairings of controlled and manipulated variables. In a later section we will see how SVD may also be used to achieve decoupling. The treatment here is brief, but more details are available in the literature[30] for those who may be interested.

Singular value decomposition is a concept from linear algebra whose properties have been exploited for multivariable control systems design. SVD analysis can be carried out with process gain matrices, as will be illustrated in this section, or with process transfer function matrices,[12] which give dynamic measures of sensitivity and interaction. The geometric representation of SVD facilitates the visualization of relationships between inputs and outputs. In multivariable systems, such as distillation columns, SVD has been applied to determine proper sensor locations. Singular value analysis has, in some instances, provided insights into the process design aspects of a problem. A large number of commercial distillation column control systems is said to have been designed successfully with SVD,[32] although some researchers[23] report examples where SVD has been unable to correctly predict proper pairing. As has been mentioned earlier, the approach to adopt is to apply the variety of tools available. If any one of them indicates a potential problem, a more detailed study, including dynamic simulation, may be carried out before a final design is selected.

To use SVD based on the process gain matrix, **K,** one performs the singular value decomposition of **K** giving three component matrices according to

$$\mathbf{K} = \mathbf{U} \, \mathbf{\Sigma} \, \mathbf{V}^{\mathbf{T}} \tag{17.23}$$

where **K** is an $n \times m$ matrix. **U** is an $n \times n$ matrix whose *columns* are called the *left singular vectors*; **V** is an $m \times m$ matrix, the *columns* of which are called the *right singular vectors*; and Σ is an $n \times m$ diagonal matrix of *scalars* called *singular values*, which are organized in descending order such that $\sigma_1 \geq \sigma_2 \geq \ldots \sigma_m \geq 0$. The following step-by-step procedure may be used to determine **U**, Σ and **V**.

1. Write out the transpose of the matrix **K**.
2. Determine the product $\mathbf{K} \, \mathbf{K}^{\mathbf{T}}$.
3. Determine the eigen values, **P**, of $\mathbf{K} \, \mathbf{K}^{\mathbf{T}}$. This requires the solution of $|\mathbf{K} \, \mathbf{K}^{\mathbf{T}} - \mathbf{PI}| = 0$.
4. $\sigma_i = \sqrt{P_i}$
5. Determine the matrix **U** by solving $(\mathbf{K} \, \mathbf{K}^{\mathbf{T}} - P_1) \, \mathbf{U}_1 = 0$, $(\mathbf{K} \, \mathbf{K}^{\mathbf{T}} - P_2) \, \mathbf{U}_2 = 0$, etc., where $\mathbf{U}_1$, $\mathbf{U}_2$, etc., are columns of the matrix **U**.
6. Determine the matrix **V** by solving $(\mathbf{K}^{\mathbf{T}}\mathbf{K} - P_1)\mathbf{V}_1 = 0$,

$(\mathbf{K^T K} - \mathbf{P}_2)\mathbf{V}_2 = 0$, etc., where $\mathbf{V}_1$, $\mathbf{V}_2$, etc., are columns of the matrix $\mathbf{V}$.

Let us take an example: The process gain matrix is given as

$$
\begin{array}{cc}
& \quad C_1 \qquad\quad C_2 \\
\mathbf{K} = \begin{array}{c} M_1 \\ M_2 \end{array}\!\!\left[\begin{array}{cc} -1.092 & 0.203 \\ -0.628 & -0.203 \end{array}\right]
\end{array}
$$

1.

$$
\mathbf{K^T} = \left[\begin{array}{cc} -1.092 & -0.628 \\ 0.203 & -0.203 \end{array}\right]
$$

2.
$$
\mathbf{K\,K^T} = \left[\begin{array}{cc} -1.092 & 0.203 \\ -0.628 & -0.203 \end{array}\right]\left[\begin{array}{cc} -1.092 & -0.628 \\ 0.203 & -0.203 \end{array}\right]
$$

$$
= \left[\begin{array}{cc} 1.234 & 0.6445 \\ 0.6445 & 0.4356 \end{array}\right]
$$

3.
$$
|\mathbf{K\,K^T} - \mathbf{PI}| = \left|\begin{array}{cc} (1.234 - P) & 0.6445 \\ 0.6445 & (0.4356 - P) \end{array}\right| = 0
$$

which gives

$$
(1.234 - P)(0.4356 - P) - (0.6445)^2 = 0
$$

or

$$
P_1 = 1.5927; \quad P_2 = 0.07654
$$

4.
$$
\sigma_1 = \sqrt{1.5876} = 1.26; \quad \sigma_2 = \sqrt{0.07654} = 0.2767
$$

5.
$$
(\mathbf{KK^T} - \mathbf{P}_1)U_1 = \left(\left[\begin{array}{cc} 1.234 & 0.6445 \\ 0.6445 & 0.4536 \end{array}\right]\right.
$$

$$
\left. - \left[\begin{array}{cc} 1.5927 & 0 \\ 0 & 1.5927 \end{array}\right]\right)\left[\begin{array}{c} U_{11} \\ U_{21} \end{array}\right] = 0
$$

which gives

$$
\left[\begin{array}{cc} -0.3590 & 0.6445 \\ 0.6445 & -1.157 \end{array}\right]\left[\begin{array}{c} U_{11} \\ U_{12} \end{array}\right] = 0
$$

Solving simultaneously gives

$$U_{11} = 0.8736 \quad \text{and} \quad U_{21} = 0.4866$$

Similarly, the solution of $(KK^T - P_2)U_2$ gives the U_2 vector. Thus,

$$\left(\begin{bmatrix} 1.234 & 0.6445 \\ 0.6445 & 0.4536 \end{bmatrix} - \begin{bmatrix} 0.07654 & 0 \\ 0 & 0.07654 \end{bmatrix} \right) \begin{bmatrix} U_{12} \\ U_{22} \end{bmatrix} = 0$$

Solving simultaneously gives

$$U_{12} = -0.4866 \quad \text{and} \quad U_{22} = 0.8736$$

Thus, the **U** matrix is

$$\mathbf{U} = \begin{bmatrix} U_{11} & U_{12} \\ U_{21} & U_{22} \end{bmatrix} = \begin{bmatrix} 0.8736 & -0.4866 \\ 0.4866 & 0.8736 \end{bmatrix}$$

6. By applying the same procedure but working with $\mathbf{K}^T \mathbf{K}$ gives the **V** matrix as

$$\mathbf{V} = \begin{bmatrix} V_{11} & V_{12} \\ V_{21} & V_{22} \end{bmatrix} = \begin{bmatrix} -0.9985 & -0.0621 \\ 0.0622 & -0.9961 \end{bmatrix}$$

Thus,

$$\mathbf{K} = \mathbf{U} \, \Sigma \, \mathbf{V}^T$$

or

$$\begin{bmatrix} -1.092 & 0.203 \\ -0.628 & -0.203 \end{bmatrix}$$

$$= \begin{bmatrix} 0.8736 & -0.4866 \\ 0.4866 & 0.8736 \end{bmatrix} \begin{bmatrix} 1.26 & 0 \\ 0 & 0.2767 \end{bmatrix} \begin{bmatrix} -0.9985 & 0.0622 \\ -0.0621 & -0.9961 \end{bmatrix}$$

Now let us return to the problem of variable pairing. The pairing that will yield the least open-loop interaction is one in which the output associated with the largest magnitude element (without regard to sign) of the U_1 vector is paired with manipulated variable associated with the largest magnitude element (again, without regard to sign) of the V_1 vector. The output associated with the largest magnitude element of U_2 is paired with the manipulated variable associated with the largest magnitude element of V_2, and so on. For the example under scrutiny, the **U** and **V** vectors are

$$\mathbf{U} = \begin{bmatrix} \mathbf{0.87} & -0.48 \\ -0.48 & \mathbf{0.87} \end{bmatrix} \begin{matrix} C_1 \\ C_2 \end{matrix}$$

and

$$\mathbf{V} = \begin{bmatrix} \mathbf{-0.99} & -0.062 \\ 0.062 & \mathbf{-0.99} \end{bmatrix} \begin{matrix} M_1 \\ M_2 \end{matrix}$$

The largest values of $\mathbf{U}_1$ and $\mathbf{V}_1$, shown in boldface, suggest that C_1 should be controlled by manipulating M_1. Similarly, the largest values in $\mathbf{U}_2$ and $\mathbf{V}_2$ suggest that C_2 should be controlled by manipulating M_2.

The foregoing SVD procedure utilized steady-state information, and, therefore, the results may or may not be valid under dynamic conditions. Lau et al.[12] have extended the SVD analysis that includes the effects of dynamics. It then becomes possible to assess the performance of the control system over the entire frequency band of the characteristic disturbances. To use this approach, one performs the singular value decomposition of the process transfer function matrix, and plots of the singular values, σ_i, alignment, θ_i, interaction, θ, and condition number, γ, as a function of frequency are prepared. Equations for the computation of these parameters are available.[12] Depending upon the magnitudes of these parameters, conclusions may be drawn about the workability of the pairings utilized, whether decoupling will improve performance, or whether alternate pairings (if feasible) or process design changes should be considered.

Lau et al.[12] have shown how dynamic decouplers (or compensators) may be designed in the framework of SVD in order to achieve interaction compensation.

17.2 Design of SISO Controllers in Multivariable Environment

If the interaction analysis suggests that only a modest amount of interaction is present, then it may be sufficient to employ SISO controllers. In this section a relatively simple procedure for tuning PID-type SISO controllers in a multivariable environment is described.[9]

A block diagram of the control system is shown in Figure 17.7. For simplicity, PI controllers will be used. Over 90% of the industrial controllers are of the PI type anyway, due to the problems caused by process noise, and, therefore, their use here is not an overly restrictive proposition. The objective is to determine the gains and integral times of these controllers such that the multivariable system gives good set point and load responses. It is also desirable that the system have *integrity,* meaning that the system must remain stable in the event one or more controllers are switched from automatic to manual or vice versa. In the following development it is assumed that the system is of dimension $n \times n$ and it is controlled by n PI controllers such that an output C_i is controlled by an input M_i. The multivariable process is assumed to be open-loop stable.

The method is based on the use of Nyquist stability criteria. The method is not to be confused with inverse and direct Nyquist arrays[20] for multivariable control, which give a measure of interaction and can be used to design a multi-input multi-output control algorithm having interaction compensation properties. For a description of INA and DNA methodology, the reader is referred to Deshpande[8] or to the original works of Rosenbrock.[20,21]

Let us review the stability properties of SISO control systems. A

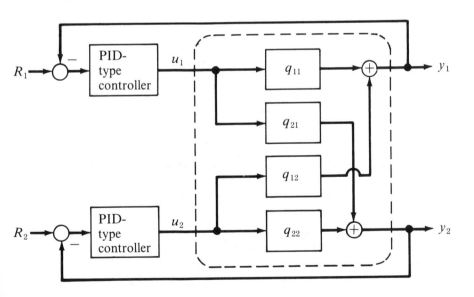

Figure 17.7
Block Diagram of a 2 × 2 Multiloop System

closed-loop control system can be shown to be stable if the roots of the characteristic equation

$$1 + D_c(s)\, G_p(s) = 0 \tag{17.24}$$

or

$$1 + G(s) = 0$$

where

$$G(s) = D_c(s)\, G_p(s)$$

$$D_c(s) = \text{controller transfer function}$$

$$G_p(s) = \text{process transfer function}$$

all lie in the left half s-plane. For a given $G_p(s)$ the controller gain, K_c, and integral time, τ_I, are selected such that the stability condition is met.

The Nyquist stability criteria are developed around a Nyquist plot. A Nyquist plot depicts the real part of $G(j\omega)$ as the x axis and the imaginary part of $G(j\omega)$ as the y axis. To prepare the Nyquist plot one substitutes $j\omega$ for s in $G(s)$ and computes $Re\,\{G(j\omega)\}$ and $Im\,\{G(j\omega)\}$. Substitution of a particular ω gives a point on the Nyquist plot corresponding to the chosen frequency. By varying ω between 0 and ∞ the complete Nyquist plot can be prepared. A typical Nyquist plot is shown

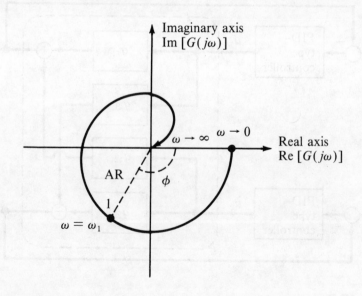

Figure 17.8
Typical Nyquist Plot

in Figure 17.8. One can easily show that a point on the Nyquist plot at a particular frequency gives the corresponding amplitude ratio, AR, and phase angle, ϕ, for that frequency, and, thus, Nyquist diagrams can be prepared from the familiar Bode plots. The Nyquist stability criterion states that a feedback control system will be unstable if the Nyquist plot of $G(s)$ encircles the point $(-1,0)$ on the negative real axis as the frequency goes from zero to infinity. The number of encirclements corresponds to the number of roots of the characteristic equation that lie in the right-half s plane assuming that the process is open-loop stable.

With the appropriate choice of K_c and τ_I the Nyquist plot can be located sufficiently away from the point $(-1,0)$, thus imparting a sufficient margin of stability on the control loop. A measure of the distance of the $G(j\omega)$ contour from the point $(-1,0)$ is given by

$$L_c = 20 \log \left| \frac{G}{1 + G} \right| \tag{17.25}$$

A specification of $+2$ dB for the maximum closed-loop log modulus, L_c^{max} has been suggested. To use this procedure for tuning a controller of an SISO system, one would plot L_c as a function of frequency, ω, for trial values of K_c and τ_I. If the maximum value of L_c so determined exceeds $+2$ dB, new values of K_c and τ_I are selected, and the procedure is repeated until satisfactory results are achieved. A Nichols chart shown in Figure 16.7 can also be used to determine L_c from $G(j\omega)$.

To use this procedure for tuning SISO controllers in a multivariable environment, we would begin with the following system equations:

$$\mathbf{C} = \mathbf{G_p M} \tag{17.26}$$

and

$$\mathbf{M} = \mathbf{G_c (R - C)} \tag{17.27}$$

where $\mathbf{C}$ is an $n \times 1$ vector of controlled variables, $\mathbf{M}$ is an $n \times 1$ vector of manipulated variables, $\mathbf{G_c}$ is an $n \times n$ diagonal matrix of PI controller transfer functions, $\mathbf{G_p}$ is an $n \times n$ matrix containing the process transfer functions relating the n controlled variables to the n manipulated variables, and $\mathbf{R}$ is a $n \times 1$ vector of set point values of the n controlled variables. Combining Equations (17.26) and (17.27) gives

$$\mathbf{C} = (\mathbf{I} + \mathbf{G_p G_c})^{-1} \mathbf{G_p G_c R} \tag{17.28}$$

Since the determinant of a matrix winds up in the denominator in the process of evaluating a matrix inverse, the characteristic equation of the multivariable system is the scalar equation

$$\det (\mathbf{I} + \mathbf{G_p}\mathbf{G_c}) = 0 \tag{17.29}$$

If the left-hand side of Equation (17.29) is plotted against frequency, the encirclements of the *origin* would indicate that the system is unstable. If a new function w is defined as

$$w = -1 + \det (\mathbf{I} + \mathbf{G_p}\mathbf{G_c}) \tag{17.30}$$

and plotted as a function of frequency, then the encirclement of the point $(-1,0)$ would indicate instability.

Now, synonymous with the SISO controller design procedure, a multivariable closed-loop log modulus is defined as

$$L_{c_m} = 20 \log \left| \frac{w}{1 + w} \right| \tag{17.31}$$

Based on a study of 10 multivariable systems, Luyben suggests that

$$(L_{c_m})^{\max} = 2n \tag{17.32}$$

where n is the dimension of the multivariable system.

The tuning procedure for n SISO controllers is as follows:

1. Calculate Ziegler-Nichols settings of n PI controllers. To do this calculation numerically requires the finding of frequency $\omega = \omega_{co}$ for which the phase angle is exactly $-180°$. The reciprocal of the gain corresponding to this frequency is K_u. Then the Ziegler-Nichols settings are

$$K_c = \frac{1}{2.2} K_u \tag{17.33a}$$

$$\tau_I = \frac{2.0\pi}{1.2\omega_{co}} \tag{17.33b}$$

2. Assume a factor F; typical values are said to vary between 2 and 5.
3. Calculate new values of controller constants by the relationships

$$K_{c_i} = K_{c_i} \Big|_{Z-N} /F \tag{17.34a}$$

$$\tau_{I_i} = F \cdot \tau_{I_{i_{Z-N}}} \tag{17.34b}$$

$i = 1, \ldots, n$

The factor F may be considered as a detuning factor; as F increases, the stability margin increases, but then the system becomes more sluggish and vice versa. The procedure suggested here is meant to give a reasonable compromise between stability and performance.

4. Compute $w = -1 + \det (\mathbf{I} + \mathbf{G_p G_c})$ (17.30)

For a 2×2 system, for example,

$$\det (\mathbf{I} + \mathbf{G_p G_c}) = 1 + G_{c_1} G_{11} + G_{c_2} G_{22} + \tag{17.35}$$
$$(G_{c_1} G_{c_2})(G_{11} G_{22} - G_{12} G_{22})$$

5. Determine

$$L_{c_m} = 20 \log \left| \frac{w}{1 + w} \right| \tag{17.31}$$

6. If $L_{c_m}^{\max} > 2n$, select a new value of F and return to step 2.

Luyben refers to this procedure as biggest log-modulus tuning (BLT). He has applied the procedure successfully to ten multivariable systems ranging from 2×2 to 4×4 processes. The results for the 2×2 systems are shown in Figure 17.9. The system transfer functions are shown in Table 17.2 and the tuning constants are shown in Table 17.3. It has been suggested that this procedure should be viewed as giving preliminary settings. The procedure guarantees stability with all controllers in automatic, also with individual controllers in automatic and the rest in manual. Further checks on stability may have to be made for other combinations of manual/automatic operations. We should point out that PID-type control is not the only solution for multiloop control. IMC and SMPC, the strategies covered in an ensuing section, are also potential candidates.

17.3 Decoupling for Noninteracting Control

Decoupling in the Framework of RGA

If the relative gains are numerically close to each other, interaction ("fighting loops") in a multivariable control system is likely to be a problem, particularly if the response times of the loops are comparable. In cases in which cross coupling between loops is severe, the system can become unstable, and decoupling will be required.

To successfully implement a decoupler, the model or frequency response of the process should be known to a high degree of accuracy.

A decoupler is a device that eliminates interaction between manipulated and controlled variables by, in effect, changing all the manipulated variables in such a manner that only the desired controlled variable changes.

For example, in a two-variable interacting system, shown in Figure 17.10, we can write:

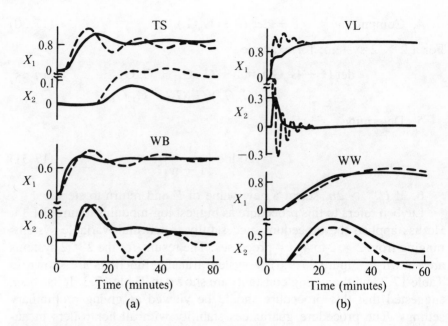

Figure 17.9
a X_1 **Set Point Responses of TS and WB**
b X_1 **Set Point Response of VL and WW**
(Reprinted by permission from an article by W.L. Luyben in *Ind.*
***Eng. Chem. Proc. Des. Dev.*, vol. 25, no. 3, pp. 654–660, © 1986,**
American Chemical Society.)

$$\Delta C_1(s) = G_{11}(s)\Delta M_1(s) + G_{12}(s)\Delta M_2(s)$$
$$\Delta C_2(s) = G_{21}(s)\Delta M_1(s) + G_{22}(s)\Delta M_2(s) \qquad (17.36)$$

The decoupler inputs are two new manipulated variables u_1 and u_2, and its outputs are the original manipulated variables M_1 and M_2, as shown in Figure 17.11. We want to design the decoupler so that u_1 affects only C_1 and u_2 affects only C_2.

From Figure 17.11 it can be seen that the decoupler equations can be written as

$$\Delta M_1(s) = D_{11}(s)\Delta u_1(s) + D_{12}(s)\Delta u_2(s)$$
$$\Delta M_2(s) = D_{21}(s)\Delta u_1(s) + D_{22}(s)\Delta u_2(s) \qquad (17.37)$$

For ease in building the decoupler, let us specify that $D_{11}(s) = D_{22}(s) = 1$. In this case, the decoupler equations become:

$$\Delta M_1(s) = \Delta u_1(s) + D_{12}(s)\Delta u_2(s)$$
$$\Delta M_2(s) = D_{21}(s)\Delta u_1(s) + \Delta u_2(s) \qquad (17.38)$$

Table 17.2
Process Open-Loop Transfer Functions of 2 × 2 Systems

	TS (Tyreus stabilizer)	WB (Wood and Berry)	VL (Vinante and Luyben)	WW (Wardle and Wood)
G_{11}	$\dfrac{-0.1153(10S+1)e^{-0.1S}}{(4S+1)^3}$	$\dfrac{12.8e^{-S}}{16.7S+1}$	$\dfrac{-2.2e^{-S}}{7S+1}$	$\dfrac{0.126e^{-6S}}{60S+1}$
G_{12}	$\dfrac{0.2429e^{-2S}}{(33S+1)^2}$	$\dfrac{-18.9e^{-3S}}{21S+1}$	$\dfrac{1.3e^{-0.3S}}{7S+1}$	$\dfrac{-0.101e^{-12S}}{(48S+1)(45S+1)}$
G_{21}	$\dfrac{-0.0887e^{-12.6S}}{(43S+1)(22S+1)}$	$\dfrac{6.6e^{-7S}}{10.9S+1}$	$\dfrac{-2.8e^{-1.8S}}{9.5S+1}$	$\dfrac{0.094e^{-8S}}{38S+1}$
G_{22}	$\dfrac{0.2429e^{-0.17S}}{(44S+1)(20S+1)}$	$\dfrac{-19.4e^{-3S}}{14.4S+1}$	$\dfrac{4.3e^{-0.35S}}{9.2S+1}$	$\dfrac{0.12e^{-8S}}{35S+1}$

Table 17.3
2 × 2 Systems

	TS (Tyreus stabilizer)	WB (Wood and Berry)	VL (Vinante and Luyben)	WW (Wardle and Wood)
RGA	4.35	2.01	1.63	2.69
NI	+0.229	+0.498	+0.615	+0.372
empirical				
K_c	−30, 30	0.2, −0.04	−2.38, 4.39	18, −24
τ_I	∞	4.44, 2.67	3.16, 1.15	19, 24
L_c	1.74	10.1	13.3	8.4
$Z - N$				
K_c	−166.2, 706	0.96, −0.19	−2.40, 4.45	59, −28.5
τ_I	2.06, 8.01	3.25, 9.20	3.16, 1.15	19.3, 24.6
L_c	unstable	unstable	13.3	18.5
BLT				
F	10	2.55	2.25	2.15
K_c	−16.6, 70.6	0.375, −0.075	−1.07, 1.97	27.4, −13.3
τ_I	20.6, 80.1	8.29, 23.6	7.1, 2.58	41.4, 52.9

Substituting Equation (17.38) into Equation (17.36) gives the equations for the process-decoupler combination:

$$\Delta C_1 = G_{11}(\Delta u_1 + D_{12}\Delta u_2) + G_{12}(D_{21}\Delta u_1 + \Delta u_2)$$
$$= (G_{11} + G_{12}D_{21})\Delta u_1 + (G_{11}D_{12} + G_{12})\Delta u_2 \quad (17.39)$$
$$\Delta C_2 = G_{21}(\Delta u_1 + D_{12}\Delta u_2) + G_{22}(D_{21}\Delta u_1 + \Delta u_2)$$
$$= (G_{21} + G_{22}D_{21})\Delta u_1 + (G_{21}D_{12} + G_{22})\Delta u_2$$

For complete decoupling, we want ΔC_1 to be affected *only* by Δu_1 and ΔC_2 *only* by Δu_2; that is,

$$\Delta C_1 = H_1 \Delta u_1 \quad (17.40)$$
$$\Delta C_2 = H_2 \Delta u_2$$

Comparing Equations (17.39) and (17.40) gives four equations, which can be solved for D_{12}, D_{21}, H_1, and H_2:

$$G_{11} + G_{12}D_{21} = H_1$$
$$G_{11}D_{12} + G_{12} = 0$$
$$G_{21} + G_{22}D_{21} = 0$$
$$G_{21}D_{12} + G_{22} = H_2$$

From which,

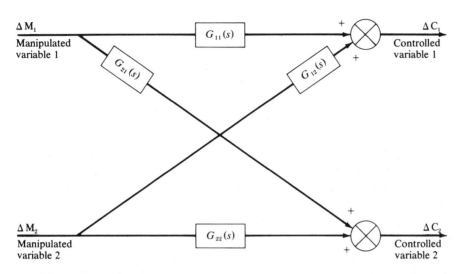

Figure 17.10
2 × 2 Multivariable System

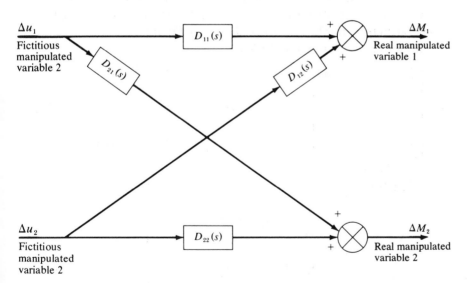

Figure 17.11
Decoupler for 2 × 2 System

$$D_{12}(s) = -\frac{G_{12}(s)}{G_{11}(s)}$$

$$D_{21}(s) = -\frac{G_{21}(s)}{G_{22}(s)}$$ (17.41)

$$H_1(s) = G_{11}(s) - \frac{G_{12}(s)G_{21}(s)}{G_{22}(s)}$$

$$H_2(s) = G_{22}(s) - \frac{G_{12}(s)G_{21}(s)}{G_{11}(s)}$$

From Equation (17.41) it can be seen why good process models are necessary to effectively decouple a system. If the models are inaccurate, then the cross term of the process-decoupler combination will not be zero.

In theory, at least, a decoupler can be built with analog hardware, particularly when simple (first-order) models are used for the process. However, success is not likely with such simple models—success is much more likely if a digital computer is used where some form of algorithm is available to make use of actual on-line performance to update the process models.

The design equations for a general decoupler for a $n \times n$ system are conveniently summarized using matrix notation. The block diagram of a general multivariable system is shown in Figure 17.12.

The terms are defined as follows:

$$\mathbf{G} = \begin{bmatrix} G_{11}(s) \cdots G_{1n}(s) \\ \cdot \\ \cdot \\ \cdot \\ G_{n1}(s) \cdots G_{nn}(s) \end{bmatrix} \qquad \mathbf{D} = \begin{bmatrix} D_{11}(s) \cdots D_{1n}(s) \\ \cdot \\ \cdot \\ \cdot \\ D_{n1}(s) \cdots D_{nn}(s) \end{bmatrix}$$ (17.42)

Transfer   Decoupler
function matrix   matrix

$$\mathbf{H} = \begin{bmatrix} H_{11}(s) & \cdot & \cdot & \cdot & 0 \\ \cdot & & & & \\ \cdot & & H_{22}(s) & & \\ 0 & & & & \\ \cdot & & & & \\ \cdot & & & & H_{nn}(s) \end{bmatrix}$$

Diagonal matrix
of process decoupler
combination; i.e.,
C = Hu

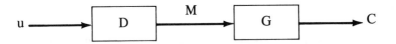

Figure 17.12
General *n* × *n* Multivariable System

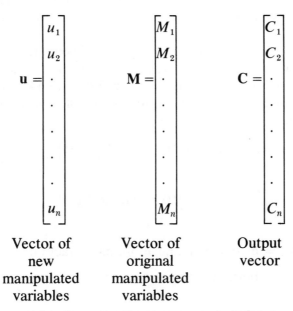

$\mathbf{u} = \begin{bmatrix} u_1 \\ u_2 \\ \cdot \\ \cdot \\ \cdot \\ \cdot \\ u_n \end{bmatrix}$	$\mathbf{M} = \begin{bmatrix} M_1 \\ M_2 \\ \cdot \\ \cdot \\ \cdot \\ \cdot \\ M_n \end{bmatrix}$	$\mathbf{C} = \begin{bmatrix} C_1 \\ C_2 \\ \cdot \\ \cdot \\ \cdot \\ \cdot \\ C_n \end{bmatrix}$
Vector of new manipulated variables	Vector of original manipulated variables	Output vector

From Figure 17.12 we can write:

$$\mathbf{C} = \mathbf{G}\,\mathbf{M}$$

$$\mathbf{M} = \mathbf{D}\,\mathbf{u}$$

Thus,

$$\mathbf{C} = \mathbf{G}\,\mathbf{D}\,\mathbf{u}$$

But we want

$$\mathbf{C} = \mathbf{H}\,\mathbf{u} \qquad\qquad (17.43)$$

Thus,

$$\mathbf{G}\,\mathbf{D} = \mathbf{H}$$

or

$$\mathbf{D} = \mathbf{G}^{-1}\mathbf{H}$$

which defines the decoupler.

Example 3

Luyben[33] has presented an application of decoupling to a 2×2 distillation control system. In this application the objectives are to control the bottoms and overhead composition of the binary column by manipulating the reboiler heat duty and the reflux flow, respectively. Since a change in either reboiler heat duty or reflux flow upsets both compositions, we have an interacting system. The purpose is to design decouplers so as to achieve noninteracting feedback control of the multivariable system.

The block diagram of the control system without the decoupling elements is shown in Figure 17.13. The block diagram of the control system with the decoupling elements is shown in Figure 17.14. The open-loop transfer functions relating the controlled variables to the manipulated variables are:

$$X_D(s) = G_{11}(s) R(s) + G_{12}(s) V_B(s) \qquad (17.44)$$

and

$$X_B(s) = G_{21}(s) R(s) + G_{22}(s) V_B(s) \qquad (17.45)$$

where

$$X_D = \text{distillate composition, mole fraction}$$

$$X_B = \text{bottoms composition, mole fraction}$$

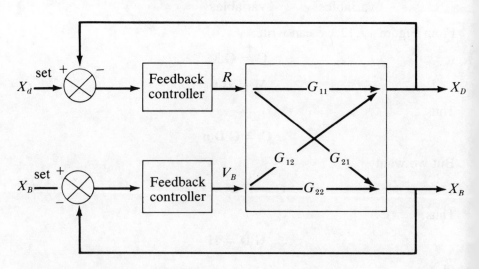

Figure 17.13
Distillation Column Control System without Decoupler

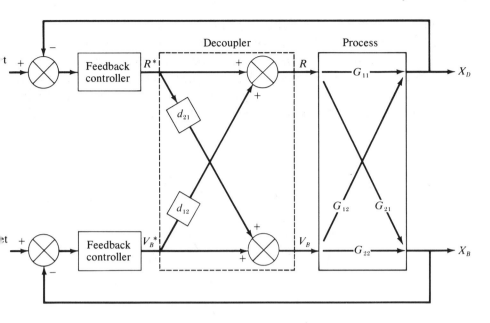

Figure 17.14
Distillation Control Scheme with Decouplers

$$V_B = \text{vapor boil up rate, moles/min}$$
$$\text{(indicative of reboiler heat duty)}$$

$$R = \text{reflux flow rate, moles/min}$$

$$G_{11}(s) = \text{open-loop transfer function relating } X_D \text{ to } R$$

$$G_{12}(s) = \text{open-loop transfer function relating } X_D \text{ to } V_B$$

$$G_{21}(s) = \text{open-loop transfer function relating } X_B \text{ to } R$$

$$G_{22}(s) = \text{open-loop transfer function relating } X_B \text{ to } V_B$$

In accordance with Equation (17.41), the decoupler design equations are

$$D_{12}(s) = -\frac{G_{12}(s)}{G_{11}(s)} \tag{17.46}$$

$$D_{21}(s) = -\frac{G_{12}(s)}{G_{22}(s)} \tag{17.47}$$

In this application, the open-loop transfer functions were available in the frequency domain in the form of Bode plots. Thus graphical manipulations via Equations (17.46) and (17.47) resulted in the Bode plot of $D_{12}(s)$ and $D_{21}(s)$ to which transfer functions were fitted. Since the decouplers turned out to be lead-lag networks, conventional hardware could be used to implement them. The computer simulations of the control system resulted in the closed-loop responses shown in Figures 17.15 and 17.16. These figures show the benefits of decoupling the multivariable system.

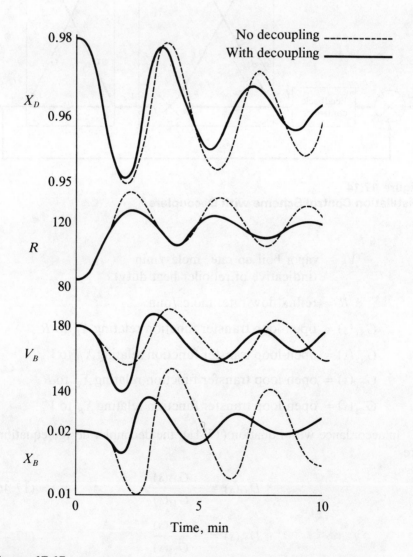

Figure 17.15
Transient Response ($X_B = 0.02$, $X_D = 0.98$) with X_D^{set} Disturbance (Reproduced with Permission from Ref. 6)

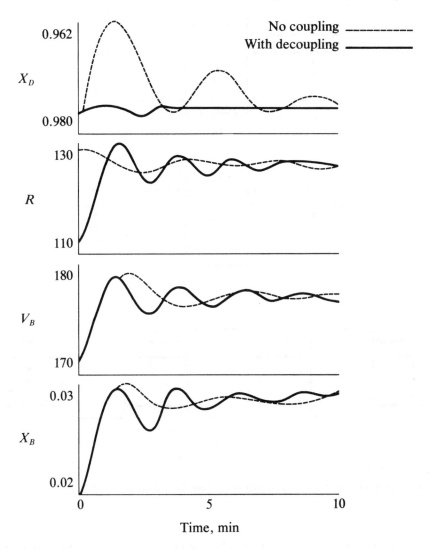

Figure 17.16
**Transient Response ($X_B = 0.02, X_D = 0.98$) with X_B^{set} Disturbance
(Reproduced with Permission from Ref. 6)**

Decoupling with SVD[8]

Recall that the singular value decomposition of the process gain matrix is of the form

$$\mathbf{C} = \mathbf{KM} = \mathbf{U} \, \Sigma \, \mathbf{V}^T \mathbf{M} \qquad (17.48)$$

Equation (17.48) is shown in block diagram form in Figure 17.17. In principle, all **M**s can affect all **C**s. The objective in this instance is to

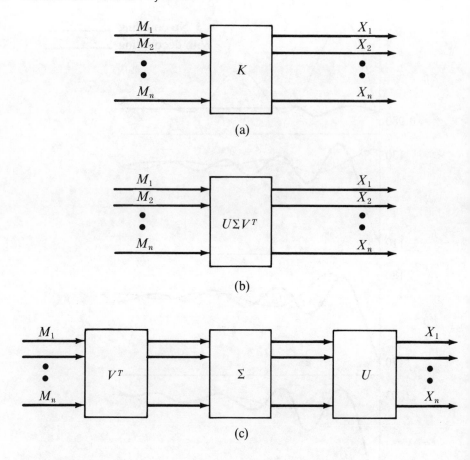

Figure 17.17
Three Equivalent Representations of the Original System in Terms of SVD[8]

compute pseudo-manipulated variables and pseudo-controlled variables such that each pseudo-manipulated variable affects only one pseudo-controlled variable. This can be achieved by utilizing the orthonormal property of U and $V(U^T U = I; V^T V = I)$, and performing the following matrix multiplications to obtain

$$\Sigma = U^T(U \Sigma V^T)V \tag{17.49}$$

Since Σ is a diagonal matrix, each input to it will affect only one output, and thus steady-state decoupling has been achieved. The singular values σ_i of Σ are the open-loop gains of the decoupled system. A block diagram of the decoupled system is shown in Figure 17.18.

The equivalence of the three representations shown in Figure 17.17

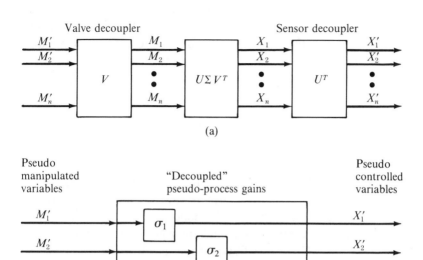

(a)

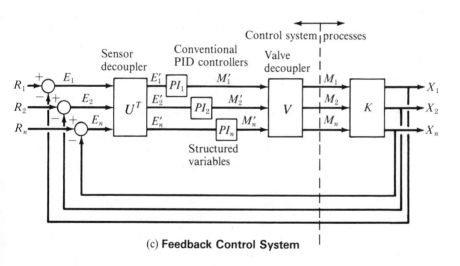

(b) **Noninteracting Pseudo Process**

(c) **Feedback Control System**

Figure 17.18
Decoupling in the Framework of SVD[8]

can be easily shown by writing out the system of equations depicted by each block and combining the result while remembering that $\mathbf{U}$, $\mathbf{\Sigma}$, and $\mathbf{V}$ are matrices. A feedback control scheme built around the decoupled system is shown in Figure 17.18. Again, due to the matrix nature of $\mathbf{U}$, $\mathbf{\Sigma}$, and $\mathbf{V}$, the matrix $\mathbf{V}$ appears ahead of the process block.

17.4 Model Predictive Control

In a model predictive control strategy a mathematical model is utilized for predicting the process outputs and in control calculations. The methods under this category include model algorithmic control[2] (MAC), dynamic matrix control[3] (DMC) and internal model control[4,5,6] (IMC). Early work on MAC and DMC was done in the seventies in France and the United States, respectively.

At the time, both were developed on the basis of heuristics. The two methods use a similar approach but utilize different optimization routines and input parameter specifications. They take full advantage of the computational power of control computers. Constraint handling is an important facility available in these methods, which was hitherto unavailable with PID-type control systems. Equally important is that these methods need only open-loop step responses for design; also, no assumptions need be made about the order of the system. Parametric process models are not required for implementation in contrast to traditional methods such as decoupling control. Interaction compensation and dead time compensation are inherent in these algorithms. Inverse response characteristics and time delays pose no special problems, but an open-loop unstable plant must be stabilized with conventional feedback before these methods can be applied. To date, these techniques have been applied successfully to such diverse systems as a crude column, a fluid catalytic cracker, a distillation column, a green house, an F-16 jet engine, and a power plant.

Motivated by a discussion of Brosilow,[33] Garcia and Morari[4,5,6] developed a new control strategy called internal model control (IMC) for SISO and multivariable systems. There are two approaches to designing IMC. The first approach is based on the factorization of the process model, while the second requires the formulation of a predictive control problem. Garcia and Morari[4,5,6] have shown that such classical techniques as dead time compensation and feedforward control could be derived in the framework of IMC. Furthermore, they have also shown that MAC and DMC are special cases of IMC.

We begin this section by describing the multivariable IMC factorization concepts and showing how this technique is implemented. In the presence of transmission zeroes in the vicinity of the inside of the unit circle in the Z-plane and in the presence of moving constraints, the factorization method may not be suitable. In such cases the IMC predictive strategy would be employed. Here, we provide only a brief treatment of the IMC predictive strategy. It is felt that the details are beyond the scope of this text, but they are available in a text on multivariable control[8] for those who may be interested. Following the IMC predictive control strategies we describe simplified model predictive control (SMPC) for multivariable systems and present some simulation results. SMPC retains some of the advantages of the IMC predictive control strategies and is easier to design and implement. SMPC may hold promise for some types of multivariable processes.

IMC Factorization Method

The block diagram of a multivariable IMC system is shown in Figure 17.19. Note that the transfer functions, inputs, and outputs are matrices or vectors. Otherwise, the multivariable IMC structure shown in Figure

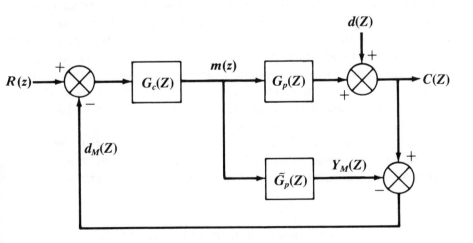

Figure 17.19
Multivariable IMC Structure

17.19 is identical with that of an SISO system shown in Figure 11.22. Following the discussion for SISO systems, the input and output system transfer functions are

$$m(Z) = [I + G_c(Z)(G_p(Z) - \tilde{G}_p(Z))]^{-1} G_c(Z)(R(Z) - d(Z)) \quad (17.50)$$

and

$$C(Z) = d(Z) + G_p(Z)[I + G_c(Z)(G_p(Z) \\ - \tilde{G}_p(Z))]^{-1} G_c(Z)(R(Z) - d(Z)) \quad (17.51)$$

In the absence of plant-model mismatch, these equations reduce to

$$m(Z) = G_c(Z)(R(Z) - d(Z)) \quad (17.52)$$

and

$$C(Z) = G_p(Z)G_c(Z)(R(Z) - d(Z)) + d(Z) \quad (17.53)$$

Thus, the stability of both the controller $G_c(Z)$ and the process $G_p(Z)$ is sufficient to ensure overall system stability in the absence of modeling errors.

In the absence of modeling errors, perfect control can be achieved by selecting

$$G_c(Z) = \tilde{G}_p(Z)^{-1} \quad (17.54)$$

However, as with SISO systems, perfect control cannot be achieved if (1) $\tilde{G}_p(Z)$ contains time delays and/or (2) if $\tilde{G}_p(Z)$ contains transmission zeroes outside the unit circle. Even in the absence of these problems, it is undesirable to select the perfect controller since (1) its use may result in rippling behavior of manipulated variables, resulting in output oscillations owing to the presence of transmission zeroes of $\tilde{G}_p(Z)$ close to the unit circle in the Z plane, and (2) a system equipped with the perfect controller is very sensitive to modeling errors.

To overcome these difficulties associated with the use of the perfect controller, the approach that may be adopted is to split the process transfer function matrix into two parts according to the factorization

$$\tilde{G}_p(Z) = \tilde{G}_+(Z) \, \tilde{G}_-(Z) \quad (17.55)$$

where $\tilde{G}_+(Z)$ accommodates any time delays present such that $\tilde{G}_-(Z)^{-1}$ is *realizable*. If $\tilde{G}_-(Z)^{-1}$ is also *stable*, then, $G_c(Z)$ is selected according to

$$G_c(Z) = \tilde{G}_-(Z)^{-1} \quad (17.56)$$

If $\tilde{G}_p(Z)$ contains transmission zeroes outside the unit circle, then $\tilde{G}_-(Z)^{-1}$ will not be stable, in which case further factorization will have to be carried out as will be explained later. But first let us consider the problem of finding $\tilde{G}_+(Z)$.

Factorization of time delays.[35] To determine $\tilde{G}_+(Z)$, the multiple time delays must be factored out. The procedure is different depending on whether time delays are equal, unequal, or nonexistent as will be described.

If the process transfer function matrix contains no time delays, then

$$\tilde{G}_+(Z) = Z^{-1}I \tag{17.57}$$

where I is an identity matrix. The term Z^{-1} accounts for the delay that is inherent in the sampling process.

If the multivariable system contains equal time delay of τ sampling instants in all the elements of the transfer function matrix, then,

$$\tilde{G}_+(Z) = Z^{-(\tau+1)I} \tag{17.58}$$

which ensures the realizability of the controller and is the optimal $\tilde{G}_+$ since it yields the least possible delay in the response of each output.

When the transfer function matrix contains unequal time delays, then a diagonal factorization matrix $\tilde{G}_+(\mathbf{Z})$ may be selected according to

$$\tilde{G}_+(\mathbf{Z}) = \text{diag}\,[Z^{-(\tau_1^+ +1)},\, Z^{-(\tau_2^+ +1)},\, \ldots Z^{-(\tau_r^+ +1)}] \tag{17.59}$$

$$\text{with } \tau_j^+ = \max\quad \max(0,\, \tilde{\tau}_{ij})\quad j = 1,\, \ldots\, r$$

where $Z^{\tilde{\tau}_{ij}+1}\bar{g}_{ij}(Z)$ are elements of the process inverse matrix $\tilde{G}_p(Z)^{-1}$ such that $\bar{g}_{ij}(Z)$ are semiproper. A ratio of polynomials $\bar{g}_{ij}(Z)$ is semiproper in Z means that the order of the numerator of $\bar{g}_{ij}(Z)$ is less than or equal to the order of its denominator.

Now, from Figure 17.19 the response of the system in the absence of modeling errors is given by

$$C(Z) = \tilde{G}_p(Z)G_c(Z)[R(Z) - d(Z)] + d(Z) \tag{17.60}$$

or in view of Equation (17.56)

$$C(Z) = \tilde{G}_p(Z)\tilde{G}_-(Z)^{-1}[R(Z) - d(Z)] + d(Z) \tag{17.61}$$

which according to Equation (17.55) gives

$$C(Z) = \tilde{G}_+(Z)[R(Z) - d(Z)] + d(Z) \tag{17.62}$$

If $\tilde{G}_+(Z)$ is diagonal, closed-loop decoupled responses will result in accordance with Equation (17.62). However, if $G_c(Z) = \tilde{G}_-(Z)^{-1}$ turns out to be unstable due to the presence of transmission zeroes in $\tilde{G}_p(Z)$, then, one of the stability requirements stated in Equation (17.52) will be violated and the input sequence will be unstable. Even if $\tilde{G}_-(Z)^{-1}$ is stable, diagonal factorization is optimal in terms of minimum settling time or in terms of ISE if, and only if, the columns and/or rows of $\tilde{G}_p(Z)$ can be rearranged in such a way that minimum delay in each row occurs on the diagonal. Otherwise, better performance can be achieved by allowing interactions in the closed-loop transfer function matrix $\tilde{G}_+(Z)$. Thus, IMC reveals when complete decoupling may not be advantageous. Let us illustrate time delay factorization concepts by examples.

Example 4[5]

Given

$$\tilde{G}_p(Z) = \frac{1}{1 - 0.5Z^{-1}} \begin{bmatrix} 2Z^{-3} & Z^{-5} \\ Z^{-2} & Z^{-3} \end{bmatrix}$$

Find $\tilde{G}_+(Z)$.

For this case,

$$\tilde{G}_p(Z)^{-1} = \frac{1 - 0.5Z^{-1}}{2Z^{-6} - Z^{-7}} \begin{bmatrix} Z^{-3} & -Z^{-5} \\ -Z^{-2} & 2Z^{-3} \end{bmatrix}$$

$$\tilde{G}_p(Z)^{-1} = \frac{1}{2} \begin{bmatrix} Z^3 & -Z \\ -Z^4 & 2Z^3 \end{bmatrix}$$

Thus $\bar{\tau}_{11} = 2$, $\bar{\tau}_{12} = 0$, $\bar{\tau}_{21} = 3$, and $\bar{\tau}_{22} = 2$

Then $\tau_1^+ = \max (\max (0, \bar{\tau}_{11}), \max (0, \bar{\tau}_{21}))$

$\qquad = \max (2, 3) = 3$

Similarly, $\tau_2^+ = \max (\max (0, \bar{\tau}_{12}), \max (0, \bar{\tau}_{22})) = 2$

and

$$\tilde{G}_+(Z) = \begin{bmatrix} Z^{-4} & 0 \\ 0 & Z^{-3} \end{bmatrix}$$

For this example, a diagonal $\tilde{G}_+(Z)$ is not optimal in terms of minimum settling time, although it yields complete decoupling, since the rows and/or columns of $\tilde{G}_p(Z)$ cannot be rearranged such that minimum delay in each row occurs on the diagonal. A faster response can be obtained by choosing a different $\tilde{G}_+$, for example, as

$$G'_+(Z) = \begin{bmatrix} Z^{-3} & 0 \\ 1/2Z^{-2}(1-Z^{-1}) & Z^{-3} \end{bmatrix} \qquad (17.63)$$

which also leads to a realizable $\tilde{G}_-^{-1}(Z)$. The increased speed of response with $\tilde{G}'_+$ has to be paid with dynamic interactions. This type of triangular factorization may be obtained in the following manner. The designer specifies the diagonal elements of $\tilde{G}_+$ and postulates a form of the off-diagonal elements so they vanish for $Z = 1$. By invoking the condition that $\tilde{G}_p^{-1}$, $\tilde{G}_+$ has to be realizable, the exact form of the off-diagonal elements can be determined.[36] Further research aimed at the development of a general procedure for the construction of arbitrary $\tilde{G}_+$ matrices is in progress.

Example 5[37]

Given

$$\tilde{G}_p(s) = \begin{bmatrix} \dfrac{12.8e^{-s}}{16.7s + 1} & \dfrac{-18.9e^{-3s}}{21s + 1} \\ \dfrac{6.6e^{-7s}}{10.9s + 1} & \dfrac{-19.4e^{-3s}}{14.4s + 1} \end{bmatrix} = \begin{bmatrix} G_{11}(s) & G_{12}(s) \\ G_{21}(s) & G_{22}(s) \end{bmatrix}$$

Find $\tilde{G}_+(Z)$.
For this problem, taking $Z\{G_{ij}(s)G_{h_0}(s)\}$ (where G_{h_0} represents the zero-order hold) gives

$$\tilde{G}_p(Z) = \begin{bmatrix} Z^{-2}\bar{g}_{11}(Z) & Z^{-4}\bar{g}_{12}(Z) \\ Z^{-8}\bar{g}_{21}(Z) & Z^{-4}\bar{g}_{22}(Z) \end{bmatrix}$$

The inverse of $\tilde{G}_p(Z)$ is

$$\tilde{G}_p(Z)^{-1} = \dfrac{1}{Z^{-6}\bar{g}_{11}(Z)\bar{g}_{22}(Z) - Z^{-12}\bar{g}_{12}(Z)\bar{g}_{21}(Z)} \begin{bmatrix} Z^{-4}\bar{g}_{22}(Z) & Z^{-4}\bar{g}_{12}(Z) \\ -Z^{-8}\bar{g}_{21}(Z) & Z^{-2}\bar{g}_{11}(Z) \end{bmatrix}$$

$$= \dfrac{1}{\bar{g}_{11}(Z)\bar{g}_{22}(Z) - Z^{-6}\bar{g}_{12}(Z)\bar{g}_{21}(Z)} \begin{bmatrix} Z^2\bar{g}_{22}(Z) & -Z^2\bar{g}_{12}(Z) \\ -Z^{-2}\bar{g}_{21}(Z) & Z^4\bar{g}_{11}(Z) \end{bmatrix}$$

Thus,

$$\bar{\tau}_{11} = 1 \quad \bar{\tau}_{12} = 1 \quad \bar{\tau}_{21} = -3 \quad \bar{\tau}_{22} = 3$$

$$\tau_1^+ = \text{max max } (0, 1), \text{max } (0, -3) = 1$$

$$\tau_2^+ = \text{max max } (0, 1), \text{max } (0, 3) = 3$$

and

$$\tilde{G}_+(Z) = \begin{bmatrix} Z^{-2} & 0 \\ 0 & Z^{-4} \end{bmatrix}$$

The foregoing time delay factorization yields a $\tilde{G}_+(Z)$ and a realizable $G_c(Z) = \tilde{G}_-(Z)^{-1}$ but the stability requirements demand that $\tilde{G}_-(Z)^{-1}$ must also be *stable*. If $\tilde{G}_-(Z)^{-1}$ turns out to be unstable, $\tilde{G}_-(Z)$ must be further split into two parts, $\tilde{G}_{+1}(Z)$ and $\tilde{G}_{-1}(Z)$, giving the factorization

$$\tilde{G}_p(Z) = \tilde{G}_+(Z)\tilde{G}_{+1}(Z)\tilde{G}_{-1}(Z) \tag{17.64}$$

where $\tilde{G}_+(Z)$ is chosen to account for the time delay elements, as has been described, and $\tilde{G}_{+1}(Z)$ is chosen to account for the zeroes outside the unit circle as will be described in the following section.

Factorization of Transmission Zeroes Outside the Unit Circle.[34] Before proceeding with the description of a method for factorization of zeroes, some comments on the definition and properties of SISO and MIMO (multi-input multi-output) zeroes are in order.

For SISO systems, the zeroes are the roots of the numerator polynomial of the process transfer function, $\tilde{G}_p(s)$. A right-half plane (RHP) zero is a zero whose real part is ≥ 0. For example, consider the continuous transfer function

$$\tilde{G}_p(s) = \frac{-0.5s + 1}{(0.25s + 1)(0.3s + 1)} \tag{17.65}$$

has a zero at $s = 2$. For SISO systems, the process having an odd number of right-half plane zeroes in its transfer function will exhibit inverse response to a step input. If there are an even number of RHP zeroes, the response will start in the proper direction, change directions, eventually reversing its sign, and then move back toward the steady state. An important property of SISO zeroes is that, unlike poles, it is not feasible to shift zeroes to different locations. They must simply be tolerated much like dead time elements. Furthermore, the zeroes are the poles of the inverse of the plant transfer function, and, therefore, they yield an unstable IMC controller. This, combined with the inability to move zeroes, implies that perfect control is not feasible for a SISO system with RHP zeroes. In discrete systems RHP zeroes are zeroes outside the unit circle in the Z-plane.

For multivariable systems the presence of RHP zeroes in the individual elements does not necessarily result in inverse response of one or more outputs. For example, although the transfer function

$$\tilde{G}_p(s) = \begin{bmatrix} \dfrac{-s+2}{s+2} & \dfrac{-s+5}{s+4} \\[3mm] \dfrac{-s+1}{s+2} & \dfrac{-s+3}{s+4} \end{bmatrix} \qquad (17.66)$$

contains a RHP zero in each of the elements, the system outputs do not exhibit inverse response. For multivariable systems, the zeroes that give rise to inverse response of one or more outputs are called *transmission zeroes*.

Returning to the problem of finding $\tilde{G}_{+1}(Z)$, the objective is to find a $\tilde{G}_{-1}(Z)$ that has a stable inverse. From Equation (17.64) this inverse may be computed as

$$\tilde{G}_{-1}(Z)^{-1} = \tilde{G}_p(Z)^{-1}\tilde{G}_+(Z)\tilde{G}_{+1}(Z) \qquad (17.67)$$

If the multivariable system is open-loop stable, the determinant of its transfer function can be shown to contain all the transmission zeroes of the system as factors. Then $\tilde{G}_{+1}(Z)$ must be chosen to include all these unstable zeroes so that $\tilde{G}_{-1}(Z)^{-1}$ will be stable.

If $\tilde{G}_{+1}(Z)$ is *required* to be diagonal, ISE will be minimized by

$$\tilde{G}_{+1}(Z) = \text{diag}\left[\prod_{i=1}^{m}\left(\frac{Z-\nu_i}{Z-\nu_i}\right)\left(\frac{1-\bar{\nu}_i}{1-\nu_i}\right), \ \prod_{i=1}^{m}\left(\frac{Z-\nu_i}{Z-\bar{\nu}_i}\right)\left(\frac{1-\bar{\nu}_i}{1-\nu_i}\right), \right.$$
$$\left. \cdots \prod_{i=1}^{m}\left(\frac{Z-\nu_i}{Z-\bar{\nu}_i}\right)\left(\frac{1-\bar{\nu}_i}{1-\nu_i}\right) \right] \qquad (17.68a)$$

where ν_i are RHP transmission zeroes and

$$\bar{\nu}_i = \nu_i \text{ for } |\nu_i| \le 1 \qquad (17.68b)$$
$$\bar{\nu}_i = 1/\nu_i \text{ for } |\nu_i| > 1$$

It can be shown that diagonal factorization of RHP transmission zeroes is not optimal in terms of ISE and better nondiagonal factorizations can be found. Thus, for systems with RHP transmission zeroes, complete decoupling is not as good an approach as some form of interactive control.

An alternative approach to diagonal factorization would be to shift the effect of the RHP transmission zeroes onto an output, which may perhaps be the least important output. Then, $\tilde{G}_{+1}(Z)$ may be designed for perfect response of all but the first output, all but the second output, and so on, and ISE for the various cases compared to decide which $\tilde{G}_{+1}(Z)$ will be chosen for computing $\tilde{G}_{-1}(Z)^{-1}$.

Let us illustrate the factorization concepts for RHP transmission zeroes by an example. Given an open-loop stable system having the transfer function

$$\tilde{\mathbf{G}}_p = \begin{bmatrix} \dfrac{0.6Z^{-5}}{Z - 0.4} & \dfrac{0.5Z^{-4}}{Z - 0.5} \\ \dfrac{0.6Z^{-3}}{Z - 0.5} & \dfrac{0.6Z^{-2}}{Z - 0.4} \end{bmatrix}$$

Find $\tilde{G}_+(Z)$ and if $\tilde{G}_-(Z)^{-1}$ is unstable, find $\tilde{G}_{+1}(Z)$.

Using the factorization

$$\tilde{G}_p(Z) = \tilde{G}_+(Z)\tilde{G}_-(Z) \qquad (17.55)$$

$\tilde{G}_+(Z)$ will be selected according to Equation (17.59) to factor out time delays. This computation requires the knowledge of $\tilde{G}_p(Z)^{-1}$.

$$\tilde{G}_p(Z)^{-1} = \frac{1}{\det \tilde{G}_p} \begin{bmatrix} \dfrac{0.6Z^{-2}}{Z - 0.4} & \dfrac{-0.5Z^{-4}}{Z - 0.5} \\ \dfrac{-0.6Z^{-3}}{Z - 0.5} & \dfrac{0.6Z^{-5}}{Z - 0.4} \end{bmatrix}$$

$$\det \tilde{G}_p = \frac{(0.6)(0.6)Z^{-7}}{(Z - 0.4)^2} - \frac{(0.5)(0.6)Z^{-7}}{(Z - 0.5)^2}$$

$$= \frac{0.06Z^{-7} (Z^2 - 2Z + 0.7)}{(Z - 0.4)^2 (Z - 0.5)^2}$$

or

$$\tilde{G}_p(Z)^{-1} = \frac{(Z - 0.4)^2 (Z - 0.5)^2}{0.06Z^{-7} (Z^2 - 2Z + 0.7)} \begin{bmatrix} \dfrac{0.6Z^{-2}}{Z - 0.4} & \dfrac{-0.5Z^{-4}}{Z - 0.5} \\ \dfrac{-0.6Z^{-3}}{Z - 0.5} & \dfrac{0.6Z^{-5}}{Z - 0.4} \end{bmatrix}$$

Now, let $f(Z) = \dfrac{(Z - 0.5)^2 (Z - 0.4)^2}{(0.06) (Z^2 - 2Z + 0.7)}$

Then

$$\bar{g}_{11}(Z) = \frac{0.6Z^{-1}f(Z)}{(Z - 0.4)} \; ; \bar{g}_{12}(Z) = \frac{-0.5Z^{-1}f(Z)}{(Z - 0.5)}$$

$$\bar{g}_{21}(Z) = \frac{-0.6Z^{-1}f(Z)}{(Z - 0.4)} \; ; \bar{g}_{22}(Z) = \frac{0.6Z^{-1}f(Z)}{(Z - 0.4)}$$

Note that $\tilde{g}_{ij}(Z)$ are semiproper in Z. With these expressions the process inverse can be written as

$$\tilde{G}_p(Z)^{-1} = \begin{bmatrix} Z^6\tilde{g}_{11} & Z^4\tilde{g}_{12} \\ Z^5\tilde{g}_{21} & Z^3\tilde{g}_{22} \end{bmatrix}$$

Applying Equation (17.59) to $G_p(Z)^{-1}$ gives

$$\tau_1^+ = 5, \ \tau_2^+ = 3$$

Thus,

$$\tilde{G}_+ = \begin{bmatrix} Z^{-6} & 0 \\ 0 & Z^{-4} \end{bmatrix}$$

Then, from Equation (17.55)

$$\tilde{G}_-(Z) = \begin{bmatrix} \dfrac{0.6Z}{(Z-0.4)} & \dfrac{0.5Z^2}{(Z-0.5)} \\ \dfrac{0.6Z}{(Z-0.5)} & \dfrac{0.6Z^2}{(Z-0.4)} \end{bmatrix}$$

It can be shown that $\tilde{G}_-(Z)^{-1}$ is unstable, indicating the presence of zeroes that must be factored out. The next step is to use the factorization

$$\tilde{G}_p(Z) = \tilde{G}_+(Z)\tilde{G}_{+1}(Z)\tilde{G}_{-1}(Z)$$

where $\tilde{G}_+(Z)$ has just been determined and

$$\tilde{G}_{+1}(Z)\tilde{G}_{-1}(Z) = \begin{bmatrix} \dfrac{0.6Z}{Z-0.4} & \dfrac{0.5Z^2}{Z-0.5} \\ \dfrac{0.6Z}{Z-0.5} & \dfrac{0.6Z^2}{Z-0.4} \end{bmatrix}$$

Now the objective is to find $\tilde{G}_{+1}(Z)$ so that $\tilde{G}_{-1}(Z)^{-1}$ will be stable and realizable.

As has been mentioned earlier, the transmission zeroes of $\tilde{G}_p(Z)$ are the roots of

$$\det\{\tilde{G}_{+1}(Z)\,\tilde{G}_{-1}(Z)\} = 0$$

or

$$= \frac{0.06Z^3\,(Z^2 - 2Z + 0.7)}{(Z-0.4)^2\,(Z-0.5)^2} = 0$$

Thus,

$$Z_1 = 1.547 \,;\, Z_2 = 0.453$$

Since one of the system zeroes is outside the unit circle, $\{\tilde{G}_{+1}(Z)\ \tilde{G}_{-1}(Z)\}^{-1}$ will be unstable.

If diagonal factorization is required, then, as per Equation (17.68a), we have

$$\tilde{G}_{+1}(Z) = \begin{bmatrix} \dfrac{-Z + 1.547}{1.547Z - 1} & 0 \\[2ex] 0 & \dfrac{-Z + 1.547}{1.547Z - 1} \end{bmatrix}$$

As in the case of time delay factorization, it is always possible to select a diagonal $\tilde{G}_{+1}(Z)$, and if $\tilde{G}_+(Z)$ is also diagonal, decoupled responses will result according to

$$C(Z) = \tilde{G}_+(Z)\tilde{G}_{+1}(Z)(R(Z) - d(Z)) + d(Z) \qquad (17.69)$$

However, in this example, the use of the selected $\tilde{G}_{+1}(Z)$ can be shown to lead to inverse response of both outputs. If one-way coupling is allowed, the factorization matrix

$$\tilde{G}_{+1}(Z) = \begin{bmatrix} 1 & 0 \\[2ex] 3.095(1 - Z^{-1}) & \dfrac{-Z + 1.547}{1.547Z - 1} \end{bmatrix}$$

can be shown to produce a stable $\tilde{G}_{-1}(Z)^{-1}$. In this instance the first output exhibits deadbeat response and the second ouput exhibits interaction and inverse response. It would also have been possible to choose $\tilde{G}_{+1}(Z)$ such that the second output exhibited deadbeat response and the first, inverse response and interaction. Thus, the designer has the option to insist on complete decoupling at the expense of poor response of both outputs or tolerate one-way coupling and obtain good response of at least one output. IMC reveals the trade-offs and allows the designer to make a choice. Due to the numerical difficulties encountered in determining the zeroes exactly, diagonal factorization of unstable zeroes of $\tilde{G}_p(Z)$ is recommended only when complete decoupling is desired.

Filter Design. As with SISO systems, a filter is introduced into the multivariable IMC structure to preserve stability in the presence of a plant/model mismatch. The filter may be introduced in the feedback path as in Figure 17.20(a), or, if the function of reference trajectory is also desired, it may be inserted ahead of the controller block as shown in Figure 17.20(b). A diagonal first-order filter matrix of the form

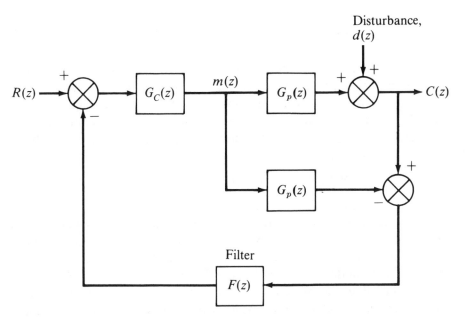

Figure 17.20a
IMC with Filter

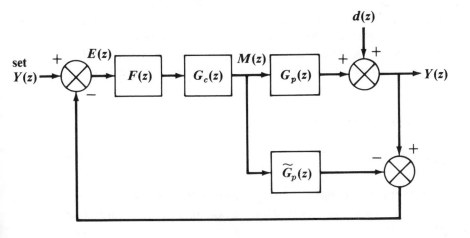

Figure 17.20b
Complete IMC Structure with Combined Filter
and Reference Trajectory Blocks

$$F(Z) = \text{diag} \left\{ \frac{1 - \alpha_F}{1 - \alpha_F Z^{-1}} \right\} \qquad 0 \le \alpha_F < 1 \qquad (17.70a)$$

may be sufficient to ensure stability in the presence of a plant/model mismatch. An important theorem relating to filter design states that if the diagonal filter given in Equation (17.70a) is selected and $G_c(Z) = \tilde{G}_-(Z)^{-1}$, there exists an $\alpha*$ ($0 \le \alpha* < 1$) such that the system is closed-loop stable for all α_F in the open interval $\alpha* \le \alpha_F < 1$ if, and only if, $G(Z)$ and $\tilde{G}(Z)$ satisfy

$$Re\{\lambda_i\{G(1)\, \tilde{G}(1)^{-1}\}\} > 0 \qquad i = 1, \ldots r \qquad (17.70b)$$

where $\lambda_i\{A\}$ denotes the ith eigenvalue of A.

This theorem indicates that, as long as the steady-state plant and model gains satisfy the relationship given in Equation (17.70b), a stable closed-loop system with offset-free performance can be obtained for any arbitrary plant/model mismatch by simply making α_F sufficiently large.

It is possible that by removing the constraints on the filter structure and introducing off-diagonal elements (which allow interactions) and more general filter elements, similar robustness with less performance degradation can be achieved if the model uncertainties are structured. Structured model uncertainties refer to the uncertainties associated with the values of the parameters in the transfer function matrix. For unstructured model uncertainties, a diagonal filter is optimal. Thus, in addition to the detrimental effects of dead time and transmission zeroes outside the unit circle, robustness requirements may serve as another reason for tolerating interactions in order to obtain better overall system performance. This discussion emphasizes the point that in the presence of nonminimum phase elements and structured model uncertainties (and only under these circumstances) closed-loop performance of a multivariable system may have to be sacrificed by subjecting it to a decoupling constraint.

Simulation Examples

Example 6.[5]

Use the factorization concepts on the Wood and Berry column[37] and determine the closed-loop responses of the overhead and bottoms composition to:

    (a) a set point change of +0.75 mol% in overhead composition,
    (b) a set point change of +0.5 mol% in bottoms composition,
    (c) a load change of 0.34 lb/min in feed flow.

The sampling period is 1 minute. Assume perfect modeling.
The experimentally determined column model is

$$
\begin{bmatrix} y_1(s) \\ y_2(s) \end{bmatrix} = \begin{bmatrix} \dfrac{12.8e^{-s}}{16.7s + 1} & \dfrac{18.9e^{-3s}}{21s + 1} \\ \dfrac{6.6e^{-7s}}{10.9s + 1} & \dfrac{-19.4e^{-3s}}{14.4s + 1} \end{bmatrix} \begin{bmatrix} M_1(s) \\ M_2(s) \end{bmatrix} + \begin{bmatrix} \dfrac{3.8e^{-8s}}{14.9s + 1} \\ \dfrac{4.9e^{-3s}}{13.2s + 1} \end{bmatrix} F'(s)
$$

where time is minutes. The physical meaning and the steady-state values
of the variables are as follows:

$$y_1 = \text{overhead composition, 96.25 mol\%}$$

$$y_2 = \text{bottoms composition, 0.5 mol\%}$$

$$m_1 = \text{reflux flow rate, 1.95 lb/min}$$

$$m_2 = \text{steam flow rate, 1.71 lb/min}$$

$$F' = \text{feed flow rate, 2.45 lb/min}$$

The plant transfer function matrix $G(Z)$ is of the form

$$
\tilde{G}(Z) = \begin{bmatrix} Z^{-2}\tilde{g}_{11}(Z) & Z^{-4}\tilde{g}_{12}(Z) \\ Z^{-8}\tilde{g}_{21}(Z) & Z^{-4}\tilde{g}_{22}(Z) \end{bmatrix}
$$

The term $Z^{-2}\tilde{g}_{11}(Z)$, for example, of the plant transfer function matrix,
$\tilde{G}(Z)$, can be obtained as

$$
Z^{-2}\tilde{g}_{11}(Z) = Z\left\{ \frac{1 - e^{-sT}}{s} \cdot \frac{12.8e^{-s}}{16.7s + 1} \right\}
$$

Similar procedure yields the other elements of $\tilde{G}(Z)$. As we have seen
earlier, the time delay factorization procedure yields

$$
\tilde{G}_+(Z) = \begin{bmatrix} Z^{-2} & 0 \\ 0 & Z^{-4} \end{bmatrix}
$$

For this system there are no transmission zeroes, and, therefore, the
resulting controller $G_c(Z) = \tilde{G}_-(Z)^{-1} = \tilde{G}_p(Z)^{-1}\tilde{G}_+(Z)$ is stable. Now,

$$
y(Z) = \tilde{G}_+(Z) F(Z) (y_d(Z) - d(Z)) + d(Z)
$$

where

$$
d(Z) = \begin{bmatrix} d_1(Z) \\ d_2(Z) \end{bmatrix} = \begin{bmatrix} Z\{G_L(s) F'(s)\} \\ Z\{G_L(s) F'(s)\} \end{bmatrix}
$$

Thus,

$$
\begin{bmatrix} y_1(Z) \\ \\ y_2(Z) \\ \underset{(2 \times 1)}{} \end{bmatrix}
=
\begin{bmatrix} Z^{-2} & 0 \\ \\ 0 & Z^{-4} \\ \underset{(2 \times 2)}{} \end{bmatrix}
\begin{bmatrix} \dfrac{1 - \alpha_F}{1 - \alpha_F Z^{-1}} & 0 \\ \\ 0 & \dfrac{1 - \alpha_F}{1 - \alpha_F Z^{-1}} \\ \underset{(2 \times 2)}{} \end{bmatrix}
$$

$$
\begin{bmatrix} y_d(Z) - d_1(Z) \\ \\ y_d(Z) - d_2(Z) \\ \underset{(2 \times 1)}{} \end{bmatrix}
-
\begin{bmatrix} d_1(Z) \\ \\ d_2(Z) \\ \underset{(2 \times 1)}{} \end{bmatrix}
$$

Part (a).

$$y_{d_1}(t) = 0.75u(t)$$

$$y_{d_1}(s) = 0.75/s, \; y_{d_1}(Z) = \frac{0.75}{1 - Z^{-1}}$$

$$y_{d_2}(Z) = 0$$

$$d_1(Z) = d_2(Z) = 0$$

Therefore,

$$
\begin{bmatrix} y_1(Z) \\ \\ y_2(Z) \end{bmatrix}
=
\begin{bmatrix} Z^{-2} & 0 \\ \\ 0 & Z^{-4} \end{bmatrix}
\begin{bmatrix} \left(\dfrac{1 - \alpha_F}{1 - \alpha_F Z^{-1}}\right) & 0 \\ \\ 0 & \left(\dfrac{1 - \alpha_F}{1 - \alpha_F Z^{-1}}\right) \end{bmatrix}
\begin{bmatrix} \dfrac{0.75}{1 - Z^{-1}} \\ \\ 0 \end{bmatrix}
$$

In the absence of the filter, these equations reduce to

$$y_1(Z) = \frac{Z^{-2}\,0.75}{1 - Z^{-1}}$$

$$y_2(Z) = 0$$

Thus, the overhead composition reaches the new set point of 96 mol% after a delay of 2 sampling instants, while the bottoms composition remains at its initial set point. Introducing a filter, with $\alpha_F = 0.35$ makes the overhead composition response sluggish as expected. These results are shown in Figure 17.21(a).

Part (b).

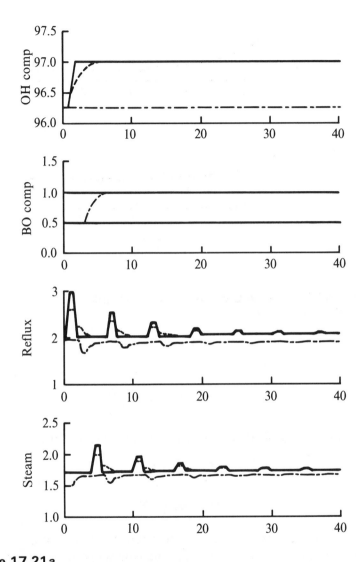

Figure 17.21*a*
Wood/Berry Column; Response to Set Point Changes;
(⎯⎯⎯) IMC with $\alpha = 0$, $(y_{1s}, y_{2s}) = (97.0, 0.5)$,
(⎯ ⎯ ⎯ ⎯) IMC with $\alpha = 0.35$, $(y_{1s}, y_{2s}) = (97.0, 0.5)$,
(⎯ · ⎯) IMC with $\alpha = 0.35$, $(y_{1s}, y_{2s}) = (96.25, 1.0)$
(Reprinted by permission from "Internal Model Control," Garcia and Morari, *Ind. Eng. Chem. Proc. Des. Dev.*, vol. 25, pp. 475–76, 479–82, © 1985, American Chemical Society.)

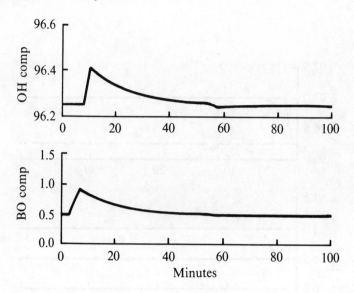

Figure 17.21*b*
Wood/Berry Column
Rejection to Disturbance Change of +0.34 lb/min;
(———) IMC with $\alpha = 0$

$$y_{d_1}(t) = 0$$

$$y_{d_2}(t) = 0.5; \quad y_{d_2}(Z) = \frac{0.5}{1 - Z^{-1}}$$

By reasoning similar to that in Part (a), in the absence of the filter,

$$y_{d_1}(Z) = 0$$

$$y_{d_2}(Z) = \frac{Z^{-4}0.5}{1 - Z^{-1}}$$

In this instance the overhead composition remains at its initial set point, while the bottoms composition reaches the new set point of 1 mol% after a delay of 4 sampling instants. Introducing a filter, $\alpha_F = 0.35$, renders the bottoms composition loop sluggish. These results are also shown in Figure 17.21(a). In Parts (a) and (b) perfect decoupling is evident.

    Part (c).

$$F'(s) = 0.34u(t)$$

$$F'(Z) = \frac{0.34}{1 - Z^{-1}}$$

$$d(Z) = \begin{bmatrix} d_1(Z) \\ \\ d_2(Z) \end{bmatrix} = \begin{bmatrix} Z\left\{ \dfrac{(3.8)(0.34)e^{-8s}}{s(14.9s + 1)} \right\} \\ \\ Z\left\{ \dfrac{(4.9)(0.34)e^{-3s}}{s(13.2s + 1)} \right\} \end{bmatrix}$$

For $\alpha_F = 0$ the response equations are

$$\begin{bmatrix} y_1(Z) \\ y_2(Z) \end{bmatrix} = \begin{bmatrix} Z^{-2} & 0 \\ 0 & Z^{-4} \end{bmatrix} \begin{bmatrix} 0 - d_1(Z) \\ 0 - d_2(Z) \end{bmatrix} - \begin{bmatrix} d_1(Z) \\ d_2(Z) \end{bmatrix}$$

Thus,

$$y_1(Z) = -Z^{-2} d_1(Z) - d_1(Z)$$

$$y_2(Z) = -Z^{-4} d_2(Z) - d_2(Z)$$

Substitution and inversion yields the load responses shown in Figure 17.21(b).

Example 7.[5]

This example involves a seventh-order linear model of a fixed bed reactor[38]. A schematic of the oxygen-hydrogen experimental reactor is shown in Figure 17.22(a). The steady-state operating data are shown in Table 17.4. In this process the controlled variables are reactor outlet temperature and oxygen concentration, and the manipulated variables are the flow rate and the temperature of a hydrogen quench stream. The open-loop responses of the controlled variables to step changes in the manipulated variables are shown in Figure 17.22(b). Determine the closed-loop set point responses, and study the effect of modeling errors. The sampling period, T, is 20 seconds.

Part 1. Set Point Responses

In the absence of dead time

$$\tilde{G}_+(Z) = Z^{-1}I$$

The resulting $G_c(Z) = \tilde{G}_-^{-1}$ is found to be stable. The responses of temperature and outlet oxygen concentration to a 1% step change in the latter and the associated manipulated variable moves are shown in Figure 17.23(a). With no filtering, the outputs show oscillations, although the amplitude of the oscillations is small. Adding filtering reduces the amplitude of the oscillations in outlet temperature and yields almost perfect decoupling. The associated manipulated variable moves are also not excessive.

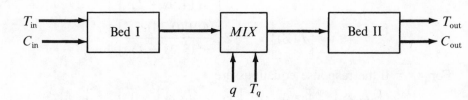

Figure 17.22a
Schematic of Reactor System
(Reprinted by permission from "Internal Model Control," Garcia and
Morari, *Ind. Eng. Chem. Proc. Des. Dev.*, vol. 25, pp. 475–76, 479–
82, © 1985, American Chemical Society.)

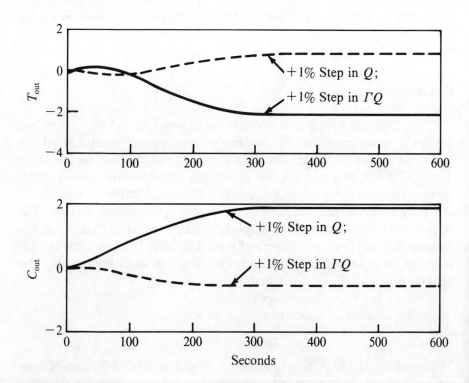

Figure 17.22b
Fixed-bed Reactor, Unit Step Responses
(Reprinted by permission from "Internal Model Control," Garcia and
Morari, *Ind. Eng. Chem. Proc. Des. Dev.*, vol. 25, pp. 475–76, 479–
82, © 1985, American Chemical Society.)

Table 17.4
Oxygen-Hydrogen Reactor Operating Conditions

Variable	Symbol	Steady-State Values	Model Ref. Units
Outlet temperature	T_{out}	454 K	167.4° C
Outlet O_2 concentration	C_{out}	0.211 mol%	1 mol%
Quench stream flow rate	q	2850 cc/min	13900 cc/min
Quench stream temperature	T_q	388 K	167.4° C
Inlet temperature	T_{in}	380 K	167.4° C
Inlet O_2 concentration	T_{out}	1.26 mol%	1 mol%

Part 2. Robustness in the Presence of Modeling Errors

To study the effect of modeling errors, the high-order dynamics are deliberately approximated by dead time elements, yielding the model

$$\tilde{G}(Z) = \begin{bmatrix} Z^{-2}\, \bar{g}_{11}(Z) & Z^{-2}\, \bar{g}_{12}(Z) \\ Z^{-1}\, \bar{g}_{21}(Z) & Z^{-3}\, \bar{g}_{22}(Z) \end{bmatrix}$$

For this model the time delay factorization

$$\tilde{G}_{+}(Z) = \begin{bmatrix} Z^{-2} & 0 \\ 0 & Z^{-1} \end{bmatrix}$$

yields a stable realizable controller, $G_c(Z) = \tilde{G}_{-}(Z)^{-1}$.

The responses of the simulated system with no filtering are shown in Fig. 17.23(b). The controller attempts to bring the system to set point at the sampling instants, but modeling errors propagate and eventually destabilize the system. For this structural plant/model mismatch, an exponential filter with $\alpha_1 = \alpha_2 = 0.5$ does an excellent job, as is evident from Figure 17.23(b).

Now let us consider the effect of a plant/model gain mismatch. The reactor true gain matrix is

$$G(1) = \begin{bmatrix} -2.27 & 0.73 \\ 1.85 & -0.65 \end{bmatrix}$$

Let us assume that the following gain matrix containing slight modeling errors is used to tune the IMC controller.

$$\tilde{G}(1) = \begin{bmatrix} -2.253 & 0.77 \\ 1.85 & -0.63 \end{bmatrix}$$

The eigenvalues of $G(1)\,\tilde{G}(1)^{-1}$ are 1.0025 and -24.398, and, therefore, the condition stipulated in Equation (17.70b) is not satisfied. Hence, no stabilizing $\alpha(0 \leq \alpha < 1)$ exists, although the modeling errors are small. This example illustrates that for inherently sensitive systems, like the

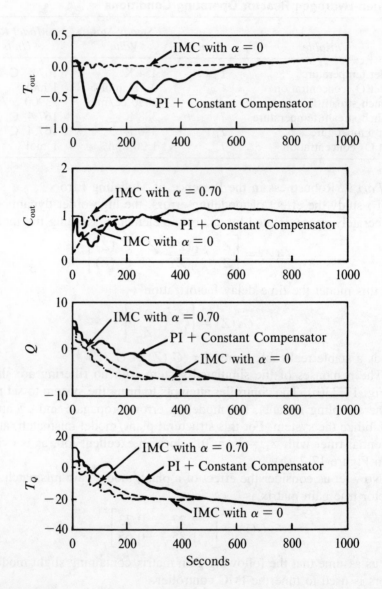

Figure 17.23*a*
Fixed-bed Reactor, +1% Step in C_{out} Set Point
(Reprinted by permission from "Internal Model Control," Garcia and Morari, *Ind. Eng. Chem. Proc. Des. Dev.*, vol. 25, pp. 475–76, 479– 82, © 1985, American Chemical Society.)

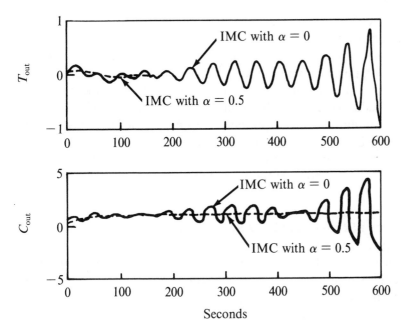

Figure 17.23b
Fixed-bed Reactor, Robustness to Modeling Errors
(Reprinted by permission from "Internal Model Control," Garcia and
Morari, *Ind. Eng. Chem. Proc. Des. Dev.*, vol. 25, pp. 475–76, 479–
82, © 1985, American Chemical Society.)

hydrogen-oxygen reactor of this example, good control is difficult to achieve in practice regardless of what type of control methodology is used.

IMC Predictive Control Formulation[5,6,8]

The analytical factorization method can become complex for all but simple cases. Furthemore, in industrial practice, constraints on inputs and outputs are often present. In such applications it becomes important to know ahead of time when constraint violations are likely to occur so that appropriate actions may be taken. To meet these needs, a predictive control problem must be formulated and solved. This type of formulation leads to the IMC predictive algorithm and, as a subset, model algorithmic control[2] (MAC) and dynamic matrix control[3] (DMC).

In this section we present the IMC predictive control algorithm for single-input/single-output systems. We will show how the selection of particular choices of parameters in the IMC predictive control algorithm leads to MAC and DMC. The derivation of the multivariable IMC predictive algorithm is similar in concept to its SISO counterpart, although certain complexities arise in the mathematical manipulations due to the fact that the impulse response models, $G_{ij}(Z)$, may contain unequal dead time elements. Again, for certain choices of parameters, the multivariable IMC predictive control algorithm gives MAC and DMC. To maintain sufficient simplicity, we present here only a brief treatment of the multivariable IMC predictive control algorithm. The detailed derivation of the IMC predictive algorithm for SISO and multivariable systems is available for those who may be interested.[6,8]

IMC Predictive Strategy for SISO Systems. In the predictive IMC strategy an impulse response model of the process is used for prediction and control calculations. The design procedure consists of the following steps:

1. Predict the process output over P future sampling instants by the impulse response model as

$$Y_M(k + \tau + i) = h_1 m(k + i - 1) \tag{17.71}$$
$$+ \, h_2 m(k + i - 2) + \ldots + h_N m(k + i - N)$$

with $i = 1, 2, \ldots P$

2. Correct the predicted output for the possible presence of modeling errors and load disturbances according to

$$Y(k + \tau + i) = Y_M(k + \tau + i) + Y(k) - Y_M(k) \tag{17.72}$$

3. Specify the desired response of the process output, for example, as a first-order trajectory

$$Y_d(k + \tau + i) = \alpha_f Y_d(k + \tau + i - 1) + (1 - \alpha_f)R \tag{17.73}$$

with

$$Y_d(k + \tau) = Y(k)$$

where

α_f = tuning parameter that determines the speed of response ($0 < \alpha_f < 1$)

R = set point

4. Calculate a set of control actions such that the corrected future output given in Equation (17.72) follows the desired set point trajectory given in Equation (17.73). Now, if Equations (17.72) and (17.73) are equated, a set of future controller outputs can be calculated that will force the corrected future outputs to match the desired trajectory. How-

ever, this procedure is not used since the controller outputs so calculated would probably not be physically realizable. Instead, a classical optimization problem is formulated whose objective is to compute a set of values $m(k + i - 1)$, $i = 1, 2, \ldots M$ (where $M \leq P$) such that the following performance index is minimized:

$$
J = \sum_{i=1}^{P} \{ \gamma_i^2 \, [Y_d(k + \tau + i) - Y(k + \tau + i)]^2 \\
+ \beta_i^2 \, [m(k + i - 1)]^2 \}
\tag{17.74}
$$

subject to

$$
m(k + M) = m(k + M + 1) = \ldots = m(k + P - 1)
$$

where

γ_i = weights on the output, $i = 1, 2, \ldots P$

β_i = weights on the input, $i = 1, 2, \ldots P$

   $= 0$ for $i > M$

The solution of this least squares optimization problem is obtained by differentiating J with respect to m, setting the derivative equal to zero and solving for m. The result is

$$
m(k) = -\sum_{i=1}^{N-1} \phi_i \, m(k - i) + \sum_{i=1}^{P} \nu_i \{ Y_d(K + \tau + i) \\
- [Y(k) - Y_M(k)] \}
\tag{17.75}
$$

where

ϕ_i, $i = 1, 2, \ldots P$ are elements of first row of $s\Gamma_p\Omega$

ν_i, $i = 1, 2, \ldots P$ are elements of first row of $s\Gamma_p$

The terms s, Γ_p, and Ω represent various matrices and vectors and are defined in Appendix A.

Equation (17.75) is the position form of the IMC predictive control algorithm. The control law can be expressed in velocity form to facilitate bumpless transfer from manual to automatic as

$$
\Delta m(k) = -\sum_{i=1}^{N-1} \phi_i \, \Delta m(k - i) + \sum_{i=1}^{P} \nu_i \{ Y_d(k + \tau + i) \\
- \Delta[Y(k) - Y_M(k)] \}
\tag{17.76}
$$

where

$$
\Delta x(k) = x(k) - x(k - 1)
$$

It can be shown that the use of Equations (17.75) or (17.76) can lead to a steady-state offset when β_i is not equal to zero. The offset is eliminated by an offset compensator—a scalar—given by

$$\Phi = \frac{1 + \sum_{i=1}^{N-1} \phi_i}{\sum_{i=1}^{P} \nu_i \cdot \sum_{i=1}^{N} h_i} \tag{17.77}$$

Finally, robustness in the presence of modeling errors may be maintained by a filter having the transfer function

$$F(Z) = \frac{1 - \alpha_F}{1 - \alpha_F Z^{-1}} \tag{17.78}$$

A block diagram of the IMC predictive control law along with the offset compensator and the filter is shown in Figure 17.24.

Garcia and Morari[4] have shown that model algorithmic control[2] and dynamic matrix control[3] are very closely related to IMC. In IMC one computes $m(k + i - 1)$, $i = 1, 2, \ldots M$, with $m(k + M) = m(k + M + 1) = \ldots = m(k + P - 1)$. By contrast, in MAC, $M = P$ and the inputs $m(k + i - i)$, $i = 1, 2, \ldots P$, are calculated with the restriction that several intermediate inputs are "blocked," i.e., held constant. For example, three inputs, m_1, m_2, m_3, may be computed to define five future control actions as $m(k) = m_1$; $m(k + 1) = m(k + 2) = m_2$; and $m(k + 3) = m(k + 4) = m_3$.

In DMC the manipulated variable moves are penalized. The performance index given in Equation (17.74) is modified according to

$$J = \sum_{i=1}^{P} \{\gamma_i^2 [Y_d(k + \tau + i) - Y(k + \tau + i)]^2 + \beta_i^2 [\Delta m(k + i - 1)]^2\} \tag{17.79}$$

subject to

$$\Delta m(k + M) = \Delta m(k + M + 1) = \ldots = \Delta m(k + P - 1) = 0$$

The solution of the optimization problem in this instance is

$$\Delta m(k) = \sum_{i=1}^{N-1} \bar{\phi}_i \, \Delta m(k - i) + \sum_{i=1}^{P} \bar{\nu}_i \{Y_d(k + \tau + i) - [Y_d(k) - Y_M(k) - Y_M(k + \tau)]\} \tag{17.80}$$

where

$\bar{\phi}_i$, $i = 1, 2, \ldots n - 1$ are elements of $S\bar{\Omega}$

$\bar{\nu}_i$, $i = 1, 2, \ldots P$ are elements of S

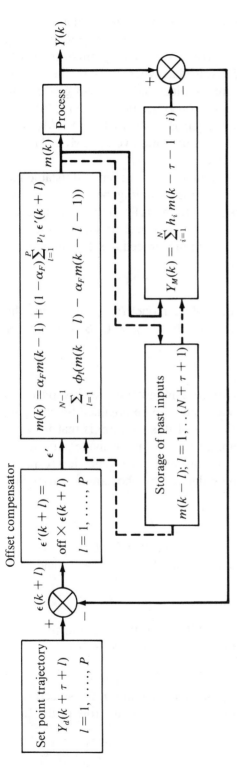

Notes:

1. $\text{Off} = \dfrac{\left(1 + \displaystyle\sum_{l=1}^{N-1} \phi_l\right)}{\left(\displaystyle\sum_{l=1}^{P} \nu_l\right)\left(\displaystyle\sum_{i=1}^{N} h_i\right)}$

2. $\phi_l, l = 1, \ldots, (N - 1)$ are the elements of the first row of $s\Gamma_p\Omega$.

3. $\nu_l, l = 1, \ldots, P$ are the elements of the first row of $s\Gamma_p$.

4. s is defined in Appendix A; Ω and Γ_p are defined in Appendix A.

Figure 17.24
IMC Predictive Control Scheme for SISO Systems with Offset Compensator and Filter — Computational Details

The terms s and $\bar{\Omega}$ represent various matrices and vectors and are defined in Appendix A. In DMC an offset compensator is not required since manipulated variable moves are penalized. A block diagram of the DMC system is shown in Figure 17.25. Note that when $\beta_i = 0$, the performance indices of IMC and DMC are identical and both will give rise to the same control law, and Equations (17.75) (or Equation (17.76)) and (17.80) will give identical results. When $\beta_i \neq 0$, IMC and DMC will give different results even if the filter is not employed in IMC. Finally, note that DMC does not utilize a filter.

The IMC predictive control algorithm contains several adjustable parameters, viz., γ_i, β_i, M, P, α_F, and the sampling period, T, whose values must be selected by the designer. In the following paragraphs we consider several stability theorems that have a bearing on the selection of these parameters and then outline tuning procedures for finding the suitable values of these parameters.[4]

Theorem 1. If the zeroes of the delay-free model $H(Z)$ lie outside the unit circle, the IMC control law is unstable for $\gamma_i \neq 0$, $\beta_i = 0$, and $M = P \leq N$.

Theorem 2. If a system exhibits a discrete monotonic step response and $\gamma_i = 1$, $\beta_i = 0$, $P = N$, a stable control law can be obtained by choosing a sufficiently small M. The significance of this theorem is that for nonminimum phase systems, the sampling period can be made sufficiently large so that all $h_i > 0$, resulting in a discrete monotonic step response for which a stable control law can be derived by choosing a sufficiently small M.

Theorem 3. There exists a finite $\beta^* > 0$ such that for $\beta_i \geq \beta^*$ ($i = 1, \ldots M$) the IMC control law is stable for all $M \geq 1$, $P \geq 1$, and $\gamma_i \geq 0$.

Theorem 4. For $\gamma_i = 1$, $\beta_i = 0$, the IMC controller is stable for a sufficiently small M and a sufficiently large $P > N + M - 1$.

Tuning guidelines. The following guidelines will aid in the selection of the parameter γ_i, β_i, P, and M. Some comments on the role of the sampling period are also offered.

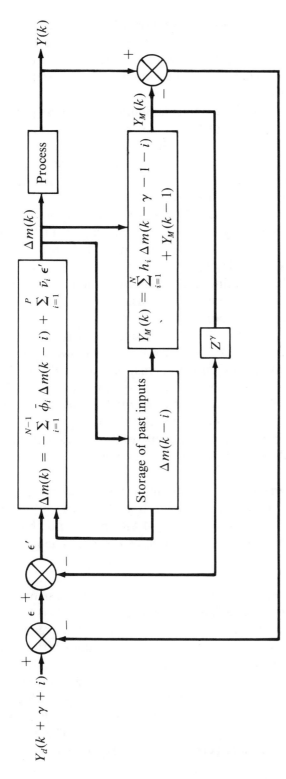

Figure 17.25
DMC Control System for SISO Systems — Computational Details

Sampling Period (T). Unlike conventional feedback control, the stability of IMC is not affected by the sampling period. Larger values of T generally lead to less extreme excursions of the manipulated variable, but they have a detrimental effect on the ability of the system to handle frequent disturbances.

Frequent sampling of a continuous nonminimum process produces an $H(Z)$ with roots outside the unit circle. However, if the sampling period is sufficiently increased, these roots can be stabilized and a stable controller can be derived. Another alternative is to increase T just enough to ensure $h_i > 0$ ($i = 1, 2, \ldots, N$) and decrease M until a stable controller results in accordance with theorem 2.

Input Suppression Parameter (M). Any system with $h_i > 0$ ($i = 1, 2, \ldots N$) can be stabilized by choosing a sufficiently small M. For minimum phase systems having stable $H(Z)^{-1}$, perfect control ($M = P = N$) can be specified, but this usually leads to severe excursions of the manipulated variables, resulting in oscillations of the process output between sampling instants. Reducing M decreases input variations and leads to a more desirable plant response.

Input Penalty Parameter (β_i). For nonminimum phase systems, a stable approximation to $H(Z)^{-1}$ may always be obtained by increasing β_i sufficiently. For minimum phase systems, increasing β_i will decrease manipulated variable moves and make the system more sluggish. Furthermore, $\beta_i \neq 0$ will in general lead to an offset. This is corrected for by the offset compensator shown in Figure 17.24.

Output Penality Parameter (γ_i). The use of these parameters is meaningful only when they are time varying. Otherwise, γ_i should be set equal to 1.

Optimization Horizon (P). Nonminimum phase systems can be stabilized by selecting a sufficiently small M and a sufficiently large P. For minimum phase systems, the length of the optimization horizon has virtually no effect on performance as soon as P exceeds about twice the system order.

The foregoing discussion results in a set of guidelines for a system of order n based on the fact that an nth order system can be brought precisely to rest in no more than n discrete steps. This suggests a predictive control law that will vary the inputs over the next n steps and the inputs will be zero afterwards. The inputs will be determined such that the outputs at $n, n + 1, n + 2, \ldots 2n - 1$ are zero. This strategy suggests the following choice of tuning parameters for a system of order n: $P = 2n - 1$; $M = n + 1$; $\beta_1 = \beta_2 = \ldots = \beta_n = 0.$; $\beta_{n+1} = $ large; $\gamma_1 = \gamma_2 = \ldots = \gamma_{n-1} = 0$; $\gamma_n = \gamma_{n+1} = \gamma_{2n-1} = 1$. These values reportedly resulted in excellent performance in test simulations.[4]

IMC Predictive Strategy for Multivariable Systems.[6] To derive the IMC predictive control law for multivariable systems, one must begin with the appropriate extension of the SISO performance measure given in Equation (17.74). The extension of Equation (17.74) to multivariable systems is complicated due to the fact that the individual elements of the process transfer function matrix may contain unequal time delays. The difficulties associated with the possible presence of unequal time delays are alleviated by using the concept of *balance* in the time delays. The time delays are said to be balanced if

$$\tau_i^{min} = \tau_i^+ \quad \text{for } i = 1, 2, \ldots, r \tag{17.81}$$

where

$$\tau_i^{min} = \min_j (\tau_{ij})$$

and

τ_i^+ = minimum number of sampling instants that are required for the output i ($i = 1, 2, \ldots, r$) to settle at the new steady-state (see Equation (17.59))

A measure of the imbalance is given by

$$\tau_o = \max (\tau_i^+ - \tau_i^{min}) \tag{17.82}$$

In the light of these definitions, the performance measure for multivariable systems can be written as

$$J = \min_{\substack{m_i(k) \ldots m_i(k + M_i - 1) \\ i=1,r}} \sum_{l=1}^{P+\tau_o} \{[y_d^+(k + l - \tau_o)$$

$$- y^+(k + l - \tau_o)]^T \, \Gamma_l^T \, \Gamma_l [y_d^+ (k + l - \tau_o) - y^+(k + l - \tau_o)] \tag{17.83a}$$

$$+ [m(k + l - 1)^T \, \mathbf{B}_l^T \mathbf{B}_l m(k + l - 1)]\}$$

subject to

$$y^+(k) = y_M^+(k) + d^+(k) = \sum_{j=1}^{N+\tau_o} \mathbf{H}_j \, m(k + \tau_o - j) + d^+(k) \tag{17.83b}$$

and

$$m_i(k + M_i - 1) = \ldots = m_i(k + P + \tau_o - 1) \tag{17.83c}$$
$$\text{for } i = 1, \ldots r.$$

where $Z\,Y^+(Z)$ represents a vector of delayed outputs given by

$$Z\,Y^+(Z) = \overset{*}{G}_+(Z)^{-1}\, Y(Z) \tag{17.84}$$

and

$$\overset{*}{G}_+(Z) = \text{diag } \{Z^{-(\tau_1^+ +1)}, Z^{-(\tau_2^+ +1)}, \ldots, Z^{-(\tau_r^+ +1)}\} \qquad (17.59)$$

Similar definitions apply to Y_d^+ and d^+. Making use of the matrix partitioning concepts, the cost functional can be written in the form of $E^T E$. Setting the gradient of the objective function equal to zero leads to the IMC control law. (The terms are defined in Appendix A.)

$$\mathbf{U}_M(\bar{k}) = \boldsymbol{\Delta}^{-1} \, \mathbf{T}_M^T \, \boldsymbol{\Lambda}^T \, \boldsymbol{\Gamma}^T \, \boldsymbol{\Gamma} \, [\boldsymbol{\epsilon}_p \, (\bar{k} + 1) - \boldsymbol{\psi} \, \mathbf{V} \, (k - 1)] \quad (17.85\text{a})$$

where

$$\boldsymbol{\epsilon}(\bar{k} + l) = Y_d^+(\bar{k} + l) - d^+(\bar{k} + l) \qquad (17.85\text{b})$$

$$\boldsymbol{\Delta} = \mathbf{T}_M^T \, \boldsymbol{\Lambda}^T \, \boldsymbol{\Gamma}^T \, \boldsymbol{\Gamma} \, \boldsymbol{\Lambda} \, \mathbf{T}_M + \mathbf{T}_M^T \, \mathbf{B}^T \, \mathbf{B} \, \mathbf{T}_M \qquad (17.85\text{c})$$

In general, $\boldsymbol{\Delta}$ can be rank-deficient, in which case the inverse will not exist. The following cases will cover all the possibilities.

Case 1. Rank $[\mathbf{B}_j] = r$ for $j = 1, \ldots, P + \tau_o$, where r is the number of inputs, it can be shown that[39]

$$\text{rank} \left[\frac{\boldsymbol{\Gamma} \, \boldsymbol{\Lambda} \, \mathbf{T}_M}{\mathbf{B} \mathbf{T}_M} \right] = M \qquad (17.86)$$

for all $\boldsymbol{\Gamma}$, $\boldsymbol{\Lambda}$, and $\mathbf{T}_M$. Therefore, det $\boldsymbol{\Delta} \neq 0$. Then, Equation (17.85a) will yield a unique solution.

Case 2. If $\mathbf{B} = 0$ and rank $[\boldsymbol{\Gamma} \, \boldsymbol{\Lambda} \, \mathbf{T}_M] = M$, then det $\boldsymbol{\Delta} \neq 0$ and a unique $\mathbf{U}_M(k)$ can be determined from Equation (17.85a) as in Case 1.

Case 3. If $\mathbf{B} = 0$ and rank $[\boldsymbol{\Gamma} \, \boldsymbol{\Lambda} \, \mathbf{T}_M] = K < M$, then det $\boldsymbol{\Delta} = 0$ and a unique $\mathbf{U}_M (k)$ does not exist. In this instance one approach is to modify the least squares problem. It is possible to find nonsingular matrices $\boldsymbol{\Omega}$ and $\mathbf{A}$ by performing Gaussian elimination such that

$$\boldsymbol{\Omega} \, \boldsymbol{\Gamma} \, \mathbf{T}_M = \left[\frac{\bar{\mathbf{A}}}{0} \right] \qquad (17.87)$$

where $\bar{\mathbf{A}}$ $(K \times M)$ is of full rank K.

Now, define

$$\mathbf{R}' = [\mathbf{I}_K \vdots \mathbf{0}] \qquad (17.88)$$

The use of "minimum norm solution," which implies a minimization of input deviations, then yields a control law

$$\mathbf{U}_M(k) = \bar{\mathbf{A}}^T \, (\bar{\mathbf{A}} \, \bar{\mathbf{A}}^T)^{-1} \, \mathbf{R}' \, \boldsymbol{\Omega} \, \boldsymbol{\Gamma} \, [\boldsymbol{\epsilon}_P \, (\bar{k} + 1) - \boldsymbol{\psi} \, \mathbf{V} \, (\bar{k} - 1)] \quad (17.89)$$

The choice of $\bar{\mathbf{A}}$ and $\boldsymbol{\Omega}$ is generally not unique.

Returning to Equation (17.85a), define

$$\mathbf{S} = \boldsymbol{\Delta}^{-1} \mathbf{T}_M^T \boldsymbol{\Lambda}^T \boldsymbol{\Gamma}^T \boldsymbol{\Gamma} \tag{17.90}$$

Then Equation (17.85a) becomes

$$\mathbf{U}_M(k) = \mathbf{S}\,\boldsymbol{\epsilon}_p(k+1) - \mathbf{S}\,\boldsymbol{\psi}\,\mathbf{V}(k-1) \tag{17.91}$$

Now, let

$$\mathbf{b}^T = [\mathbf{I}_r \vdots \mathbf{0}] \tag{17.92}$$

Consequently,

$$\begin{aligned}
m(k) &= \mathbf{b}^T \mathbf{U}_M(k) \\
&= \mathbf{b}^T \mathbf{S}\,\boldsymbol{\epsilon}_p(k+1) - \mathbf{b}^T \mathbf{S}\,\boldsymbol{\psi}\,\mathbf{V}(k-1)
\end{aligned} \tag{17.93}$$

Next, let $\mathbf{D}_1, \mathbf{D}_2, \ldots, \mathbf{D}_{N+\tau_o-1}$ be the elements of $\mathbf{b}^T \mathbf{S}\,\boldsymbol{\psi}$

and $\mathbf{E}_1, \mathbf{E}_2, \ldots, \mathbf{E}_p + \tau_o$ be the elements of $\mathbf{b}^T \mathbf{S}$

Then, Equation (17.93) may be reorganized to give

$$m(k) + \sum_{l=1}^{N+\tau_o-1} \mathbf{D}_l\, m(k-l) = \sum_{l=1}^{P+\tau_o} \mathbf{E}_l\,\boldsymbol{\epsilon}(k+l-\tau_o)$$

or, in view of Equation (17.85b),

$$\begin{aligned}
m(k) + \sum_{l=1}^{N+\tau_o-1} \mathbf{D}_l\, m(k-l) &= \sum_{l=1}^{P+\tau_o} \mathbf{E}_l\{y_d^+(k+l-\tau_o) \\
&\qquad\quad - d^+(k+l-\tau_o)\} \\
&= \sum_{l=1}^{\tau_o} \mathbf{E}_l\{y_d^+(k+l-\tau_o) \\
&\qquad\quad - d^+(k+l-\tau_o)\} \\
&\quad + \sum_{l=\tau_o+1}^{P+\tau_o} \mathbf{E}_l\{y_d^+(k+l-\tau_o) \\
&\qquad\quad - d^+(k+l-\tau_o)\}
\end{aligned} \tag{17.94}$$

or

$$\begin{aligned}
m(k) + \sum_{l=1}^{N+\tau_o-1} \mathbf{D}_l\, m(k-l) &= \sum_{l=1}^{\tau_o} \mathbf{E}_l\{y_d^+(k+l-\tau_o) - d^+(k+l-\tau_o)\} \\
&\quad + \sum_{l=1}^{P} \mathbf{E}_{l+\tau_o}\{y_d^+(k+l) - d^+(k)\}
\end{aligned}$$

Note that $d^+ (k + l) = d^+ (k)$. This is the position form of the IMC multivariable predictive control law shown in the block diagram form in Figure 17.26a. Taking Z-transforms of both sides of Equation (17.94) gives

$$m(Z) + \sum_{l=1}^{N+\tau_o-1} \mathbf{D}_l Z^{-l} m(Z) = \sum_{l=1}^{\tau_o} \mathbf{E}_l Z^{l-\tau_o} \{y_d^+(Z) - d^+(Z)\}$$
$$+ \sum_{l=1}^{P} \mathbf{E}_{l+\tau_o} \{Z^l y_d^+(Z) - d^+(Z)\} \tag{17.95}$$

Define

$$D_c(Z) = I_r + \sum_{l=1}^{N+\tau_o-1} \mathbf{D}_l Z^{-l}$$

$$N_p(Z) = \sum_{l=1}^{\tau_o} \mathbf{E}_l Z^{l-\tau_o} \tag{17.96}$$

$$N_F(Z) = \sum_{l=1}^{P} \mathbf{E}_{l+\tau_o} Z^l$$

Substituting these expressions into Equation (17.95) we obtain

$$D_c(Z) m(Z) = \{N_p(Z) + N_F(Z)\} y_d^+ (Z)$$
$$- \{N_p(Z) + N_F(1)\} d^+(Z) \tag{17.97}$$

or

$$m(Z) = D_c(Z)^{-1} \{\{N_p(Z) + N_F(Z)\} y_d^+ (Z)$$
$$- \{N_p(Z) + N_F(1)\} d^+(Z)\} \tag{17.98}$$

The block diagram representation of the multivariable control law given in Equation (17.98) is shown in Figure 17.26b.

For a unit step change in set point and for any asymptotically constant disturbance, under the assumption of perfect modeling, the final value of the output is

$$\lim_{t\to\infty} Y(t) = \lim_{Z\to 1} (1 - Z^{-1})[\tilde{G}(Z)D_c(Z)^{-1} \{N_p(Z)$$
$$+ N_F(Z)\} y_d^+(Z)] \tag{17.99}$$

Zero offset performance is achievable when

$$\tilde{G}(1) D_c(1)^{-1} \{N_p(1) + N_F(1)\} = I_r$$

For certain choices of tuning parameters, e.g., when $\mathbf{B} \neq 0$, $\tilde{G}(1)D_c(1)^{-1}\{N_p(1) + N_F(1)\} \neq I_r$, leading to an offset. This difficulty is corrected by introducing an offset compensator, G_{off}, given by

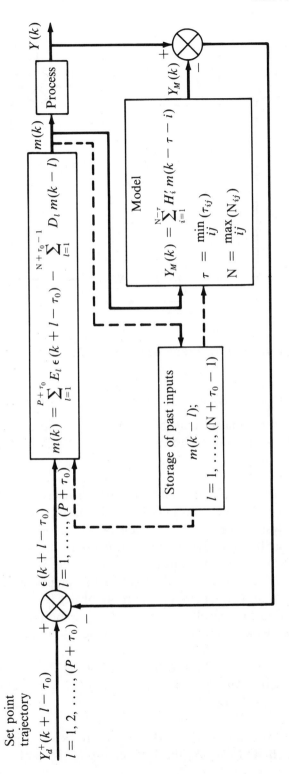

$Y(k)$

Process

$m(k)$

Model

$$Y_M(k) = \sum_{i=1}^{N-\tau} H'_i \, m(k - \tau - i)$$

$$\tau = \min_{ij} (\tau_{ij})$$

$$N = \max_{ij} (N_{ij})$$

$Y_M(k)$

$$m(k) = \sum_{l=1}^{P+\tau_0} E_l \, \epsilon(k + l - \tau_0) - \sum_{l=1}^{N+\tau_0-1} D_l \, m(k - l)$$

Storage of past inputs

$m(k - l)$;

$l = 1, \ldots, (N + \tau_0 - 1)$

Set point
trajectory

$Y_d^+(k + l - \tau_0)$

$l = 1, 2, \ldots, (P + \tau_0)$

$\epsilon(k + l - \tau_0)$

$l = 1, \ldots, (P + \tau_0)$

Note:

1. $D_1, \ldots, D_{N+\tau_0-1}$ are the elements of $\mathbf{b}^T \mathbf{S} \boldsymbol{\psi}$

   $E_1, \ldots, E_{P+\tau_0}$ are the elements of $\mathbf{b}^T \mathbf{S}$

2. $\mathbf{S}$ is defined in Equation (17.90)

   $\boldsymbol{\psi}$ is defined in Appendix A

Figure 17.26a
**Multivariable IMC Predictive Control Scheme —
Computational Details**

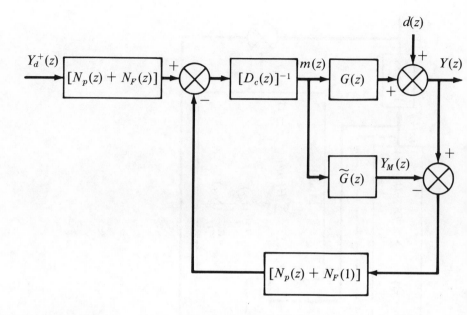

Figure 17.26b
Block Diagram of Multivariable IMC Predictive Control Scheme

$$G_{\text{off}} = \{N_p(1) + N_F(1)\}^{-1} D_c(1) \, \tilde{G}(1)^{-1} \qquad (17.100)$$

As in SISO systems, a filter matrix may be introduced in the control scheme, giving the block diagrams shown in Figures 17.26(c) and 17.26(d).

Stability Theorems. It has been pointed out that in the absence of modeling errors the stability of the IMC controller is sufficient to ensure the stability of the control system, provided the plant is open-loop stable. In terms of the multivariable predictive control formulation, this means that the roots of det $D_c(Z) = 0$ must lie inside the unit circle in the Z-domain. Garcia and Morari[6] have presented the following theorems, which show the effect of the tuning parameters on the roots of det $[D_c(Z)] = 0$. These theorems will lead to controller design and tuning procedures. For proof, the interested reader may refer to their article.

Theorem 1. The choice of $\Gamma_l = I$, $\mathbf{B}_l = 0$, $l = 1, \ldots P + \tau_o$; $M_j = P + \tau_o$, $j = 1, \ldots r$ in the problem formulation, Equation (17.83), yields the perfect control law, $G_c = G\underset{\sim}{*}(Z)^{-1}$.

Theorem 2. Assume det $\mathbf{B}_l \neq 0$, $l = 1, \ldots P + \tau_o$. There exists a finite $\beta^* > 0$ such that, if $\|\mathbf{T}_M^T \mathbf{B}^T \mathbf{B} \mathbf{T}_M\| > \beta^*$, the IMC controller is stable.

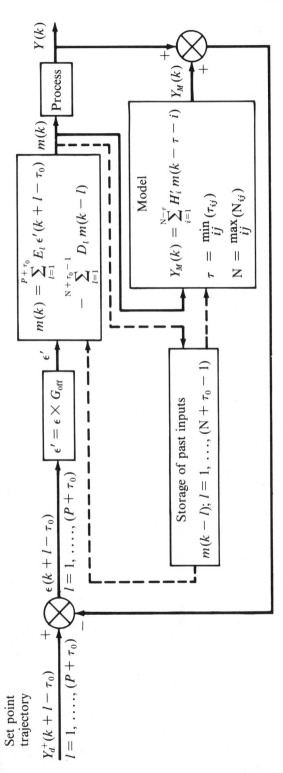

Set point
trajectory

$Y_d^+(k+l-\tau_0)$

$l = 1, \ldots, (P+\tau_0)$

$\epsilon(k+l-\tau_0)$

$l = 1, \ldots, (P+\tau_0)$

$\epsilon' = \epsilon \times G_{\text{off}}$

ϵ'

$m(k) = \sum_{l=1}^{P+\tau_0} E_l\,\epsilon'(k+l-\tau_0)$
$\qquad - \sum_{l=1}^{N+\tau_0-1} D_l\,m(k-l)$

$m(k)$

Process

$Y(k)$

Model

$Y_M(k) = \sum_{i=1}^{N-\tau} H_i\,m(k-\tau-i)$

$\tau = \min_{ij}(\tau_{ij})$

$N = \max_{ij}(N_{ij})$

$Y_M(k)$

Storage of past inputs

$m(k-l);\ l = 1, \ldots, (N+\tau_0-1)$

Notes:

1. $G_{\text{off}} = \left\{\left(\left(\sum_{i=1}^{N-\tau} H_i\right)\left(I_r + \sum_{l=1}^{N+\tau_0-1} D_l\right)^{-1}\right)^{-1}\left[\sum_{l=1}^{\tau_0} E_l + \sum_{l=1}^{P} E_{l+\tau_0}\right]\right\}^{-1}$

2. $D_1, \ldots, D_{N+\tau_0-1}$ are the elements of $\mathbf{b}^T\,\mathbf{S}\boldsymbol{\psi}$
   $E_1, \ldots, E_{P+\tau_0}$ are the elements of $\mathbf{b}^T\,\mathbf{S}$

3. $\mathbf{S}$ is defined in Equation (17.90)
   $\boldsymbol{\psi}$ is defined in Appendix A

Figure 17.26c
Multivariable IMC Predictive Control Scheme
with Offset Compensator — Computational Details

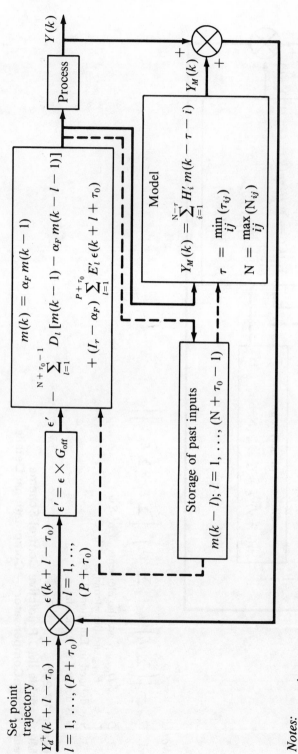

Notes:

1. $G_{\text{off}} = \left\{ \left(\sum\limits_{i=1}^{N-\tau} H_i \right) \left(I_r + \sum\limits_{l=1}^{N+\tau_0-1} D_l \right)^{-1} \left[\sum\limits_{l=1}^{\tau_0} E_l + \sum\limits_{l=1}^{P} E_{l+\tau_0} \right] \right\}^{-1}$

2. $D_1, \ldots, D_{N+\tau_0-1}$ are the elements of $\mathbf{b}^T \mathbf{S} \boldsymbol{\psi}$
   $E_1, \ldots, E_{P+\tau_0}$ are the elements of $\mathbf{b}^T \mathbf{S}$

3. $\mathbf{S}$ is defined in Equation (17.90)
   $\boldsymbol{\psi}$ is defined in Appendix A

Figure 17.26d
Multivariable IMC Predictive Control Scheme
with Filter and Offset Compensator — Computational Details

Theorem 3. Assume that $\Gamma_l = \mathbf{I}_r$, $\mathbf{B}_l = 0$, $l = 1, \ldots P + \tau_o$. Then for a sufficiently small M and a sufficiently large $P > N$ chosen subject to the condition that rank $[\Gamma \wedge \mathbf{T}_M] = M$, the IMC control law is stable.

Tuning Procedures. The tuning parameters leading to the perfect control law should be tried first. Should the perfect controller be unstable or exhibit rippling of inputs (and ensuing oscillations of outputs between sampling instants), other parameter settings (and, consequently, different approximations of $G_-(Z)^{-1}$) would have to be tried to bring the roots of det $\{D_c(Z)\} = 0$ inside the unit circle, or away from the unit circle, closer to the origin.

Garcia and Morari[6] suggest a computer routine[40] to find the determinant of $D_c(Z)$ by triangularizing it. Then, the largest root of the resulting polynomial in Z may be found by any convenient procedure.

Tuning Guidelines. The following guidelines for the selection of the tuning parameters have been suggested.

Sampling Period, T. If T is sufficiently small, a continuous system with zeroes outside the unit circle produces a $G_-(Z)$ with roots outside the unit circle. These zeroes disappear when T is sufficiently increased. The stability of IMC-based open-loop stable systems is not affected by T. However, rapid sampling produces strong variations of the manipulated variables. On the other hand, frequent sampling is preferable so that the control system can handle frequent disturbances. Thus, an optimal value of T exists that represents a compromise between these two conflicting requirements.

Input Suppression Parameters (M_i). Small M_i, relative to P, produce a stable controller for nonminimum phase systems. For minimum phase systems, extreme excursions of the manipulated variables are reduced by decreasing values of M_i.

Input Penalty Matrices, $\mathbf{B}_l$. For systems with zeroes outside the unit circle, nonsingular and sufficiently large $\mathbf{B}_l$ ensures a stable IMC controller. For open-loop stable processes a sluggish controller response is produced by increasing $\mathbf{B}_l$. In addition, if $D_c(1)^{-1} [N_P(1) + N_F(1)] \neq G_*^*(1)^{-1}$, an offset results, which may be removed by the multivariable offset compensators.

Output penalty parameter, Γ_l. Guidelines for the optimal selection of Γ_l are currently not available. In general terms, individual output responses may be effected directly by the suitable selection of the Γ_l matrices. For example, a large relative weight on a particular output decouples its response from the other outputs. As in the case of SISO systems, time varying Γ_l can improve the responses.

Optimization Horizon, P. Any system with transmission zeroes outside the unit circle can be stabilized by increasing P sufficiently. For open-loop stable systems, the IMC controller becomes more sluggish as P is increased.

Application of the IMC Predictive Algorithm to SISO and MIMO Systems.

Three simulation examples are presented in this section that show the capability and performance of the IMC predictive control algorithm.

Example 8[5]

$$G_p(s) = \frac{Y(s)}{m(s)} \frac{e^{-20s}}{100s^2 + 12s + 1}$$

$T = 10$. Thus, $\tau = 2$.

Determine the closed-loop response of the system to a step change of 0.5 $u(t)$ in d introduced at $t = T$. Use model inverse controller $G_c(Z) = H(Z)^{-1}$ and the various inverse approximations. Use the tenth-order step response model derived from the given transfer function.

The tenth order step response model of the process is

$$y(k + 1) = 0.322\, m(k - 2) + 0.461\, m(k - 3) + 0.255\, m(k - 4)$$

$$+ 0.056\, m(k - 5) - 0.034\, m(k - 6) - 0.043\, m(k - 7)$$

$$- 0.023\, m(k - 8) - 0.004\, m(k - 9) - 0.003\, m(k - 10)$$

$$- 0.002\, m(k - 11)$$

Effect of M. With $\gamma_l = 1$, $\beta_l = 0$, $P = 10$, the model inverse controller $G_c(Z) = H(Z)^{-1}$ is obtained by setting $M = 10$. The results shown in Figure 17.27 indicate perfect set point compensation after $\tau + 1 = 3$ sampling periods but at the expense of severe input oscillations. The use of $M = 2$ and 1 reduces the oscillations, although the set point is not satisfied exactly at the sampling instants.

Effect of Penalty on Input m. With $M = P = 10$, and $\beta_l = 5$, a significant offset is observed, as shown in Figure 17.28. Introducing the offset compensator, a smooth response with good input dynamics is obtained, as shown in Figure 17.28.

Effect of Penalty on Δm. With $\beta_l = 0.5$, a penalty on input *changes* yields a sluggish controller, as shown in Figure 17.29. The choice of $\beta_l = 1.5$ gives better response, also shown in Figure 17.29.

Effect of Reference Trajectory. Figure 17.30(a) shows the closed-

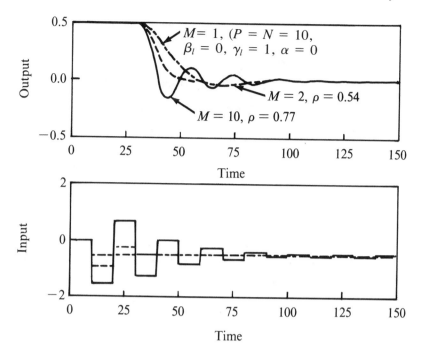

Figure 17.27
Effect of $M < N$ for Example 8
(Reprinted by permission from "Internal Model Control," Garcia and Morari, *Ind. Eng. Chem. Proc. Des. Dev.*, vol. 21, pp. 316-16, 319-20, © 1982, American Chemical Society.)

loop response for increasing values of α. The exponential (discrete) approach to the set point, in contrast to a step change in set point, reduces input amplitudes; however, a smoother response may be obtained by reducing M or by using input penalties.

Effect of Penalty on Outputs. For the second-order system, choosing $P = M = 3$; $\beta_1 = \beta_2 = 0$; $\beta_3 = $ large; $\gamma_1 = 0$; $\gamma_2 = \gamma_3 = 1$ gives excellent response for different reference trajectories, as shown in Figure 17.30(b).

Example 9[5]

For the nonminimum phase system,

$$G_p(s) = \frac{1 - 5s}{(10s + 1)^2}\, e^{-20s}$$

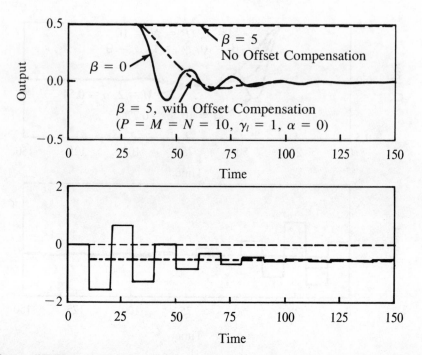

Figure 17.28
Effect of Penalty on $m(k)$ For Example 8
(Reprinted by permission from "Internal Model Control," Garcia and
Morari, *Ind. Eng. Chem. Proc. Des. Dev.*, vol. 21, pp. 316-16, 319-20,
© 1982, American Chemical Society.)

whose step response is shown in Figure 17.31, determine the closed-loop response.

In order to guarantee stability for a sufficiently small M, a sampling period of 10 is chosen, which produces a discrete monotonic response (see Figure 17.31). With $T = 10$ and $N = 10$, the largest root of $H(Z)$ has a magnitude of 3.97 and, therefore, the model inverse controller is unstable. Figure 17.32 shows the stable responses when M is reduced. The authors mention that use of $T = 5$, $M = 1$ results in $h_1 < 0$ and a stable controller. Also for $T = 20$ the magnitude of the largest root of $H(Z)$ reduces to 0.62 and, thus, the exact model inverse controller could be used.

Example 10[6]

This example involves the Wood and Berry binary distillation column[37] to which we applied the factorization concepts in the previous section.

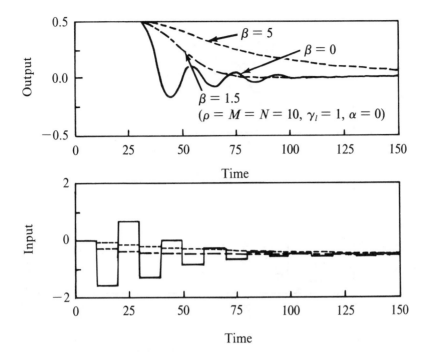

Figure 17.29
Effect of Penalty on $\Delta m(k)$ **For Example 8**
(Reprinted by permission from "Internal Model Control," Garcia and
Morari, *Ind. Eng. Chem. Proc. Des. Dev.*, vol. 21, pp. 316-16, 319-20,
© 1982, American Chemical Society.)

Recall that the diagonal time delay factorization is optimal for this example in terms of the sum of squared errors or in terms of the settling time since $\tau_o = 0$. The diagonal time delay factorization matrix for this system is

$$G_+(Z) = \begin{bmatrix} Z^{-2} & 0 \\ 0 & Z^{-4} \end{bmatrix}$$

which gives perfectly decoupled responses.

In terms of the predictive algorithm the perfect controller for this system involves the settings $P = M_j = 1$; $B_1 = 0$; $\Gamma_1 = I$. Figure 17.33 shows the set point response of the system to a step change in the overhead composition with the perfect controller and an exponential filter with $\alpha_i = 0.8$ ($i = 1, 2$). The perfect controller yields an exponential approach to the set point and perfectly decoupled responses since $F(Z)$ and $G(Z)$ are both diagonal.

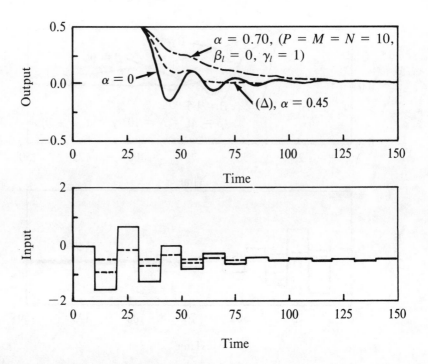

Figure 17.30a
Reference Trajectory For Example 8
(Reprinted by permission from "Internal Model Control," Garcia and Morari, *Ind. Eng. Chem. Proc. Des. Dev.*, vol. 21, pp. 316-16, 319-20, © 1982, American Chemical Society.)

A smooth approach to the set point may also be obtained with an input penalty matrix $\mathbf{B}_l$, instead of the filter matrix, $F(Z)$. Figure 17.33 shows the responses with $P = M_j = 1$; $\Gamma_1 = I$; $B = \text{diag}(1.25, 3.0)$. However, since input moves are restricted, decoupling is lost. An alternative to penalizing the inputs, the number of input moves can be reduced in comparison with the horizon to achieve a similar effect. The results with $P = 10$, $M_j = 2$, $B_i = 0$, $\Gamma_i = I$ shown in Figure 17.33 confirm that the overhead composition reaches the set point fast but, due to the input restriction, closed-loop interactions arise.

This example illustrates that it is possible to shape the closed-loop responses using the various tuning constants of the predictive algorithm. However, this procedure is not recommended unless it is absolutely necessary, for instance, when severe rippling behavior is present or if the inverse is unstable.[6] In all other cases, the use of the filter matrix is suggested. The filter time constant is a much simpler and direct tool to shape the closed-loop response and to dampen severe input oscillations.[6]

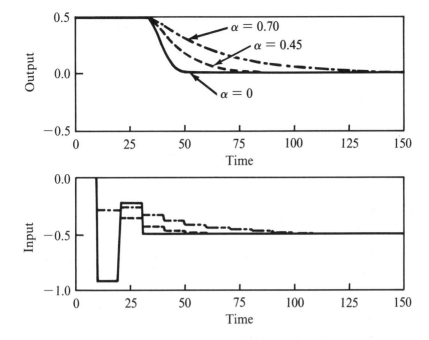

Figure 17.30*b*
Deadbeat Controller with Reference Trajectory for Example 8

Simplified Model Predictive Control

Simplified model predictive control (SMPC) is the last of the model predictive control methods to be presented here. The method,[41,42] which also utilizes easily available step responses for implementation, is easy to understand and implement. It requires a digital computer for implementation, but the memory requirements are modest. Some interaction compensation is inherent in the multivariable version of SMPC, and the robustness issues can be addressed in the design stage. The algorithm also has some constraint-handling capacity that is important in industrial applications.

The block diagram of a multivariable feedback control system is shown in Figure 17.34. Vectors and matrices are indicated by bold letters. The method is illustrated for a 2 × 2 process, although it can be readily extended to processes of higher dimensions. It is assumed that the process is open-loop stable.

The normalized open-loop response of the multivariable system is

$$\mathbf{C} = \mathbf{G(K)}^{-1}\mathbf{M} \qquad (17.101)$$

where $\mathbf{(K)}^{-1}$ is the inverse of process gain matrix.

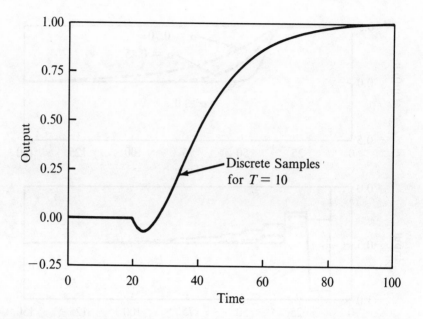

Figure 17.31
Unit Step Response for Example 9
(Reprinted by permission from "Internal Model Control," Garcia and Morari, _Ind. Eng. Chem. Proc. Des. Dev._, vol. 21, pp. 316-16, 319-20, © 1982, American Chemical Society.)

The closed-loop pulse transfer function of the system may be developed as usual by writing out the system of equations indicated in Figure 17.34. The result is

$$\mathbf{C} = (\mathbf{I} + \mathbf{GD})^{-1}\,\mathbf{GDR} \qquad (17.102)$$

The Z-transform operator has been omitted in Equations (17.101) and (17.102) to maintain clarity.

The ratio $\mathbf{C}/\mathbf{R}$ does not exist for multivariable systems since $\mathbf{C}$ and $\mathbf{R}$ are matrices. However, it is still possible to define the closed-loop transfer function matrix as[43]

$$\mathbf{P} = (\mathbf{I} + \mathbf{GD})^{-1}\,\mathbf{GD} \qquad (17.103)$$

Then, the closed-loop response may be evaluated by the equation

$$\mathbf{C} = \mathbf{PR} \qquad (17.104)$$

Similarly, the normalized open-loop transfer function matrix may be defined as

$$\mathbf{Q} = \mathbf{G}(\mathbf{K})^{-1} \qquad (17.105)$$

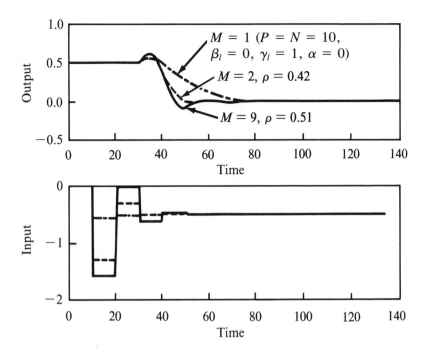

Figure 17.32
Stable Controllers for NMP System for Figure 17.31 by Reducing *M*
(Reprinted by permission from "Internal Model Control," Garcia and
Morari, *Ind. Eng. Chem. Proc. Des. Dev.*, vol. 21, pp. 316-16, 319-20,
© 1982, American Chemical Society.)

Then, the normalized open-loop response may be obtained by the equation

$$\mathbf{C} = \mathbf{QM} \qquad (17.106)$$

Now an argument is made that it should always be possible to design
a control algorithm that will give a set point response having the same
dynamics as that of the open-loop response. The effect of this argument
is that Equations (17.103) and (17.105) may be equated, giving

$$(\mathbf{I} + \mathbf{GD})^{-1}\,\mathbf{GD} = \mathbf{G(K)}^{-1} \qquad (17.107)$$

or

$$\mathbf{D} = (\mathbf{K} - \mathbf{G})^{-1} \qquad (17.108)$$

From Figure (17.34) we may write

$$\mathbf{M} = \mathbf{DE} \qquad (17.109)$$

or in the view of Equation (17.108)

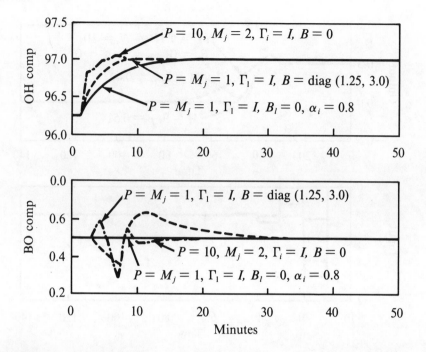

Figure 17.33
Wood/Berry Column. Change in Overhead Composition Set Point
(Reprinted by permission from "Internal Model Control," Garcia and
Morari, *Ind. Eng. Chem. Proc. Des. Dev.*, vol. 24, no. 2, pp. 490–491,
© 1985, American Chemical Society.)

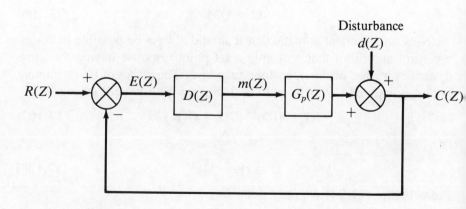

Figure 17.34
Typical Sampled-data Control System

$$\mathbf{M} = (\mathbf{K} - \mathbf{G})^{-1}\mathbf{E} \qquad (17.110)$$

Equation (17.110) may be simplified to obtain

$$\mathbf{M} = (\mathbf{K})^{-1}\mathbf{E} + (\mathbf{K})^{-1}\mathbf{GM} \qquad (17.111)$$

or letting $(K)^{-1} = k$ for convenience

$$\mathbf{M} = \mathbf{kE} + \mathbf{kGM} \qquad (17.112)$$

Introducing the Z-transform operator in Equation (17.112) gives

$$\mathbf{M}(\mathbf{Z}) = \mathbf{kE}(\mathbf{Z}) + \mathbf{kG}(\mathbf{Z})\,\mathbf{M}(\mathbf{Z}) \qquad (17.113)$$

The expanded version of Equation (17.113) for 2×2 systems is

$$M_1(Z) = k_{11}E_1(Z) + k_{12}E_2(Z) + k_{11}[G_{11}(Z)\,M_1(Z) \qquad (17.114a)$$
$$+\ G_{12}(Z)\,M_2(Z)] + k_{12}\,[G_{21}(Z)\,M_1(Z)$$
$$+\ G_{22}(Z)\,M_2(Z)]$$

and

$$M_2(Z) = k_{21}E_1(Z) + k_{22}E_2(Z) + k_{21}\,[G_{11}(Z)\,M_1(Z) \qquad (17.114b)$$
$$+\ G_{12}(Z)\,M_2(Z)] + k_{22}\,[G_{21}(Z)\,M_1$$
$$+\ G_{22}(Z)\,M_2(Z)]$$

The process transfer functions, $G_{ij}(Z)$, can be represented with the aid of impulse response coefficients as

$$G_{ij}(Z) = h_{ij}^1 Z^{-1} + h_{ij}^2 Z^{-2} + \ldots + h_{ij}^N Z^{-N} = \sum_{l=1}^{N} h_{ij}^l Z^{-l} \qquad (17.115)$$

In SMPC the multivariable process is represented by its open-loop step responses. The relationship between the impulse response coefficients and the step response coefficients is

$$h_{ij}^l = a_{ij}^l - a_{ij}^{l-1} \qquad (17.116)$$

Substitution of Equation (17.115) into Equation (17.114) and inversion gives the result

$$M_1^n = k_{11}E_1^n + k_{12}E_2^n + k_{11}\sum_{l=1}^{N}(h_{11}^l M_1^{n-l} + h_{12}^l M_2^{n-l}) \qquad (17.117a)$$
$$+\ k_{12}\sum_{l=1}^{N}(h_{21}^l M_1^{n-l} + h_{22}^l M_2^{n-l})$$

and

$$M_2^n = k_{21}E_1^n + k_{22}E_2^n + k_{21} \sum_{l=1}^{N} (h_{11}^l M_1^{n-l} + h_{12}^l M_2^{n-l})$$

$$+ k_{22} \sum_{l=1}^{N} (h_{21}^l M_1^{n-l} + h_{22}^l M_2^{n-1}) \tag{17.117b}$$

Equations (17.117a) and (17.117b) yield closed-loop responses having open-loop dynamics. The algorithm can be speeded up by introducing a matrix of gains α giving

$$M_1^n = \alpha_{11}E_1^n + \alpha_{12}E_2^n + k_{11} \sum_{l=1}^{N} (h_{11}^l M_1^{n-l} + h_{12}^l M_2^{n-l})$$

$$+ k_{12} \sum_{l=1}^{N} (h_{21}^l M_1^{n-l} + h_{22}^l M_2^{n-l}) \tag{17.118a}$$

and

$$M_2^n = \alpha_{21}E_1^n + \alpha_{22}E_2^n + k_{21} \sum_{l=1}^{N} (h_{11}^l M_1^{n-l} + h_{12}^l M_2^{n-l})$$

$$+ k_{22} \sum_{l=1}^{N} (h_{21}^l M_1^{n-l} + h_{22}^l M_2^{n-l}) \tag{17.118b}$$

Equation (17.118) is the final form off the algorithm. The α's are tuning parameters of the algorithm that are determined by an off-line optimization program such that a suitable performance index is minimized. One possible type of performance index is

$$I = \Sigma (E_1^2 + E_2^2) \tag{17.119}$$

In general, the objective of optimization for set point changes should be to ensure good transient response of one variable consistent with zero offset and zero steady-state error for the other. For load changes the objective would be to ensure fast recovery combined with minimum overshoot and no steady-state error for both variables.

Theoretical Properties of the SMPC Algorithm. The SMPC algorithm has the following important properties:

1. *Open-loop Characteristics.* For first-order plus dead time-type process models, the open-loop step response of the algorithm is similar to that of a PI controller; the controller output, following a step change in error, suddenly increases (or decreases), which is followed by a ramp.

An illustrative plot of the controller output for an illustrative process is shown in Figure 17.35.

2. *No Reset Problem.* An examination of the system equations will show that even in the presence of modeling errors the algorithm guarantees offset-free performance. An illustrative plot of the response of an arbitrary SISO control system to a step change in disturbance, d, in the presence of plant-model gain mismatch is shown in Figure 17.36.

3. *Controller Tuning.* The algorithm contains N^2 parameters, where N is the dimension of the multivariable system. Thus, the technique in its current form is unsuitable for large dimensional systems. For 2×2 and 3×3 systems, the determination of the α matrix does not present any difficulties.

4. *Robustness of the Algorithm.* Since SMPC can be expressed in IMC format, the robustness tools available to IMC are also available to SMPC.

5. *Constraint Handling.* The constraints on inputs and/or outputs are incorporated by suitably modifying the performance index. The suggested procedure is to solve the unconstrained optimization problem first, then examine the resulting output responses and the manipulated variable moves. If constraints are violated, the optimization problem must be solved again with the performance index modified to include appropriate weights on inputs and/or outputs. In the presence of constraints, the performance index, Equation (17.119), may be modified according to

$$I = \Sigma \, (\gamma_1 E_1^2 + \gamma_2 E_2^2 + \beta_1 M_1^2 + \beta_2^2 M_2^2) \qquad (17.120)$$

where

β_i = input penalty parameters

γ_i = weights on outputs

The objective of the optimization effort would be to determine the α matrix subject to the minimization of the function given in Equation (17.120).

6. *Interaction Compensation.* The use of normalized step responses as basis for constrol systems design gives some compensation for interaction.

Applications of SMPC to Multivariable Systems. The algorithm has been tested on three simulated processes: a hot water-cold water blending process,[44] a binary distillation column,[42] and a 3×3 side-draw distillation column.[45] The following paragraphs review the simulation results pertaining to one of these applications.[42]

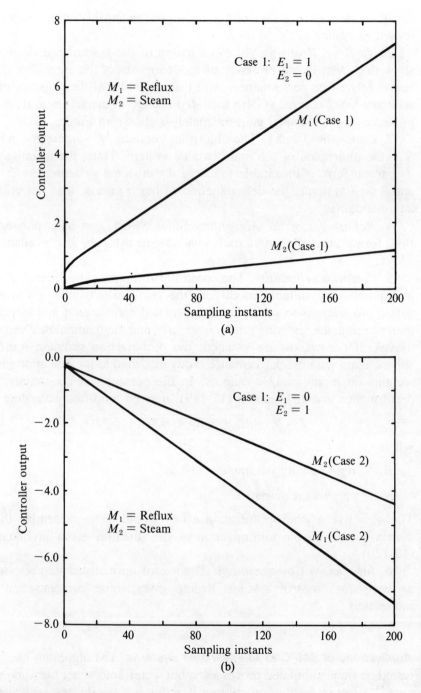

Figure 17.35
Open-loop Response of SMPC Algorithm for Wood-Berry Column
(Arulalan and Deshpande, 1986)

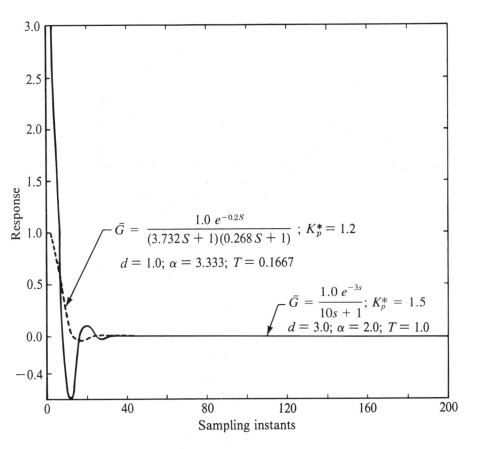

The plot contains the following labels:

$$\tilde{G} = \frac{1.0\,e^{-0.2S}}{(3.732\,S + 1)(0.268\,S + 1)} \; ; \; K_p^* = 1.2$$

$$d = 1.0; \; \alpha = 3.333; \; T = 0.1667$$

$$\tilde{G} = \frac{1.0\,e^{-3s}}{10s + 1}; \; K_p^* = 1.5$$

$$d = 3.0; \; \alpha = 2.0; \; T = 1.0$$

Figure 17.36
**Effect of Plant-model Gain Mismatch on Load Response
of Two SISO Systems**

The simulation example involves a binary distillation column that separates a mixture of methanol and water into two relatively pure products in a column that is equipped with a total condenser and a reboiler.[37] A schematic of the column and control system is shown in Figure 17.37. The goal of the control effort is to maintain the two product compositions at set point in the presence of disturbances. The transfer functions and the steady-state operating data are shown in Table 17.5. The open-loop step responses determined from these transfer functions are used as inputs to SMPC.

The matrix of controller gains α is determined by an off-line optimization program such that the objective function in Equation (17.119) is minimized. The optimization program is based on a modified random

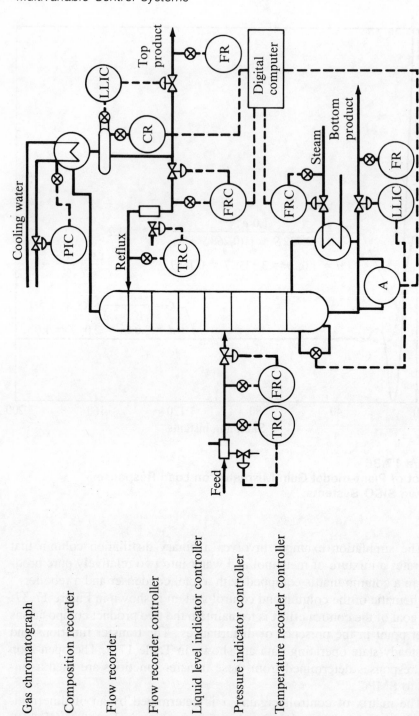

Gas chromatograph

Composition recorder

Flow recorder

Flow recorder controller

Liquid level indicator controller

Pressure indicator controller

Temperature recorder controller

Figure 17.37
Schematic of a Distillation Column

Table 17.5(a)
Column Data
Steady-State Operating Conditions
(Reprinted by permission from *Ind. Eng. Chem. Proc. Des. Dev.*, vol. 26, no. 2, pp. 352–355, © 1987, American Chemical Society.)

Stream	Flow, lb/min	Composition wt % methanol
Distillate	1.18	96
Reflux	1.95	96
Bottoms	1.27	0.5
Feed	2.45	46.5
Steam	1.71	—

Steady-state Temperature Profile

Stream/tray	Temperature, °F
Reflux	151.7
Feed	168.0
Steam	233.0
Condensate	227.5
Reboiler	209.6
Plate 1	203.6
2	194.4
3	181.2
4	172.9
5	164.1
6	156.8
7	152.1
8	148.5
Condenser	143.9

Transfer Functions

$$\begin{bmatrix} X_D(s) \\ X_B(s) \end{bmatrix} = \begin{bmatrix} \dfrac{12.80e^{-2}}{16.7s + 1} & \dfrac{-18.9e^{-3s}}{21s + 1} \\ \dfrac{6.6e^{-7s}}{10.9s + 1} & \dfrac{-19.4e^{-3s}}{14.4s + 1} \end{bmatrix} \begin{bmatrix} R(s) \\ S(s) \end{bmatrix} + \begin{bmatrix} \dfrac{3.8e^{-8.1s}}{14.9s + 1} \\ \dfrac{4.9e^{-3.4s}}{13.2s + 1} \end{bmatrix} F(s)$$

search procedure that is described elsewhere.[46] The process gain matrix and its inverse and the controller gains are shown in Table 17.5(b).

Figures 17.38(a) and 17.38(b) show the responses of the product compositions for a change in the set point of the overhead composition. The manipulated variable moves for this test are shown in Figure 17.38(c). Figure 17.39 presents similar information for a set point change in bottoms composition and Figure 17.40 for a load upset in feed flow to the

Table 17.5(b)
Controller Parameters
Process Gain Matrix and Inverse

$$K = \begin{bmatrix} 12.8 & -18.9 \\ 6.6 & -19.4 \end{bmatrix}$$

$$K^{-1} = \begin{bmatrix} 0.15698 & -0.15294 \\ 0.05341 & -0.10358 \end{bmatrix}$$

Tuning Parameters

Type of change	Magnitude of change	α_{11}	α_{12}	α_{21}	α_{22}	Minimum IAE
Set point (distillate comp.)	1%	0.5004	−0.2907	0.0509	−0.230	7.406
Set point (bottoms comp.)	1%	0.5418	−0.2463	0.1717	−0.2298	8.109
Load (feed flow)	0.34 lb/min	1.0580	−0.0822	−0.1564	−0.2640	7.17

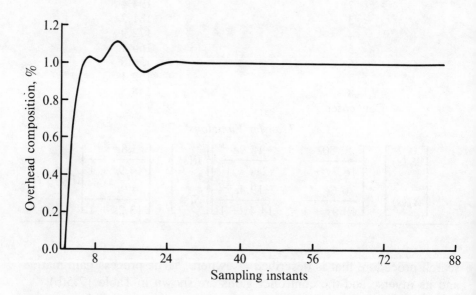

Figure 17.38*a*
Response of Overhead Composition
for a Set Point Change in Overhead Composition
(Reprinted by permission from *Ind. Eng. Chem. Proc. Des. Dev.*, vol. 26, no. 2, pp. 352–355, © 1987, American Chemical Society.)

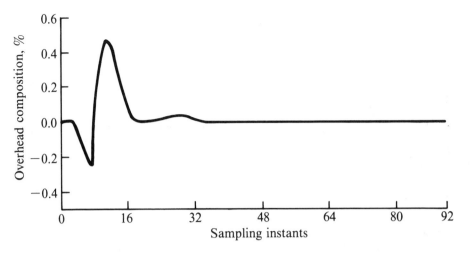

Figure 17.38b
Response of Bottoms Composition
for a Set Point Change in Overhead Composition

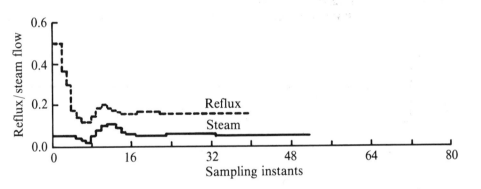

Figure 17.38c
Manipulated Variable Moves Associated
with Overhead Composition Set Point Change

column. The load responses shown in Figure 17.40 are somewhat oscillatory, although the magnitude of the oscillations is small. The oscillations can be reduced, albeit at the expense of ISE, by selecting the elements of the matrix that are smaller than α_{ISE} or by expressing SMPC as an IMC system and then utilizing a diagonal filter network in the feedback path.

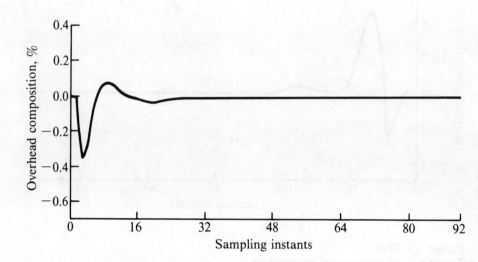

Figure 17.39a
Response of Overhead Composition
for a Set Point Change in Bottoms Composition

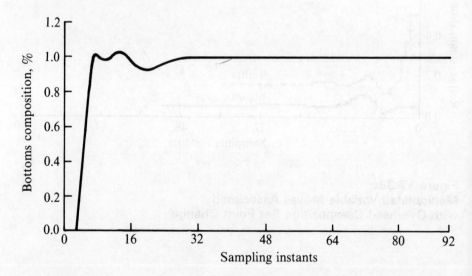

Figure 17.39b
Response of Bottoms Composition
for a Set Point Change in Bottoms Composition

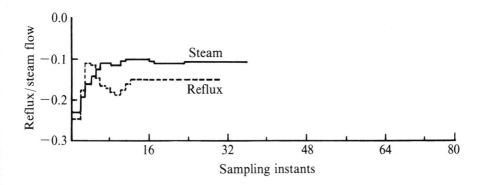

Figure 17.39c
Manipulated Variable Moves Associated with
Bottom Composition Set Point Change

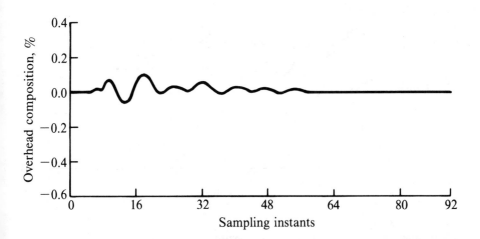

Figure 17.40a
Response of Overhead Composition
for an Upset in Feed Flow Rate
(Reprinted by permission from *Ind. Eng. Chem. Proc. Des. Dev.*, vol.
26, no. 2, pp. 352–355, © 1987, American Chemical Society.)

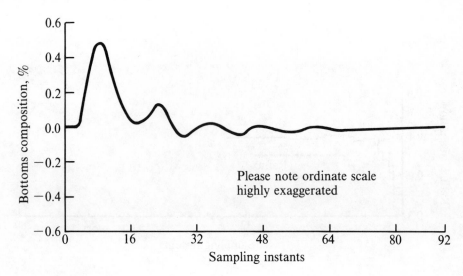

Figure 17.40*b*
Response of Bottoms Composition for an Upset in Feed Flow Rate

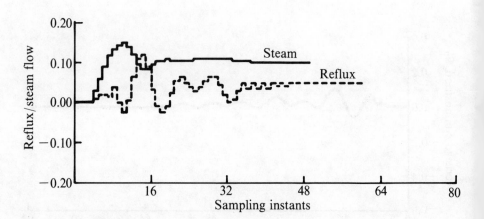

Figure 17.40*c*
**Manipulated Variable Moves Associated
with Feed Flow Disturbance Test**

As we pointed out in Chapter 11, if the matrix **k** is replaced by one
containing the elements $k_{ij} (1 - \beta_{ij}Z)/(1 - \beta_{ij})$, a new algorithm called
Conservative Model-Based Controller[47] results, which has several excellent properties. The details have been left out as an exercise for the
reader.

NOMENCLATURE

a	step response coefficients		transfer function
a_{ij}	closed loop gain defined in Equation (17.9a)	F	detuning factors in BLT tuning procedure
$\mathbf{A}$	dynamic matrix; matrix of step response coefficients	$\tilde{g}_{ij}$	elements of process inverse matrix; defined under Equation (17.59)
$\bar{\mathbf{A}}$	defined in Equation (17.87)	G	process (pulse) transfer function
$\mathbf{b}^T$	vector defined in Appendix A	G_c	IMC controller
$\mathbf{B}$	diagonal matrix $\{\mathbf{B}_l\}$; refer to Appendix A	G_p	transfer function of process
$\mathbf{B}_l$	matrix of input weights, $l = 1, \ldots, (P + \tau_o)$; see Appendix A	G_+	transfer function containing nonminimum phase elements
$\mathbf{B}_M$	diagonal matrix $\{\beta_i\}$	G_-	invertible part of $\tilde{G}_p(Z)$
$\mathbf{C}$	matrix defined in Equation (17.15)	G_{off}	Offset compensator
C_i	controlled variable	$\tilde{G}_p$	transfer function of model
C_{ij}	elements of matrix defined in Equation (17.15)	$\tilde{G}_+$	part of $\tilde{G}_p$ containing delay terms
d	disturbance variable	$\tilde{G}_{+1}$	part of $\tilde{G}_p$ containing zeroes outside unit circle
D	feedback controller algorithm	$\tilde{G}_{-1}$	invertible part of $\tilde{G}_p$
D_c	defined in Equation (17.96)	h	impulse response coefficients
$\mathbf{D}_i$	elements of $\mathbf{b}^T \mathbf{S}\boldsymbol{\psi}$	H	transfer function of the delay-free model
D_{ij}	decoupler transfer functions	H_i	gain of the ith decoupled loop in Equation (17.37)
E	error	$\mathbf{H}_j$	impulse response matrix
$\mathbf{E}_i$	elements of $\mathbf{b}^T \mathbf{S}$	$\mathbf{I}_j$	identity matrix of order j
F	transfer function of desired closed-loop response in model predictive control or filter	J	performance index, cost functional
		$\mathbf{k}$	process gain inverse matrix
		k	current sampling instant

K	process gain matrix
K_c	controller gain
K_{ij}	process gain
K_u	ultimate gain
L_c	closed loop log modulus for SISO systems
L_{c_m}	closed loop log modulus for multiloop systems
M, m	manipulated variable
N	number of sampling periods in settling time
N_F	defined in Equation (17.96)
N_p	defined in Equation (17.96)
P	optimization horizon; horizon of prediction
P_i	eigen values of $\mathbf{KK}^T$ in SVD analysis
$\mathbf{P}_i$	diagonal matrix having P_i as elements
$\mathbf{Q}$	transfer function matrix
R	set point
$\mathbf{S}$	defined in Equation (17.90)
T	sampling period
$\mathbf{T}_M$	matrix of zeroes and one; see Appendix A
$\mathbf{U}$	SVD matrix containing left singular vectors (see Equation 17.23)
$\mathbf{U}$	vector of future inputs
$\mathbf{U}_M$	vector of future inputs
u_i	pseudo-manipulated variable in Equation (17.37)
$\mathbf{V}$	SVD matrix containing right singular vectors (see Equation 17.23)

$\mathbf{V}$	vector of past inputs
y_d	desired output
Y_M	model output
Y	process output
$\mathbf{Y}^+$	defined in Equation (17.84)
Z	Z-transform variable

Greek Letters

λ_{ij}	relative gain
λ_i	i^{th} eigenvalue
$\boldsymbol{\Sigma}$	diagonal matrix of singular values (see Equation 17.23)
σ	elements of matrix of singular values
ω	frequency
ω_{co}	crossover frequency
ϕ_i	elements of first row of $s\Gamma_p\Omega$
ν_i	elements of first row of $s\Gamma_p$
Φ	offset compensator in Equation (17.77)
τ_I	integral time
α_F	filter constant, Equation (17.70a)
γ	weight on output
β	weight on input
α	SMPC tuning parameter
$\bar{\phi}_i$	elements of $S\bar{\Omega}$
$\bar{\nu}_i$	elements of S
Γ	matrix of output weights
ϕ	phase angle in Section 17.2
Λ	matrix of impulse response coefficients (h_i);

	see Appendix A		(H_i); refer to Appendix A
ν_i	zeroes of $\tilde{G}_p$; refer to Equation (17.68)	Ω	defined in Appendix A
π	product operator		
Σ	summation operator		
τ	number of integer sampling periods in dead time	**Superscripts, Subscripts or Arguments**	
τ_o	measure of imbalance in time delays; Equation (17.82)	$\tilde{}$	pertaining to model
		m	pertaining to model
		d	pertaining to set point
$\tau_i^{\min}$	minimum τ_{ij}; defined in Equation (17.81)	k	current sampling instant
τ_i^+	defined in Equation (17.59)	p	optimization horizon or horizon of prediction
$\bar{\tau}_{ij}$	defined under Equation (17.59)	M	input suppression parameter
$\boldsymbol{\Psi}$	matrix of impulse response coefficients	T	transpose of a matrix
		$+$	pertaining to delayed variables

References

1. Edgar, T. J., "Status of Design Methods for Multivariable Control," in *Chemical Process Control,* AIChE Symposium Series, A. S. Foss and Morton M. Denn, Editors, **72,** 159, 1976.
2. Richalet, J., A. Rault, J. L. Testud, and J. Papon, "Model Predictive Heuristic Control: Application to Industrial Processes," *Automatica,* **14,** 1978, p. 413–428.
3. Cutler, C. R. and B. L. Ramaker, *Dynamic Matrix Control—A Computer Control Algorithm,* Paper No. 51b, AIChE 86th National Meeting, April, 1979.
4. Garcia, C. E. and M. Morari, "Internal Model Control 1—A Unifying Review and Some New Results," *Ind. Eng. Chem. Process Des. Dev.,* **21,** 1982, p. 308–323.
5. Garcia, C. E. and M. Morari, "Internal Model Control 2—Design Procedures for Multivariable Systems," *Ind. Eng. Chem. Process Des. Dev.,* **24,** 1985, p. 472–484.
6. Garcia, C. E. and M. Morari, "Internal Model Control 3—Multivariable Control Law Computation and Tuning Guidelines," *Ind. Eng. Chem. Process Des. Dev.,* **24,** 1985, p. 484–494.

7. Mehra, R. K., Chapter on "Model Algorithmic Control," in P. Deshpande, *Distillation Dynamics and Control,* Instrument Society of America Research Triangle Park, NC, 1985.

8. Deshpande, P. B., Ed., *Multivariable Control Methods,* ISA, 1988.

9. Luyben, W. L., "Simple Method for Tuning SISO Controllers in Multivariable Systems," *Ind. Eng. Chem. Process Des. Dev.,* **25,** 3, 1986, p. 654–660.

10. Bristol, E. H., "On a New Measure of Interaction for Multivariable Process Control," *IEEE Transactions on Automatic Control,* AC-11, 1966, p. 133.

11. Bruns, D. D. and C. R. Smith, "Singular Value Analysis: A Geometrical Structure for Multivariable Processes," AIChE Winter Meeting, 1982.

12. Lau, H., et al., "Synthesis of Control Structures by Singular Value Analysis: Dynamic Measures of Sensitivity and Interaction," *AIChE J.,* **31,** 3, 1985, p. 427–439.

13. Economou, C. G. and M. Morari, "Internal Model Control: 6. Multiloop Design," *Ind. Eng. Chem. Process Des. Dev.,* **25,** 2, 1986, p. 411–419.

14. Rijnsdorp, J. E., "Interaction in Two Variable Control Systems for Distillation Columns," *Automatica,* **1,** 1965, p. 15.

15. Tung, L. S. and T. F. Edgar, "Dynamic Interaction Index and Its Application to Distillation Column Control," 16th IEEE Conf. Decision and Control, New Orleans, **1,** December, 1977, p. 107.

16. Tung, L. S. and T. F. Edgar, "Analysis of Control-Output Interactions in Dynamic Systems," 71st Annual AIChE Meeting, Miami, FL, November 1978.

17. Tung, L. S. and T. F. Edgar, "Analysis of Control-Output Interactions in Dynamic Systems," *AIChE J.,* **27,** 4, 1981, p. 690.

18. Gagnepain, J. P. and D. E. Seborg, "An Analysis of Process Interactions with Applications to Multiloop Control Systems Design," 72nd Annual AIChE Meeting, San Francisco, November, 1979.

19. Jaaksoo, O., "Interaction Analysis in Multivariable Control Systems Design," 2nd IFAC/IFIP Symposium on Software for Complete Control, Prague, A-XXV-1 (1979).

20. Rosenbrock, H. H., *State Space and Multivariable Theory,* John Wiley and Sons, N.Y., 1979.

21. Rosenbrock, H. H. and C. Storey, *Mathematics of Dynamical Systems,* John Wiley and Sons, N.Y., 1970.

22. Rosenbrock, H. H., *Computer-Aided Control Systems Design,* Academic Press, 1974.

23. Mijares, G., et al., "A New Criterion for the Pairing of Control and Manipulated Variables," *AIChE J.,* **32,** 9, 1986, p. 1439–1449.

24. Jensen, N., D. G. Fisher, and S. L. Shah, "Interaction Analysis in Multivariable Control Systems," *AIChE J.,* **32,** 6, 1986, p. 959–970.

25. Yu, C. C. and W. L. Luyben, "Design of Multiloop SISO Controllers in Multivariable Processes," *Ind. Eng. Chem. Process Des. Dev.*, **25**, 2, 1986, p. 498–503.
26. Nisenfeld, A. E. and H. M. Schultz, "Interaction Analysis Applied to Control Systems Design," *Instrumentation Technology*, April 1971.
27. Desphande, P. B., W. L. S. Laukhuf, and N. G. Patke, "Advanced Process Control Experiments," *Chemical Engineering Education*, XIV, 2, 1980.
28. Niederlinski, A., "A Heuristic Approach to the Design of Linear Multivariable Control Systems," *Automatica*, **7**, 1971, p. 691.
29. McAvoy, T. J., *Interaction Analysis*, ISA, 1983.
30. Moore, C. F., *Singular Value Decomposition in Multivariable Control Methods*, Deshpande, Ed., ISA, 1988.
31. Smith, C. R., C. F. Moore and D. D. Bruns, "A Structural Framework for Multivariable Control Applications," Proceedings of Joint American Conference, 1981.
32. Doss, J., Private Communication, Tennessee Eastman Co., Kingsport, TN, 1986.
33. Brosilow, C. B., "The Structure and Design of Smith Predictors from the Viewpoint of Inferential Control," JACC, Denver, CO, June 1979.
34. Holt, B. R. and M. Morari, "Design of Resilient Processing Plants— VI. The Effect of Right Half Plane Zeros on Dynamic Resilience," *Chem. Eng. Sc.*, **40**, 1, 1985(a), p. 59–74.
35. Holt, B. R. and M. Morari, "Design of Resilient Processing Plants— V. The Effect of Deadtime on Dynamic Resilience," *Chem. Eng. Sc.*, **40**, 7, 1985(b), p. 1229–1237.
36. Morari, M., Private Communication, Dept. of Chemical Engineering, California Institute of Technology, 1986.
37. Wood, R. K. and M. W. Berry, "Terminal Composition Control of Binary Distillation Column," *Chemical Engineering Science*, **28**, 1973, p. 1707.
38. Foss, A. S., S. M. Edmunds, and B. Kouvaritakis, "Multivariable Control Systems for Two-Bed Reactors by the Characteristic Locus Method," *Ind. Eng. Chem. Fundamentals*, **19**, 1980, p. 109.
39. Garcia, C. E., "Studies in Optimizing and Regulatory Control of Chemical Processing Systems," Ph.D. Thesis, University of Wisconsin, 1982.
40. Kucera, V., *Discrete Linear Control*, John Wiley & Sons, New York, 1979.
41. Tu, F. C. Y. and J. Y. H. Tsing, "Synthesizing a Digital Algorithm for Optimized Control," InTech., May 1979.
42. Arulalan, G. R. and P. B. Deshpande, "Simplified Model Predictive Control," *I&EC Proc. Des. Dev.*, **25**, 2, 1987.
43. Kuo, B. C., *Multivariable Control Systems*, John Wiley & Sons, New York, 1983, p. 81.

44. Arulalan, G. R. and P. B. Deshpande, "Synthesizing a Multivariable Prediction Algorithm for Noninteracting Control," American Control Conference, Seattle, Washington, June 1986(a).
45. Arulalan, G. R. and P. B. Deshpande, "A New Algorithm for Multivariable Control," *Hydrocarbon Processing*, June, 1986.
46. Ralston, P. A. S., K. R. Watson, A. A. Patwardhan, and P. B. Deshpande, "A Computer Algorithm for Optimized Control," *Ind. Eng. Chem. Process Des. Dev.*, **24**, 4, 1985, p. 1132–1136.
47. Chawla, V. K., "An Expert System for Multivariable Control Systems Design," Ph.D. Thesis, University of Louisville, Louisville, KY, 1988.

Appendix A

Matrix Definitions

I. Internal Model Control (Predictive Formulation) for SISO Systems

$$s = b^T (T_M^T \Lambda^T R_p^T \Gamma_p^2 R_p \Lambda T_M + B_M^2)^{-1} T_M^T \Lambda^T R_p^T \Gamma_p^T \tag{A.1}$$

$$\Gamma_p = \text{diagonal}\ (\gamma_1, \gamma_2, \ldots, \gamma_p) \quad \Gamma_p = P \times P\ \text{matrix} \tag{A.2}$$

$$\Omega = [I_p \ \vdots \ 0]
\begin{bmatrix}
h_2 & h_3 & \ldots & h_n \\
h_3 & h_4 & \ldots h_N & 0 \\
\cdot & & & \\
\cdot & & & \\
h_N & 0 & \ldots & 0
\end{bmatrix}
\ ;
\begin{array}{l}
\Omega = P \times (N - 1)\ \text{matrix} \\
I_P = P \times P\ \text{matrix}
\end{array}
\tag{A.3}$$

$$T_M =
\begin{bmatrix}
\begin{array}{c|c}
\multicolumn{1}{c}{} & I_M \\ \hline
& 1 \\
0 & 1 \\
& \cdot \\
& \cdot \\
& 1
\end{array}
\end{bmatrix}
\quad
\begin{array}{l}
T_M = N \times M\ \text{matrix} \\
I_M = M \times M\ \text{matrix}
\end{array}
\tag{A.4}$$

$$\Lambda = \begin{bmatrix} h_1 & 0 & \ldots & 0 \\ h_2 & h_1 & \ldots & 0 \\ \vdots & & & \\ h_N & h_{N-1} & \ldots & h_1 \\ 0 & h_N & \ldots & (h_1 + h_2) \\ \vdots & & & \\ 0 & 0 & \ldots & \sum_{i=1}^{N} h_i \end{bmatrix} \qquad \Lambda = (2N - 1) \times N \text{ matrix} \qquad (A.5)$$

$$R_p = [I_p \ \vdots \ O] \quad R_p = P \times (2N - 1) \text{ matrix} \qquad (A.6)$$

$$B_M = \text{diagonal} (\beta_1, \beta_2, \ldots \beta_M) \quad B_M = M \times M \text{ matrix} \qquad (A.7)$$

$$b^T = [1 \quad 0 \quad 0 \quad \ldots \quad 0] \quad b^T = 1 \times (M - 1) \text{ matrix} \qquad (A.8)$$

II. Dynamic Matrix Control (SISO Systems)

$$S = b^T (A^T \Gamma_p^2 A + B_M^2)^{-1} A^T \Gamma_p^2 \qquad (A.9)$$

$$B_M = \text{diagonal} (\beta_1, \beta_2, \ldots, \beta_M) \quad B_M = M \times M \text{ matrix} \qquad (A.10)$$

$$b^T = [1 \quad 0 \quad 0 \quad \ldots \quad 0] \quad b^T = 1 \times (M - 1) \text{ matrix} \qquad (A.11)$$

$$\Gamma_p = \text{diagonal} (\gamma_1, \gamma_2, \ldots \gamma_p) \quad \Gamma_p = P \times P \text{ matrix} \qquad (A.12)$$

$$A = \begin{bmatrix} a_1 & 0 & \ldots & 0 \\ a_2 & a_1 & \ldots & 0 \\ \vdots & & & \\ a_M & a_{M-1} & \ldots & a_1 \\ a_p & a_{p-1} & \ldots & a_{p-M+1} \end{bmatrix} \qquad A = P \times M \text{ matrix} \qquad (A.13)$$

$$\bar{\Omega} = \begin{bmatrix} 1 & 0 & \ldots & 0 \\ 1 & 1 & \ldots & 0 \\ \vdots & & & \vdots \\ 1 & 1 & \ldots & 1 \end{bmatrix} \ 0 \ \begin{bmatrix} h_2 & h_3 & \ldots & h_N \\ h_3 & h_4 & \ldots & 0 \\ \vdots & & & \vdots \\ h_N & 0 & & 0 \end{bmatrix} \qquad (A.14)$$

$$\bar{\Omega} = P \times (N - 1) \text{ matrix}$$

III. Internal Model Control (Multivariable Predictive Formulation)

$$d^+(k + l) = d^+(k) = Y^+(k) - Y_M^+(k)$$

$$\Lambda = \begin{bmatrix} \mathbf{H}_1 & 0 & \ldots\ldots & 0 \\ \mathbf{H}_2 & \mathbf{H}_1 & \ldots\ldots & 0 \\ \vdots & \vdots & & \vdots \\ \mathbf{H}_{\tau_0+1} & \mathbf{H}_{\tau_0} & \ldots \mathbf{H}_1 \ldots & 0 \\ \vdots & \vdots & & \vdots \\ \mathbf{H}_{\tau_0+N} & \mathbf{H}_{\tau_0+N-1} & \ldots \mathbf{H}_1 & 0 \\ 0 & \ldots \mathbf{H}_{\tau_0+N} & \ldots \mathbf{H}_2 & \mathbf{H}_1 \end{bmatrix} \qquad \gamma(P + \tau_0) \times \gamma(P + \tau_0)$$

$$\mathbf{H}_j = \begin{bmatrix} h_{11}^{\tau_1^+ - \tau_{11} - \tau_0 + j} & \cdots & h_{1r}^{\tau_1^+ - \tau_{11} - \tau_0 + j} \\ \vdots & & \\ h_{r1}^{\tau_\gamma^+ - \tau_{\gamma 1} - \tau_0 + j} & \cdots & h_{r\gamma}^{\tau_\gamma^+ - \tau_\gamma - \tau_0 + j} \end{bmatrix}$$

$$\boldsymbol{\Psi} = \begin{bmatrix} \mathbf{H}_2 & \mathbf{H}_3 & \cdots & & & \mathbf{H}_{\tau_0 + N} \\ \mathbf{H}_3 & \mathbf{H}_4 & \cdots & \mathbf{H}_{\tau_0 + N} & & 0 \\ \vdots & & & & & \\ \mathbf{H}_{\tau_0 + 2} & \cdots & \mathbf{H}_{\tau_0 + N} & 0 & \cdots & 0 \\ \vdots & & & & & \\ \mathbf{H}_{\tau_0 + N} & 0 & \cdots & 0 & \cdots & 0 \\ \vdots & & & & & \\ 0 & 0 & \cdots & 0 & \cdots & 0 \end{bmatrix}$$

$$\gamma(P + \tau_0) \times \gamma(N + \tau_0 - 1)$$

$$\mathbf{U}(\bar{k}) = \mathbf{T}_M \mathbf{U}_M(\bar{k})$$

$$\boldsymbol{\epsilon}_p(\bar{k} + 1) = \begin{bmatrix} \epsilon(\bar{k} + 1 - \tau_0) \\ \vdots \\ \epsilon(\bar{k} + P) \end{bmatrix} \qquad \text{where } \mathbf{U}(\bar{k}) = \begin{bmatrix} m(\bar{k}) \\ \vdots \\ m(\bar{k} + \tau_0 + P - 1) \end{bmatrix}$$

$$\mathbf{V}(\bar{k} - 1) = \begin{bmatrix} m(\bar{k} - 1) \\ \vdots \\ m(\bar{k} - N - \tau_0 + 1) \end{bmatrix}$$

T_M is a matrix of zeroes and one,

$$r(P + \tau_0) \times M \text{ and } M = \sum_{i=1}^{r} M_i$$

$$\mathbf{B} = \text{diag } \{\mathbf{B}_1 \ldots \mathbf{B}_{P + \tau_0}\}$$
$$\boldsymbol{\Gamma} = \text{diag}\{\boldsymbol{\Gamma}_1 \ldots \boldsymbol{\Gamma}_{P + \tau_0}\}$$
$$\mathbf{B}_l = \text{diag}\{\beta_l^1, \beta_l^2, \ldots \beta_l^r\}$$
$$\boldsymbol{\Gamma}_l = \text{diag}\{\gamma_l^1 \gamma_l^2 \ldots \gamma_l^r\}$$

IV. Table of Z Transforms

Reprinted with Permission from B.C. Kuo's *Analysis and Synthesis of Sampled-data Control Systems*, Prentice-Hall, 1963.

Laplace transform $E(s)$	Time function $e(t)$	z-transform $E(z)$	Modified z-transform $E(z,m)$
1	$\delta(t)$	1	0
e^{-nTs}	$\delta(t-nT)$	z^{-n}	z^{-n-1+m}
$\dfrac{1}{s}$	$u(t)$	$\dfrac{z}{z-1}$	$\dfrac{1}{z-1}$
$\dfrac{1}{s^2}$	t	$\dfrac{Tz}{(z-1)^2}$	$\dfrac{mT}{z-1}+\dfrac{T}{(z-1)^2}$
$\dfrac{2!}{s^3}$	t^2	$\dfrac{T^2 z(z+1)}{(z-1)^3}$	$T^2\left[\dfrac{m^2}{z-1}+\dfrac{2m+1}{(z-1)^2}+\dfrac{2}{(z-1)^3}\right]$
$\dfrac{(n-1)!}{s^n}$	t^{n-1}	$\lim_{a\to 0}(-1)^{n-1}\dfrac{\partial^{n-1}}{\partial a^{n-1}}\left(\dfrac{z}{z-e^{-aT}}\right)$	$\lim_{a\to 0}(-1)^{n-1}\dfrac{\partial^{n-1}}{\partial a^{n-1}}\left(\dfrac{e^{-amT}}{z-e^{-aT}}\right)$
$\dfrac{1}{s+a}$	e^{-at}	$\dfrac{z}{z-e^{-aT}}$	$\dfrac{e^{-amT}}{z-e^{-aT}}$
$\dfrac{1}{(s+a)(s+b)}$	$\dfrac{1}{b-a}(e^{-at}-e^{-bt})$	$\dfrac{1}{b-a}\left(\dfrac{z}{z-e^{-aT}}-\dfrac{z}{z-e^{-bT}}\right)$	$\dfrac{1}{b-a}\left(\dfrac{e^{-amT}}{z-e^{-aT}}-\dfrac{e^{-bmT}}{z-e^{-bT}}\right)$
$\dfrac{1}{s(s+a)}$	$\dfrac{1}{a}(u(t)-e^{-at})$	$\dfrac{1}{a}\left(\dfrac{(1-e^{-aT})z}{(z-1)(z-e^{-aT})}\right)$	$\dfrac{1}{a}\left(\dfrac{1}{z-1}-\dfrac{e^{-amT}}{z-e^{-aT}}\right)$
$\dfrac{1}{s^2(s+a)}$	$\dfrac{1}{a}\left(t-\dfrac{1-e^{-at}}{a}\right)$	$\dfrac{1}{a}\left[\dfrac{Tz}{(z-1)^2}-\dfrac{(1-e^{-aT})z}{a(z-1)(z-e^{-aT})}\right]$	$\dfrac{1}{a}\left[\dfrac{T}{(z-1)^2}+\dfrac{amT-1}{a(z-1)}+\dfrac{e^{-amT}}{a(z-e^{-aT})}\right]$
$\dfrac{(s+b)}{s^2(s+a)}$	$\dfrac{a-b}{a^2}u(t)+\dfrac{b}{a}t+\dfrac{1}{a}\left(\dfrac{b}{a}-1\right)e^{-at}$	$\dfrac{1}{a}\left[\dfrac{bTz}{(z-1)^2}+\dfrac{(a-b)z}{a(z-1)}+\dfrac{(1-e^{-aT})z}{a(z-1)(z-e^{-aT})}\right]$	$\dfrac{1}{a}\left[\dfrac{bT}{(z-1)^2}+\left(bmT+1-\dfrac{b}{a}\right)\dfrac{1}{z-1}+\dfrac{b-a}{a}\dfrac{e^{-amT}}{z-e^{-aT}}\right]$
$\dfrac{1}{s(s+a)(s+b)}$	$\dfrac{1}{ab}\left(u(t)+\dfrac{b}{a-b}e^{-at}-\dfrac{a}{a-b}e^{-bt}\right)$	$\dfrac{1}{ab}\left[\dfrac{z}{z-1}+\dfrac{bz}{(a-b)(z-e^{-aT})}-\dfrac{az}{(a-b)(z-e^{-bT})}\right]$	$\dfrac{1}{ab}\left[\dfrac{1}{z-1}+\dfrac{be^{-amT}}{(a-b)(z-e^{-aT})}-\dfrac{ae^{-bmT}}{(a-b)(z-e^{-bT})}\right]$
$\dfrac{1}{(s+a)^2}$	te^{-at}	$\dfrac{Tze^{-aT}}{(z-e^{-aT})^2}$	$\dfrac{Te^{-amT}[e^{-aT}+m(z-e^{-aT})]}{(z-e^{-aT})^2}$

$f(s)$	$f(t)$	Z-transform	Modified Z-transform
$\dfrac{1}{s^3(s+a)}$	$\dfrac{1}{2a}\left(t^2 - \dfrac{2}{a}t + \dfrac{2}{a^2}u(t) - \dfrac{2}{a^2}e^{-at}\right)$	$\dfrac{1}{a}\left[\dfrac{T^2z}{(z-1)^3} + \dfrac{(aT-2)Tz}{2a(z-1)^2} + \dfrac{z}{a^2(z-1)} - \dfrac{z}{a^2(z-e^{-aT})}\right]$	$\dfrac{1}{a}\left[\dfrac{T^2z}{(z-1)^3} + \dfrac{T^2(m+\frac{1}{2}) - T/a}{(z-1)^2} + \dfrac{(amT)^2/2 - amT + 1}{a^2(z-1)} - \dfrac{e^{-amT}}{a^2(z-e^{-aT})}\right]$
$\dfrac{a}{s^2+a^2}$	$\sin at$	$\dfrac{z\sin aT}{z^2 - 2z\cos aT + 1}$	$\dfrac{z\sin amT + \sin(1-m)aT}{z^2 - 2z\cos aT + 1}$
$\dfrac{s}{s^2+a^2}$	$\cos at$	$\dfrac{z(z-\cos aT)}{z^2 - 2z\cos aT + 1}$	$\dfrac{z\cos amT - \cos(1-m)aT}{z^2 - 2z\cos aT + 1}$
$\dfrac{a}{s^2-a^2}$	$\sin h\,at$	$\dfrac{z\sin haT}{z^2 - 2z\cos haT + 1}$	$\dfrac{z\sin hamT + \sin h(1-m)aT}{z^2 - 2z\cos haT + 1}$
$\dfrac{s}{s^2-a^2}$	$\cos h\,at$	$\dfrac{z(z-\cos haT)}{z^2 - 2z\cos haT + 1}$	$\dfrac{z\cos hamT - \cos h(1-m)aT}{z^2 - 2z\cos haT + 1}$
$\dfrac{a}{s(s^2+a^2)}$	$\dfrac{1}{a}(u(t) - \cos at)$	$\dfrac{1}{a}\left[\dfrac{z}{z-1} - \dfrac{z(z-\cos aT)}{z^2 - 2z\cos aT + 1}\right]$	$\dfrac{1}{a}\left[\dfrac{1}{z-1} - \dfrac{z\cos amT - \cos(1-m)aT}{z^2 - 2z\cos aT + 1}\right]$
$\dfrac{a^2}{s^2(s^2+a^2)}$	$t - \dfrac{1}{a}\sin at$	$\dfrac{Tz}{(z-1)^2} - \dfrac{1}{a}\dfrac{z\sin aT}{z^2 - 2z\cos aT + 1}$	$\dfrac{mT}{z-1} + \dfrac{T}{(z-1)^2} - \dfrac{z\sin amT + \sin(1-m)aT}{a(z^2 - 2z\cos aT + 1)}$
$\dfrac{1}{s(s+a)^2}$	$\dfrac{1}{a^2}[u(t) - (1+at)e^{-at}]$	$\dfrac{1}{a^2}\left[\dfrac{z}{z-1} - \dfrac{z}{z-e^{-aT}} - \dfrac{aTe^{-aT}z}{(z-e^{-aT})^2}\right]$	$\dfrac{1}{a^2}\left[\dfrac{1}{z-1} - \left(\dfrac{1+amT}{z-e^{-aT}} + \dfrac{aTe^{-aT}}{(z-e^{-aT})^2}\right)e^{-amT}\right]$
$\dfrac{1}{s^2(s+a)^2}$	$\dfrac{t}{a^2} - \dfrac{2}{a^3}u(t) + \left(\dfrac{t}{a^2} + \dfrac{2}{a^3}\right)e^{-at}$	$\dfrac{1}{a^3}\left[\dfrac{(aT+2)z - 2z^2}{(z-1)^2} + \dfrac{2z}{z-1} + \dfrac{aTe^{-aT}z}{(z-e^{-aT})^2}\right]$	$\dfrac{1}{a^3}\left[\dfrac{aT}{(z-1)^2} + \dfrac{amT-2}{z-1} + \left(\dfrac{aTe^{-aT}}{(z-e^{-aT})^2} - \dfrac{amT-2}{z-e^{-aT}}\right)e^{-amT}\right]$
$\dfrac{1}{(s+a)^2+b^2}$	$\dfrac{1}{b}e^{-at}\sin bt$	$\dfrac{1}{b}\left(\dfrac{ze^{-aT}\sin bT}{z^2 - 2ze^{-aT}\cos bT + e^{-2aT}}\right)$	$\dfrac{1}{b}\dfrac{e^{-amT}[z\sin bmT + e^{-aT}\sin(1-m)bT]}{z^2 - 2ze^{-aT}\cos bT + e^{-2aT}}$
$\dfrac{s+a}{(s+a)^2+b^2}$	$e^{-at}\cos bt$	$\dfrac{z^2 - ze^{-aT}\cos bT}{z^2 - 2ze^{-aT}\cos bT + e^{-2aT}}$	$\dfrac{e^{-amT}[z\cos bmT + e^{-aT}\sin(1-m)bT]}{z^2 - 2ze^{-aT}\cos bT + e^{-2aT}}$

Table of Z Transforms and Modified Z Transforms (continued)

Laplace transform $E(s)$	Time function $e(t)$	z-transform $E(z)$	Modified z-transform $E(z, m)$
$\dfrac{1}{s[(s+a)^2 + b^2]}$	$\dfrac{1}{a^2+b^2}[1 - e^{-at}\sec\phi\cos(bt+\phi)]$ $\phi = \tan^{-1}\left(\dfrac{-a}{b}\right)$	$\dfrac{1}{a^2+b^2}\left[\dfrac{z}{z-1} - \dfrac{z^2 - ze^{-aT}\sec\phi\cos(bT-\phi)}{z^2 - 2ze^{-aT}\cos bT + e^{-2aT}}\right]$	$\dfrac{1}{a^2+b^2}\left[\dfrac{1}{z-1} - \dfrac{e^{-amT}\sec\phi\{z\cos(bmT+\phi)}{z^2 - 2ze^{-aT}\cos bT + e^{-2aT}} \\ \qquad - e^{-aT}\cos[(1-m)bT - \phi]\}\right]$

Some Useful Properties and Theorems of Z Transforms

1. $Z\{f(t - kT)\} = Z^{-k}F(Z)$

2. $Z\{f_1(t) + f_2(t)\} = F_1(Z) + F_2(Z)$

3. $Z\{e^{-at}f(t)\} = \sum_{n=0}^{\infty} f(nT)\, Z^{-n}\, e^{-anT}$

4. $\displaystyle\lim_{t\to\infty} f(t) = \lim_{Z\to 1}\left(\frac{Z-1}{Z}F(Z)\right)$

5. $\displaystyle\lim_{t\to 0} f(t) = \lim_{Z\to\infty} F(Z)$

Appendix B

Exercises

Chapter 1   Review of Conventional Process Control

1. For the three transfer functions given show that K_p is indeed the steady-state gain for a unit step disturbance.

a. First-order: $G(s) = \dfrac{K_p}{\tau s + 1}$

b. Second-order overdamped: $G(s) = \dfrac{K_p}{(\tau_1 s + 1)(\tau_2 s + 1)}$

c. Second-order underdamped: $G(s) = \dfrac{K_p}{\tau^2 s^2 + 2\zeta\tau s + 1}$

Hint: Use the final value theorem.

2. (a) Determine the transient closed-loop response of the process shown in the following block diagram to a change in set point $R(t) =$

$5(1 - e^{-t/2})$. Design the PID controller using Bode plot and the Ziegler–Nichols method.

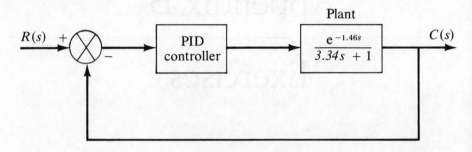

Hint: Represent the dead time by the fourth-order Pade' approximation.

$$e^{-\tau s} = \frac{e^{-\tau s/2}}{e^{\tau s/2}} \approx \frac{1 - \tau s/2 + \tau^2 s^2/12 - \tau^3 s^3/120 + \tau^4 s^4/1680}{1 + \tau s/2 + \tau^2 s^2/12 + \tau^3 s^3/120 + \tau^4 s^4/1680}$$

(b). Simulate the system shown in part (a) on the analog computer using the fourth-order Pade' approximation and the Ziegler–Nichols tuning constants and obtain the transient response to the same change in set point. If computation and simulation are done correctly, the results of parts (a) and (b) should agree.

3. Two process-control systems are to be tuned experimentally. The derivative mode on the controller is turned off, and the integral mode is set at the lowest setting. The gain of the controller is gradu-

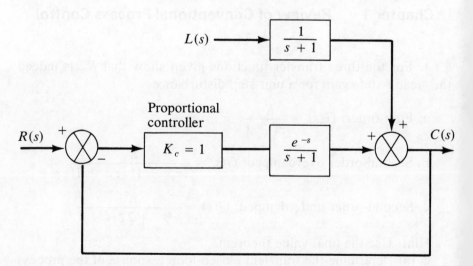

Hint: Use second-order Pade' approximation for dead time.
Figure B1. Block Diagram for Problem No. 4

ally increased (proportional band decreased) until the controlled variable begins to oscillate with constant amplitude. The final gain setting and the period of oscillation of each system is given below. Determine the two- and three-mode controller settings for each system.

System	Ultimate Gain	Ultimate Period
1	0.4	20
2	2.0	0.5

4. Determine the transient response of the controlled variable shown in the block diagram, Figure B1, to a unit step change in load L

Hint: Use second-order Pade' approximation for dead time.

5. For the overdamped second-order plus dead-time process shown in the following block diagram, how much better will the control be if the dead time were halved? Answer in terms of allowable controller gains and the associated step responses

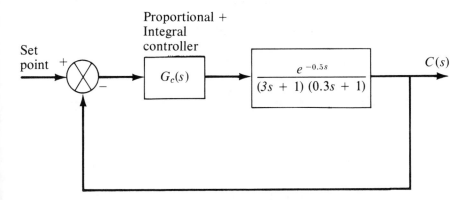

6. For the block diagram shown in the following figure (a) prepare the frequency response diagram and determine the ultimate gain and Ziegler–Nichols setting for the proportional controller; (b) apply Routh Criteria and determine the ultimate gain; and (c) determine if the closed-loop system would be stable if K_c were equal to 6.

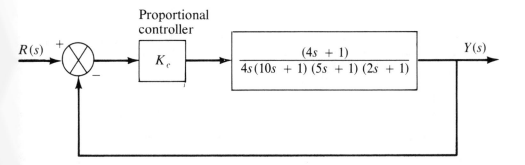

7. The block diagram of an open-loop furnace control system is shown below. Approximate this system by a first-order lag with dead time. Compare the step response of the process with that of the approximating model.

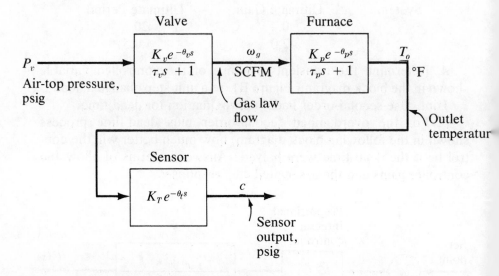

Parameters: $K_v = 100 \dfrac{\text{SCFM}}{\text{psig}}$; $K_p = 2.5 \dfrac{°\text{F}}{\text{SCFM}}$;

$K_T = 0.005 \dfrac{\text{psig}}{°\text{F}}$;

$\theta_v = 0.02$ sec;

$\theta_T = 0.13$ sec;

$\theta_p = 0.15$ sec;

$\tau_p = 0.5$ sec

$\tau_v = 0.088$ sec;

8. The block diagram of a heat exchanger control system is shown in the following block diagram

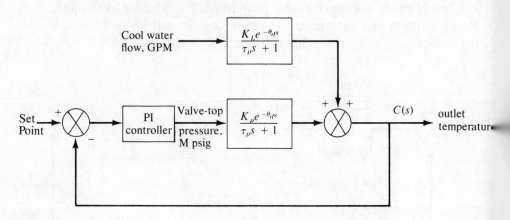

$$K_p = 2 \text{ psig/psig}; \qquad K_L = 0.5 \text{ psig/gpm}$$

Case No.	τ_p min	θ_d g_{pm}	ΔL gpm
1	2.0	1.0	+10
2	5.0	1.0	− 5
3	3.0	3.0	+ 3
4	10.0	0	−30
5	10.0	7.5	−10

(a) For each of the five cases above find
(1) Ziegler–Nichols controller gain K_c
(2) Cohen–Coon controller gain K_c
(3) ΔC_{peak} in psig if Ziegler–Nichols and Cohen–Coon controller settings are used
(4) Period of oscillation of response Pu
(5) settling time TS
(6) time to first peak TP
(b) If the controller gain is cut in half, the overshoot to a step-load disturbance, when a proportional-only controller is used, is reduced from about 50% to about 15%. When integral mode is added, the peak is about 10% higher than with proportional-only controller. For the first case, PI control, estimate ΔC_{peak} if K_c were set at half the Ziegler–Nichols gain.

9. Show that the block diagram of the control system shown in the first figure will reduce to that shown in the second figure if the load transfer function $G_L(s)$ equals the process transfer function $G_p(s)$.

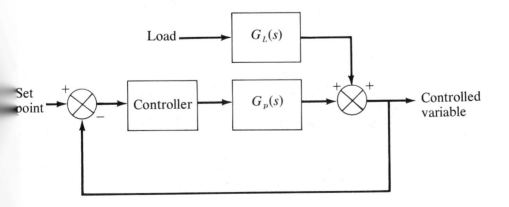

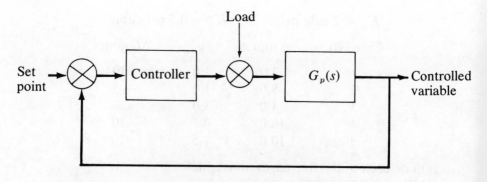

Chapter 5   *Z* Transforms

1. Determine the Z transform of the following functions by both the method of residues and from the definition of Z transforms (the first part will require outside reading).

(a) $\cos \beta t$

(b) $\dfrac{1}{s(s+1)}$

(c) $\dfrac{s}{(s+1)(s+2)}$

(d) $\dfrac{s+2}{s^2+2s+1}$

2. Determine the inverse Z transform of

(1) $\dfrac{Z^3+Z^2+Z}{(Z^2-Z+1)(Z^2-2Z+1)}$

(2) $\dfrac{Z}{Z^2-3Z+2}$

(3) $\dfrac{1}{Z^2-3Z+2}$

(4) $\dfrac{Z}{(Z-0.5)}$

(5) $\dfrac{0.5Z}{(Z-1)(Z-0.5)}$

(6) $\dfrac{1}{(Z-1)^2(Z-2)}$

by the following methods:
(a) partial fraction expansion
(b) power series expansion
(c) method of residues (this would require outside reading)
(d) using the computer program in the appendix

3. Determine the initial and final value of $F(t)$ whose Z transform is given by

$$F(Z) = \frac{2Z}{(Z-1)(1-0.4Z^{-1})}$$

4. Apply the definition of Z transforms and the translation of the function theorem to determine the Z transform of

$$F(s) = \frac{1 - e^{-sT}}{s}$$

where T is a constant.

Chapter 6   Pulse Transfer Functions

1. For the following systems determine the pulse transfer function $G(Z)$. For Parts (c) and (d) let $\theta_d = NT$ where N is an integer.

(a) $G(s) = \dfrac{K_p(1 - e^{-sT})}{s(\tau_p s + 1)}$

(b) $G(s) = \dfrac{K_p(1 - e^{-sT})}{s(\tau_1 s + 1)(\tau_2 s + 1)}$

(c) $G(s) = \dfrac{K_p e^{-\theta_d s}(1 - e^{-sT})}{s(\tau_p s + 1)}$

(d) $G(s) = \dfrac{K_p e^{-\theta_d s}(1 - e^{-sT})}{s(\tau_1 s + 1)(\tau_2 s + 1)}$

2. Determine the pulse transfer function $G(Z)$ of the following system

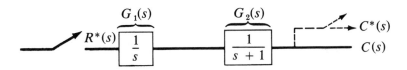

Calculate $Z\{G_1(s)\}Z\{G_2(s)\}$. Is it equal to $Z\{G_1(s)G_2(s)\}$?

Chapter 8   Open-Loop Response

1. For the sampled-data control systems shown below determine

the open-loop response to a unit step change in input, $X(t)$ at several sampling instants.

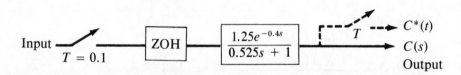

(a) $G_p(s) = \dfrac{1}{(s + 1)}$ $\qquad$ (b) $G_p(s) = \dfrac{e^{-0.2s}}{s + 1}$

(c) $G_p(s) = \dfrac{1}{(s + 1)\,(\frac{1}{2}s + 1)}$ $\qquad$ (d) $G_p(s) = \dfrac{e^{-0.2s}}{(s + 1)\,(0.2s + 1)}$

2. An approximating model for an industrial process is shown below. Determine the open-loop response to a step change in input.

Chapter 10   Stability

1. Determine the proportional gain K_c for which the following systems become unstable:

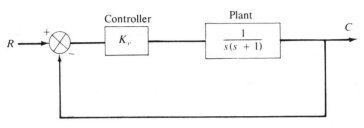

(a) Conventional control system

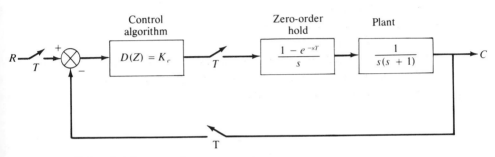

(b) Sampled-data control system

2. In the following figure is shown a second-order sampled-data control system. Using the Schur-Cohn determinants, determine the range of K values for which the system is stable.

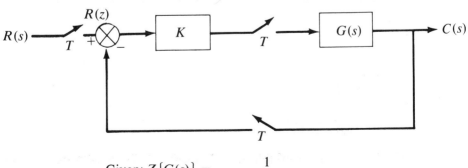

Given: $Z\{G(s)\} = \dfrac{1}{(Z - 0.5)(Z - 1)}$

Chapter 11   Control Systems Design

1. Determine and compare the closed-loop response of a process with sampled-data control system having a transfer function

$$G_p(s) = \frac{1.0e^{-s}}{\tau s + 1}$$

for (a) $\tau = 1$, (b) $\tau = 5$, for a unit step change in set point and load. Let $T = 0.2$. Test the following algorithms:
   a. Deadbeat control
   b. Dahlin algorithm (IMC with first-order filter)
   c. PI algorithm
   d. PID algorithm
   e. Smith Predictor
   f. SMPC algorithm
   g. Predictive IMC algorithm
   h. CMBC algorithm
Comment on the advantages of disadvantages of the algorithms considered.

2. Repeat Problem 1 for the case where there are ±30% modeling errors in each of the model parameters.

Chapter 12   Modified *Z* Transforms

1. Find the modified Z transforms of the following functions:

(a) $\dfrac{1}{s^2}$

(b) $\dfrac{1}{s^2(s + 1)}$

(c) $\dfrac{2}{s(s + 1)(s + 2)}$

2. Determine the complete time response of the following control system to a unit step change in set point

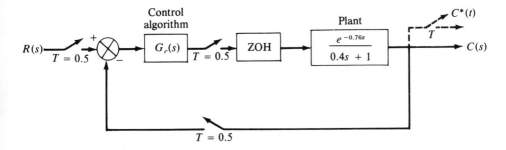

(a) Proportional control algorithm: $G_c(s) = 0.3$

(b) PI algorithm: $G_c(s) = 0.3\left(1 + \dfrac{1}{0.5s}\right)$

Chapter 13   Process Identification

1. The results from a pulse test are shown in the following table. Use the computer program from this chapter to develop the parameters of the process model, assuming that the data can be adequately described by a second-order plus dead-time model.

Time	Input	Time	Output
0.03300	1.74000	0.16700	0.04000
0.06700	1.95000	0.20000	0.07000
0.30000	1.95000	0.23300	0.13000
0.33000	1.09000	0.27000	0.17000
0.36700	0.39000	0.30000	0.23000
0.40000	0.20000	0.33000	0.27000
0.43300	0.10000	0.40000	0.37000
0.46700	0.00000	0.43000	0.43000
		0.47000	0.46000
		0.50000	0.49000
		0.53000	0.50000
		0.57000	0.50000
		0.63000	0.48000
		0.67000	0.45000
		0.73000	0.43000
		0.80000	0.39000
		0.83000	0.37000
		0.87000	0.36000
		0.93000	0.32000
		1.00000	0.30000
		1.07000	0.27000
		1.13000	0.27000
		1.30000	0.25000
		1.50000	0.23000
		1.67000	0.18000
		1.83000	0.18000
		1.97000	0.16000
		2.03000	0.14000
		2.37000	0.14000
		3.70000	0.05000
		5.03000	0.05000
		6.37000	0.00000

2. Fit an approximate transfer function to each of the frequency-response data shown in the following figure. (From W.L., Luyben, *Process Modeling, Simulation, and Control for Chemical Engineers*, McGraw-Hill, 1973 with permission.)

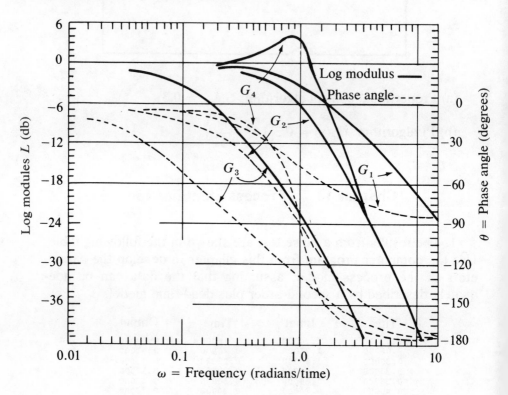

Chapter 15   Feedforward Control

1. Design a feedforward controller for the following heat exchanger system (from P. Harriott, *Process Control*, McGraw Hill, 1965 with permission).

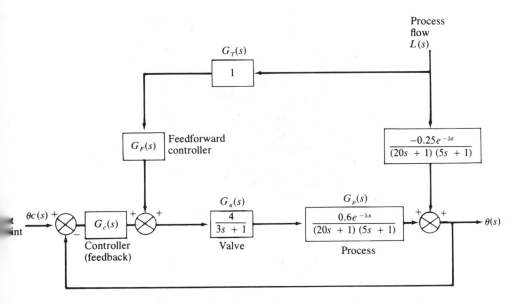

Chapter 16   Cascade Control

1. The schematic and block diagram of a reactor control system is shown below. For this control system determine (a) Ziegler–Nichols tuning of the two controllers (use three-mode master controller and a PI slave controller) of the cascade system, (b) Ziegler–Nichols setting of the equivalent single-loop PID controller, (c) determine the transient response to a unit step change in set point, and to a step change of 0.2 in L_1 and L_2 respectively, and (d) determine ITAE for each case and assess the benefits of cascade control.

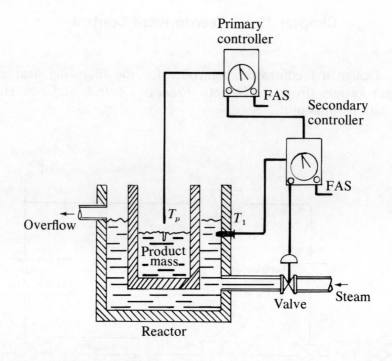

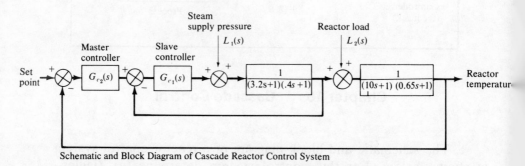

Schematic and Block Diagram of Cascade Reactor Control System

2. The block diagram of a second-order control system is shown below. Answer the questions asked in problem 1 with respect to this control system.

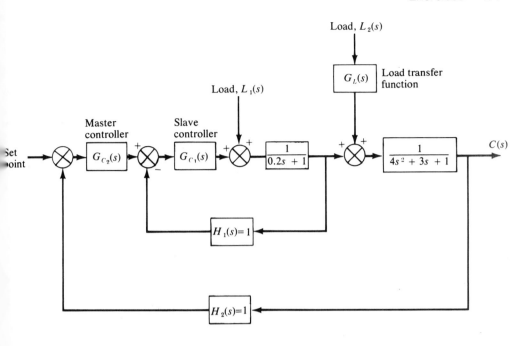

Chapter 17   Multivariable Control Systems

1. The transfer function models of the Wood-Berry column are given in Example 1. Apply the tools presented in this chapter and determine
   a. the extent of interaction by RGA and SVD methods
   b. the performance of multiloop control for set point and load change
   c. the performance of PI and SMPC control
   d. the performance of IMC

2. Repeat Problem 1 where there are ±30% modeling errors in each of the parameters.

3. Luyben (Ind. Eng. Chem. Proc. Des. Dev., 25, 31, 1986) has presented dynamic data of 4 multivariable systems. Apply the concepts in problems 1 and 2 and decide on the merits and drawbacks of each control methodology.

Appendix C1

Computer Program Listing for Z-Transform Inversion

This appendix presents a Fortran computer program for Z-transform inversion.* The program performs long division on the equation

$$C(Z) = \frac{P_0 + P_1 Z^{-1} + P_2 Z^{-2} + P_3 Z^{-3} + \ldots P_n Z^{-n}}{Q_0 + Q_1 Z^{-1} + Q_2 Z^{-2} + Q_3 Z^{-3} + \ldots Q_n Z^{-n}}$$

and outputs $C(nT)$ according to the equation

$$C(Z) = C(O) + C(T)Z^{-1} + C(2T)Z^{-2} + \ldots C(mT)Z^{-m}$$

To use the program read in the following data:

1. $P_0, P_1, P_2, \ldots P_{10}$

2. $Q_0, Q_1, Q_2, \ldots Q_{10}$

3. Set MM equal to the number of output points desired.

---

* Crosby, H. A., Peterson, D. M. Fortran Subroutine Solves Z-Transform Inversion, *Control Engineering*, August 1967, pp. 92–93. Cited with Permission of the Publisher.

Given below is the program listing and a sample printout of results.

```
file: LONDIV.FOR
Edit: DSKE:LONDIV.FOR[424,246]
*PØ:*
00100             DIMENSION PØ(10),QØ(10),CØ(100)
00200             DO 7 I=1 , 10
00300             PØ(I)=0.
00400             QØ(I)=0.
00500       7     CONTINUE
00600             TYPE 300
00700     300     FORMAT (' ENTER # OF TERMS IN NUMERATOR & DENOMINATOR.')
00800             ACCEPT 400,NNR,NDR
00900     400     FORMAT (I3)
01000             TYPE 500
01100     500     FORMAT(' ENTER COEFF. OF NUMERATOR TERM BY TERM.')
01200             DO 8 I=1,NNR
01300             ACCEPT 100,PØ(I)
01400       8     CONTINUE
01500             TYPE 600
01600     600     FORMAT(' ENTER COEFF. OF DENOMINATOR TERM BY TERM.')
01700             DO 9 I=1, NDR
01800             ACCEPT 100,QØ(I)
01900       9     CONTINUE
02000     100     FORMAT (10F8.3)
02100             MM=25
02200             CALL ZTRANS(CØ,PØ,QØ,MM)
02300             TYPE 700
02400     700     FORMAT('            SAMP.INST.            OUTPUT')
02500             DO 1 K=1,MM
02600             N=K-1
02700             TYPE 200 , N,CØ(K)
02800       1     CONTINUE
02900     200     FORMAT(10X,I3,20X,F5.3)
03000             END
03100             SUBROUTINE ZTRANS (C,P,Q,MM)
03200             DIMENSION P(10),Q(10),C(100)
03300             DO 1 I=2,10
03400             P(I)=P(I)/Q(1)
03500             Q(I)=Q(I)/Q(1)
03600       1     CONTINUE
03700             P(1)=P(1)/Q(1)
03800             Q(1)=1
03900             C(1)=P(1)
04000             DO 6 N=2,MM
04100             SIGMA =0.
04200             IF (N-10) 2,2,3
04300       2     ABC=P(N)
04400             M=N
04500             GO TO 4
04600       3     M=10
04700             ABC=0.
04800       4     DO 5 I=2,M
04900             K=(N+1)-I
05000             SIGMA=SIGMA+C(K)*Q(I)
05100       5     CONTINUE
05200             C(N)=ABC-SIGMA
05300       6     CONTINUE
05400             RETURN
05500             END
*
```

```
 EX LONDIV.FOR
FORTRAN: LONDIV
MAIN.
ZTRANS
LINK:   Loading
[LNKXCT LONDIV Execution]

ENTER # OF TERMS IN NUMERATOR & DENOMINATOR.
3
3

ENTER COEFF. OF NUMERATOR TERM BY TERM.
0.0
0.368
0.264

ENTER COEFF. OF DENOMINATOR TERM BY TERM.
1.0
-1.368
0.368
```

SAMP.INST.	OUTPUT
0	.000
1	.368
2	.767
3	.914
4	.969
5	.988
6	.996
7	.998
8	.999
9	1.000
10	1.000
11	1.000
12	1.000
13	1.000
14	1.000
15	1.000
16	1.000
17	1.000
18	1.000
19	1.000
20	1.000
21	1.000
22	1.000
23	1.000
24	1.000

```
END OF EXECUTION
CPU TIME: 0.16  ELAPSED TIME: 1:12.83
EXIT
```

Appendix C2

Closed-Loop Simulation Program for Sampled-Data Systems

This appendix presents a digital computer program, written in time-sharing Fortran for simulation of a sampled-data control system. The program can be used to obtain the closed-loop response of a first- or second-order system with dead time to a unit step change in the load variable or any specified change in the set point. The block diagram of the control system is shown in Figure C2.

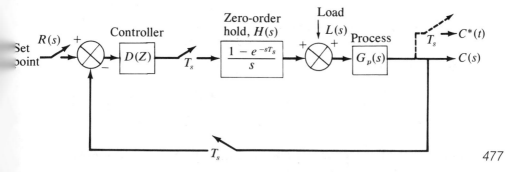

The following inputs are required for the program:

1. Order of process.

If process is first-order, enter 1; if second order, enter 2

2. Parameter of the process transfer function.

If process is first order, enter gain K_p, time constant τ_p, sampling period T_s, and dead time θ_d. If process is second order, enter gain K_p, the two process time constants τ_{p_1} and τ_{p_2}, sampling period T, and dead time θ_d.

3. Enter PID controller constants K_c, τ_I, and τ_D. (User may set τ_I equal to a high value and τ_D to zero if these modes are not required).

4. Is response required for a change in load or set point? Enter 1 if load response is desired or zero if set point response is desired.

5. If zero has been entered as answer to question number 4, enter 1 if response to a unit step change is desired or enter zero if response to some other kind of set point change is required.

6. If the answer to question number 5 is zero, enter the number of terms in the numerator and denominator of $R(Z)$ where

$$R(Z) = (a + bZ^{-1} + cZ^{-2})/(D + EZ^{-1} + FZ^{-2})$$

7. Enter the coefficients of the numerator polynomical from left to right.

8. Enter the coefficients of the denominator polynomical.

With these steps the computer will execute the program and output the response of the controlled variable at the various samples instants. The user can terminate execution of the program when the controlled variable reaches the final steady-state by typing Control C ($\uparrow$C).

The program solves the following equations by multiplication and long division

For set point changes: $C(Z) = \dfrac{D(Z)\,HG(Z)\,R(Z)}{1 + D(Z)\,HG(Z)}$

For load changes $C(Z) = \dfrac{LG(Z)}{1 + D(Z)\,HG(Z)}$

For a PID controller

$$D(Z) = \frac{K_c\left\{\left(1 + \dfrac{T_s}{\tau_I} + \dfrac{\tau_D}{T_s}\right) - \left(1 + 2\dfrac{\tau_D}{T_s}\right)Z^{-1} + \left(\dfrac{\tau_D}{T_s}\right)Z^{-2}\right\}}{1 - Z^{-1}}$$

The listing of the program and a sample printout of the results are shown on the following pages:

```
File: ZTRANS.FOR
Edit: DSKE:ZTRANS.FOR[424,246]
*P0:*


00100    C          THIS PROGRAM EVALUATES CLOSED LOOP RESPONSE OF A FIRST
00200    C     ORDER OR SECOND ORDER SYSTEM WITH DEAD TIME TO A LOAD CHANGE
00300    C     OR SET-POINT CHANGE USING Z-TRANSFORM METHOD. THE CONTROLLER
00400    C     IS IN P-I-D MODE AND THE PROGRAM TAKES CONTROLLER CONSTANTS,
00500    C     PROCESS CONSTANTS AND SAMPLING TIME AS INPUT.
00600    C     THE LOOP ASSUMES A ZERO ORDER HOLD AFTER THE CONTROLLER.
00700    C     ..........WRITTEN & EDITED BY P.C.GOPALRATNAM...........
00800    C
00900            DIMENSION GCN(20),GPN(20),GCD(20),GPD(20),RD(20),RN(20)
01000            DIMENSION NUMER(100,20),PROD1(20),PROD2(20),PROD3(20)
01100            DIMENSION PROD4(20),PROD5(20),PROD6(20),DENOM(20)
01200            DIMENSION LGN(20),LGD(20),DUM(20)
01300            REAL KC,KP,M
01400            TYPE 105
01500    105     FORMAT(' ENTER THE ORDER OF THE PROCESS.')
01600            ACCEPT 210,IORDER
01700            GO TO (150,155) ,IORDER
01800    150     TYPE 100
01900    100     FORMAT(' ENTER VALUES OF KP,TAUP1,TSAMP,THETA AS REAL#')
02000            ACCEPT 200,KP,TAUP1,TSAMP,THETA
02100    200     FORMAT(F8.4)
02200            TAUP2=0.0
02300            GO TO 180
02400    155     TYPE 101
02500    101     FORMAT(' ENTER KP,TAUP1,TAUP2,TSAMP,THETA AS REAL#:')
02600            ACCEPT 200,KP,TAUP1,TAUP2,TSAMP,THETA
02700    180     TYPE 110
02800    110     FORMAT(' ENTER KC & TAUI,TAUD OF CONTROLLER AS REAL #.')
02900            ACCEPT 200,KC,TAUI,TAUD
03000            QUO=THETA/TSAMP
03100            N=INT(QUO)
03200            M=1.-(THETA/TSAMP-N)
03300            EX1=EXP(-TSAMP/TAUP1)
03400            L=N+6
03500            EXM1=EXP(-M*TSAMP/TAUP1)
03600            IF(TAUP2.EQ.0) GO TO 21
03700            EX2=EXP(-TSAMP/TAUP2)
03800            EXM2=EXP(-M*TSAMP/TAUP2)
03900            GO TO 22
04000    21      EX2=0.0
04100            EXM2=0.0
04200    22      TYPE 181
04300    181     FORMAT(' IS THIS LOAD CHANGE OR SET-POINT CHANGE ?')
04400            TYPE 185
04500    185     FORMAT(' ENTER 1 FOR LOAD CHANGE & 0 FOR SET-POINT CHANG
E')
04600            ACCEPT 210,KONST1
04700    C
04800    C    INITIALIZE THE POLYNOMIALS.
04900    C
05000            DO 300 I=1,L
05100            DUM(I)=0.0
05200            LGN(I)=0.0
05300            LGD(I)=0.0
05400            GCN(I)=0.
05500            GPN(I)=0.
05600            GCD(I)=0.
05700            GPD(I)=0.
05800            RD(I)=0.0
05900            RN(I)=0.
06000    300     CONTINUE
06100    C
06200    C    ESTABLISH THE POLYNOMIAL FOR SET-POINT.
06300    C
06400            IF(KONST1.EQ.1) GO TO 40
06500            TYPE 120
```

```
06600    120    FORMAT(' DO YOU WANT THE RESPONSE FOR A UNIT STEP ?')
06700           TYPE 125
06800    125    FORMAT(' ENTER 1 FOR YES & 0 FOR NO.     :')
06900           ACCEPT 210,KONST
07000    210    FORMAT(I2)
07100           IF(KONST) 10,10,20
07200     10    TYPE 130
07300    130    FORMAT(' ENTER # OF TERMS IN NUMER. & DENOM. OF R(Z)')
07400           ACCEPT 220,IRN,IRD
07500    220    FORMAT(I2)
07600           IR=MAX0(IRN,IRD)
07700           L=L-1+IR
07800           TYPE 140
07900    140    FORMAT(' ENTER COEFF.OF NUMERATOR TERM BY TERM.   :')
08000           ACCEPT 230,(RN(I),I=1,IRN)
08100    230    FORMAT(F12.6)
08200           TYPE 151
08300    151    FORMAT(' ENTER COEFF.OF DENOMINATOR TERM BY TERM.        :')
08400           ACCEPT 230,(RD(I),I=1,IRD)
08500           GO TO 40
08600     20    RN(1)=1.0
08700           RD(1)=1.0
08800           RD(2)=-1.0
08900   C
09000   C    ESTABLISH THE POLYNOMIAL FOR D(Z).
09100   C
09200     40    GCN(1)=KC*(1.0+TSAMP/TAUI+TAUD/TSAMP)
09300           GCN(2)=-KC*(1.0+2.0*TAUD/TSAMP)
09400           GCN(3)=KC*TAUD/TSAMP
09500           GCD(1)=1.0
09600           GCD(2)=-1.0
09700   C
09800   C    ESTABLISH THE POLYNOMIAL FOR THE PROCESS HG(Z) AND LG(Z).
09900   C
10000           GPN(N+2)=KP*(1.+(TAUP1*EXM1-TAUP2*EXM2)/(TAUP2-TAUP1))
10100           TEMP=(TAUP1*EXM1*(1.+EX2)-TAUP2*EXM2*(1.+EX1))
10200           GPN(N+3)=-KP*(EX1+EX2+TEMP/(TAUP2-TAUP1))
10300           TEMP1=TAUP1*EXM1*EX2-TAUP2*EXM2*EX1
10400           GPN(N+4)=KP*(EX1*EX2+TEMP1/(TAUP2-TAUP1))
10500           GPD(1)=1.0
10600           GPD(2)=-1.*(EX1+EX2)
10700           GPD(3)=EX1*EX2
10800           IF(KONST1.EQ.0) GO TO 80
10900           DUM(1)=1.
11000           DUM(2)=-1.
11100           CALL MPLY(DUM,GPD,LGD,L)
11200   C
11300   C    PERFORM MULTIPLICATIONS
11400   C
11500     80    CALL MPLY(GCN,GPN,PROD1,L)
11600           CALL MPLY(GPD,GCD,PROD3,L)
11700           DO 350 I=1,L
11800           PROD4(I)=PROD1(I)+PROD3(I)
11900    350    CONTINUE
12000           IF (KONST1.EQ.0) GO TO 410
12100           CALL MPLY(GPN,PROD3,PROD2,L)
12200           CALL MPLY(PROD4,LGD,DENOM,L)
12300           GO TO 420
12400    410    CALL MPLY(PROD1,RN,PROD2,L)
12500           CALL MPLY(PROD4,RD,DENOM,L)
12600    420    DO 360 I=2,L
12700           PROD2(I)=PROD2(I)/DENOM(1)
12800           DENOM(I)=DENOM(I)/DENOM(1)
12900    360    CONTINUE
13000           PROD2(1)=PROD2(1)/DENOM(1)
13100           DENOM(1)=1.0
13200           TYPE 11
13300     11    FORMAT(' SAMP.INST.                OUTPUT VALUES.')
13400           I=0
13500           PROD2(L+1)=0.0
```

```
13600    500      RSPONS=PROD2(1)
13700             TYPE 170,I,RSPONS
13800    170      FORMAT(I5,25X,F6.3)
13900             DO 370 J=1,L
14000             PROD2(J)=PROD2(J+1)-DENOM(J+1)*RSPONS
14100    370      CONTINUE
14200             I=I+1
14300             IF (I.LE.100) GO TO 500
14400             END
14500             SUBROUTINE MPLY(A,B,C,N)
14600             DIMENSION A(20),B(20),C(20)
14700             DO 330 I=1,N
14800             C(I)=0.0
14900             DO 340 J=1,I
15000             C(I)=C(I)+A(J)*B(I+1-J)
15100    340      CONTINUE
15200    330      CONTINUE
15300             RETURN
15400             END
*
```

```
         .EX ZTRANS.FOR
         FORTRAN: ZTRANS
         MAIN.
         MPLY
         LINK:    Loading
         [LNKXCT ZTRANS Execution]

         ENTER THE ORDER OF THE PROCESS.
         1

         ENTER VALUES OF KP,TAUP1,TSAMP,THETA AS REAL#
         1.0
         0.4
         0.5
         0.76

         ENTER KC & TAUI,TAUD OF CONTROLLER AS REAL #.
         0.3
         0.5
         0.0

         IS THIS LOAD CHANGE OR SET-POINT CHANGE ?
         ENTER 1 FOR LOAD CHANGE & 0 FOR SET-POINT CHANGE
         1

         SAMP.INST.                OUTPUT VALUES.
              0                        0.000
              1                        0.000
              2                        0.451
              3                        0.843
              4                        0.833
              5                        0.592
              6                        0.314
              7                        0.123
              8                        0.035
              9                        0.016
             10                        0.024
             11                        0.030
             12                        0.028
             13                        0.020
             14                        0.011
             15                        0.005
             16                        0.002
             17                        0.001
             18                        0.001
             19                        0.001
             20                        0.001
             21                        0.001
             22                        0.000
```

```
23                              0.000
24                              0.000
25                              0.000
26                              0.000
27                              0.000
28                              0.000
29                              0.000
30                              0.000
```

```
 EX ZTRANS.FOR
LINK:    Loading
[LNKXCT ZTRANS Execution]
```

ENTER THE ORDER OF THE PROCESS.
1

ENTER VALUES OF KP,TAUP1,TSAMP,THETA AS REAL#
1.0
0.4
0.5
0.76

ENTER KC & TAUI,TAUD OF CONTROLLER AS REAL #.
0.3
0.5
0.0

IS THIS LOAD CHANGE OR SET-POINT CHANGE ?
ENTER 1 FOR LOAD CHANGE & 0 FOR SET-POINT CHANGE
0

DO YOU WANT THE RESPONSE FOR A UNIT STEP ?
ENTER 1 FOR YES & 0 FOR NO.   :
1

```
SAMP.INST.                 OUTPUT VALUES.
     0                          0.000
     1                          0.000
     2                          0.271
     3                          0.641
     4                          0.888
     5                          0.993
     6                          1.004
     7                          0.983
     8                          0.968
     9                          0.967
    10                          0.976
    11                          0.987
    12                          0.995
    13                          0.999
    14                          0.999
    15                          0.999
    16                          0.999
    17                          0.999
    18                          0.999
    19                          1.000
    20                          1.000
    21                          1.000
    22                          1.000
    23                          1.000
    24                          1.000
    25                          1.000
    26                          1.000
    27                          1.000
    28                          1.000
    29                          1.000
    30                          1.000
    31                          1.000
    32                          1.000
    33                          1.000
    34                          1.000
```

```
     35                        1.000
     36                        1.000
     37                        1.000
     38                        1.  ↑C



.EX ZTRANS.FOR
LINK:    Loading
[LNKXCT ZTRANS Execution]

ENTER THE ORDER OF THE PROCESS.
1

ENTER VALUES OF KP,TAUP1,TSAMP,THETA AS REAL#
1.0
0.4
0.5
0.76

ENTER KC & TAUI,TAUD OF CONTROLLER AS REAL #.
0.3
0.5
0.0

IS THIS LOAD CHANGE OR SET-POINT CHANGE ?
ENTER 1 FOR LOAD CHANGE & 0 FOR SET-POINT CHANGE
0

DO YOU WANT THE RESPONSE FOR A UNIT STEP ?
ENTER 1 FOR YES & 0 FOR NO.     :
0

ENTER # OF TERMS IN NUMER. & DENOM. OF R(Z)
1
2

ENTER COEFF.OF NUMERATOR TERM BY TERM.   :
1.0


ENTER COEFF.OF DENOMINATOR TERM BY TERM.        :
1.0
-1.0

SAMP.INST.                 OUTPUT VALUES.
     0                        0.000
     1                        0.000
     2                        0.271
     3                        0.641
     4                        0.888
     5                        0.993
     6                        1.004
     7                        0.983
     8                        0.968
     9                        0.967
    10                        0.976
    11                        0.987
    12                        0.995
    13                        0.999
    14                        0.999
    15                        0.999
    16                        0.999
    17                        0.999
    18                        0.999
    19                        1.000
    20                        1.000
    21                        1.000
    22                        1.000
    23          ↑C
```

```
EX ZTRANS.FOR
LINK:   Loading
[LNKXCT ZTRANS Execution]

ENTER THE ORDER OF THE PROCESS.
2

ENTER KP,TAUP1,TAUP2,TSAMP,THETA AS REAL#:
1.0
0.4
0.0
0.5
0.76

ENTER KC & TAUI,TAUD OF CONTROLLER AS REAL #.
0.3
0.5
0.0

IS THIS LOAD CHANGE OR SET-POINT CHANGE ?
ENTER 1 FOR LOAD CHANGE & 0 FOR SET-POINT CHANGE
1

SAMP.INST.                    OUTPUT VALUES.
     0                          0.000
     1                          0.000
     2                          0.451
     3                          0.843
     4                          0.833
     5                          0.592
     6                          0.314
     7                          0.123
     8                          0.035
     9                          0.016
    10                          0.024
    11                          0.030
    12                          0.028
    13                          0.020
    14                          0.011
    15                          0.005
    16                          0.002
    17                          0.001
    18                          0.001
    19                          0.001
    20                          0.001
    21                          0.001
    22                          0.000
    23                          0.000
    24                          0.000
    25                          0.000
    26                          0.000
    27                          0.000
    28                          0.000
    29                          0.000
    30                           ^C
```

Appendix D

Pulse Analysis Program

This appendix (by permission of Instrument Society of America, Research Triangle Park, North Carolina) describes a Fortran program for computing the frequency response data from the pulse test data. A listing of the program is also included. When the input and output pulse data are entered as a function of time, the program prints a table of system gain (amplitude ratio) and phase angles as a function of frequency.

Program Input Instructions

Enter the problem data on input data sheets as described in the following paragraphs (see Tables I and IV):

Related input items are grouped into the lists described below.

List 1: Problem-Set Identification—Enter this list to initialize the program or to nullify data from a preceding set of related problems. The contents of columns 11 to 70 will be printed in the heading of each output page.

List 2: Case Identification—Enter this list to print a 60 character case title below the problem-set title on each output page. The list is optional.

List 3: Response Control Data—Enter this list to specify which of the pulses defined by list 6 is the input and which is the output. Omit this list to print only the Fourier transforms of the pulses.

Lists 4 and 5: Frequency Specifications—These lists provide two alternate methods of specifying the frequencies at which the program will compute the pulse amplitude and phase angle. Do not enter both lists. If neither is entered, the program provides ten frequencies per decade from 0.001 to 100 radians per unit time.

List 6: Pulse data—Enter this list for each pulse curve to be considered in the problem.

A curve may be split into as many as ten sections with equal time intervals in each section. The section starting at zero time is the first section.

The ordinates of the curve are numbered sequentially. Ordinate number one is at zero time after excitation. The maximum number of ordinates per curve is 1000. Ordinates may be entered with any units desired. If a chart factor or other multiplier is required to change the ordinate values to actual physical units, it can be entered for each curve.

The initial steady-state value (value prior to excitation) must be given, although it is usually the same as the first ordinate.

A pulse closure effect exists when the end of the curve fails to return to the drift corrected initial steady-state value. This may be handled in one of two ways by the pulse closure code:

(a) Enter 0 if the pulse should be closed. A step change between the last ordinate and the drift corrected initial steady state value is assumed.

(b) Enter 2 if it is believed that the curve need not (or should not) return to the initial steady-state value and that the final steady-state value is indeed different from the value prior to excitation. (An example of this type appears later in this appendix.)

List 99: End of Case Control—Enter this list to mark the end of the input data for a case.

Keypunching Rules

1. Punch data items in the order in which they appear on the input data sheets. Start at the top and read from left to right.

2. Do not punch a list number if it is immediately followed by a blank entry.

3. Punch numeric items into ten-column fields with eight fields per card.

4. Right-justify each integer within its ten-column field and omit the decimal point.

5. Punch each item preceded by a double dagger (‡) into colums one to ten of a new card. Do not punch the double dagger.

Tables II and V illustrate the punched card input prepared from the Input Data Sheets.

Output

The output corresponding to typical input data are given in Tables III and VI. For the Fourier transform the real part, imaginary part, amplitude and phase angle are given for each frequency. Any pulse closure corrections made are noted. When the frequency response option is requested, the amplitude ratio, $\log_{10}$ (amplitude ratio) and phase difference are given in addition to the Fourier transform items.

A warning message is printed in the frequency table after the first frequency for which the input pulse amplitude is less than 1% of its zero frequency amplitude. If either curve persists to infinity, the area under the curve is calculated from time 0 to time of the last point.

Program Execution

Add the control cards needed to load the program into the computer to the program deck which should be prepared in accordance with the listing shown in Table VII and the data cards prepared according to the instructions just described. The control cards and the job submission procedure will depend on the specific computer installation you use.

Table I

	INPUT DATA SHEET 1 OF 3 **FREQUENCY RESPONSE FROM PULSE TEST DATA**	
		BY _____ DATE _____

LIST 1: PROBLEM-SET IDENTIFICATION

‡ *1, ISA S26 FREQUENCY RESPONSE FROM PULSE TEST DATA*
10   15   20   25   30   35   40   45   50   55   60   65   70

LIST 2: CASE IDENTIFICATION (Optional)

‡ *2 SAMPLE PROBLEM TO TEST COMPUTER PROGRAM PROPERLY EXCITED*
10   15   20   25   30   35   40   45   50   55   60   65   70

LIST 3: DYNAMIC RESPONSE CONTROL DATA (Optional)

Description	Entry
List Number	‡ _____ 3
Input Pulse Number (integer 1 through 9)	‡ _____ 1
Output Pulse Number (integer 1 through 9)	_____ 2
Response Normalization Code (Enter 0 for normalization, otherwise enter 1)	_____ 0
Transform Print Code (Enter 0 to print transforms of both pulses, Enter 1 to print Input Pulse Transform only, Enter 2 to print Output Pulse Transform only, Enter 3 to suppress printing of both transforms)	_____ 0

LIST 4: FREQUENCY RANGE (Optional *. Use List 5 if more appropriate.)

Description	Entry
List Number	‡ _____ 4
Frequency Units (Enter 0 for radians/time, Enter 1 for cycles/time) (0 normal)	‡ _____ 1
Number of Decades (5 normal, 5 max)	_____ 3
Number of Frequencies per Decade (10 normal, 10 max)	_____ 10
Initial Frequency (0.001 normal)	_____ 0.01

LIST 5: FREQUENCY POINTS (Alternate to List 4) *

Description	Entry
List Number	‡ _____ 5
Frequency Units (Enter 0 for radians/time, Enter 1 for cycles/time)	‡ _____
Number of Frequencies (50 max)	_____

Frequencies			
‡ _____	_____	_____	_____
_____	_____	_____	_____
_____	_____	_____	_____
_____	_____	_____	_____
_____	_____	_____	_____
_____	_____	_____	_____
_____	_____	_____	_____
_____	_____	_____	_____

Note: The normal values in List 4 are used if both List 4 and List 5 are omitted.

Start new card. Right justify numeric entries in 10 column fields.

Table I (*continued*)

	INPUT DATA SHEET 2 OF 3 FREQUENCY RESPONSE FROM PULSE TEST DATA	
		BY _____ DATE _____

LIST 6. PULSE DATA (Repeat this list for each pulse using a unique pulse number, k)

Description	Entry
List Number and Pulse Identification	‡ 6 *RUN 5.1.2.I.*
Pulse Number, k (1 ≤ k ≤ 9)	‡ 1 15 18
Pulse Height Prior to Excitation at Time 0⁻	0
Base Time Drift from Time 0 to T	0
Multiplier to Convert Pulse Height Units. Enter 1 if conversion not required.	1
Pulse Height Data Summary Code (0 normal. Enter 1 if printout desired.)	1
Pulse Closure Code. Enter 0 for closed pulses. Enter 2 for open pulses.	0
Number of Sections in Curve, M (1 ≤ M ≤ 10)	1

Section Number, m	1 •	2 •	3 •	4 •
Time Interval Between Points	‡ .05			
Number of Last Point in Section	18			

Number of Pulse Height Data Points, N (2 ≤ N ≤ 1000)	‡ 18

Point No.	Pulse Height Data Starting at Time 0⁺ and Ending at Time T⁻				
1-5	‡ 0	4	11	43	56
6-10	69	76	75	70	63
11-15	48	35	22	14	5
16-20	3	1	0		
21-25					
26-30					
31-35					
36-40					
41-45					
46-50					
51-55					
56-60					
61-65					
66-70					
71-75					
76-80					
81-85					
86-90					
91-95					
96-100					
101-105					
106-110					
111-115					
116-120					
121-125					

ST 99: END OF CASE CONTROL	‡ 99

unch entries below column-wise.

tart new card. Right justify numeric entries in 10 column fields.

Table I   (concluded)

INPUT DATA SHEET 3 OF 3 FREQUENCY RESPONSE FROM PULSE TEST DATA		
	BY_____	DATE_____

LIST 6: PULSE DATA (Repeat this list for each pulse using a unique pulse number, k)

Description	Entry
List Number and Pulse Identification	‡ 6 P.U.M 5.1.2.0
Pulse Number, k (1 ≤ k ≤ 9)	‡ 2
Pulse Height Prior to Excitation at Time 0⁻	C
Base Line Drift from Time 0 to T	O
Multiplier to Convert Pulse Height Units. Enter 1 if conversion not required.	1
Pulse Height Data Summary Code (0 normal. Enter 1 if printout desired.)	1
Pulse Closure Code. Enter 0 for closed pulses. Enter 2 for open pulses.	O
Number of Sections in Curve, M (1 ≤ M ≤ 10)	2

Section Number, m	1 *	2 *	3 *	4 *
Time Interval Between Points	‡ .05	0.1		
Number of Last Point in Section	.15	31		

Number of Pulse Height Data Points, N (2 ≤ N ≤ 1000)	‡ 31

Point No.	Pulse Height Data Starting at Time 0⁺ and Ending at Time T⁻				
1-5	‡ O	O	2	4	7
6-10	12	18	24	30	32
11-15	40	44	42	41	38
16-20	36	27	21	16	13
21-25	10	8	6	4.5	3.5
26-30	2.5	1.7	1.1	0.6	0.2
31-35	O				
36-40					
41-45					
46-50					
51-55					
56-60					
61-65					
66-70					
71-75					
76-80					
81-85					
86-90					
91-95					
96-100					
101-105					
106-110					
111-115					
116-120					
121-125					

LIST 99 END OF CASE CONTROL	‡ 99

* Punch entries below column-wise.

‡ Start new card. Right justify numeric entries in 10 column fields.

Table II
Printout Showing Input Data

```
1ISA - FREQUENCY RESPONSE FROM PULSE TEST DATA

   ISA S26 FREQUENCY RESPONSE FROM PULSE TEST DATA
   SAMPLE PROBLEM TO TEST COMPUTER PROGRAM PROPERLY EXCITED


   LIST 3 DATA        1.        2.       0.        0.


   LIST 4 DATA        1.        3.      10.   0.01000


   LIST 6 DATA   **********    PULSE NO. = 1  RUN S12I

      HEIGHT AT 0-   0.0000E+00  MULTIPLIER    1.000      CLOSURE CODE   0.
      BASE LINE DRIFT 0.0000E+00  SUMMARY CODE  1.        NO. SECTIONS   1.

      SECTION NUMBER        1
      TIME INTERVAL      0.500E-01
      LAST POINT NO.       18.

      PT. VALUE     PT. VALUE     PT. VALUE     PT. VALUE     PT. VALUE
       1   .000E+00   2  4.00       3  11.0       4  43.0       5  56.0
       6  69.0        7  76.0       8  75.0       9  70.0      10  63.0
      11  48.0       12  35.0      13  22.0      14  14.0      15  5.00
      16  3.00       17  1.00      18  .000E+00


   LIST 6 DATA   **********    PULSE NO. = 2  RUN S12O

      HEIGHT AT 0-   0.0000E+00  MULTIPLIER    1.000      CLOSURE CODE   0.
      BASE LINE DRIFT 0.0000E+00  SUMMARY CODE  1.        NO. SECTIONS   2.

      SECTION NUMBER        1         2
      TIME INTERVAL      0.500E-01 0.100E+00
      LAST POINT NO.       15.       31.

      PT. VALUE     PT. VALUE     PT. VALUE     PT. VALUE     PT. VALUE
       1   .000E+00   2   .000E+00   3  2.00       4  4.00       5  7.00
       6  12.0        7  18.0        8  24.0        9  30.0      10  37.0
      11  40.0       12  44.0       13  42.0       14  41.0      15  38.0
      16  36.0       17  27.0       18  21.0       19  16.0      20  13.0
      21  10.0       22  8.00       23  6.00       24  4.50      25  3.50
      26  2.50       27  1.70       28  1.10       29  .600      30  .200
      31   .000E+00
```

Table III   Results

1ISA - FREQUENCY RESPONSE FROM PULSE TEST DATA

ISA S26 FREQUENCY RESPONSE FROM PULSE TEST DATA
SAMPLE PROBLEM TO TEST COMPUTER PROGRAM PROPERLY EXCITED

PULSE NO. = 1   RUN S12I   INPUT
    TOTAL TIME SPAN =   0.8500
    PULSE AREA   =   29.75

PULSE NO. = 2   RUN S120   OUTPUT
    TOTAL TIME SPAN =   2.300
    PULSE AREA   =   33.01

STEADY STATE GAIN =   1.110

FREQUENCY RESPONSE (OUTPUT/INPUT)				FOURIER TRANSFORM OF THE INPUT CURVE				FOURIER TRANSFORM OF THE OUTPUT CURVE			
FREQUENCY CYC./TIME	AMPLITUDE RATIO	LOG AMP. RAT	PH. DIF DEGREES	REAL PART	IMAG. PART	AMPLITUDE	PH.ANG DEG.	REAL PART	IMAG. PART	AMPLITUDE	PH.ANG DEG.
1.000E-02	1.00	-0.00010	-1.52	29.7	-0.671	29.7	358.71	33.0	-1.62	33.0	357.19
1.300E-02	1.00	-0.00017	-1.97	29.7	-0.873	29.7	358.32	32.9	-2.10	33.0	356.35
1.600E-02	0.999	-0.00025	-2.42	29.7	-1.07	29.7	357.93	32.9	-2.58	33.0	355.51
2.000E-02	0.999	-0.00039	-3.03	29.7	-1.34	29.7	357.41	32.8	-3.23	33.0	354.38
2.500E-02	0.999	-0.00062	-3.79	29.7	-1.68	29.7	356.77	32.7	-4.03	33.0	352.98
3.200E-02	0.998	-0.00101	-4.85	29.7	-2.15	29.7	355.86	32.5	-5.14	32.9	351.02
4.000E-02	0.996	-0.00158	-6.06	29.6	-2.68	29.7	354.83	32.2	-6.40	32.9	348.77
5.000E-02	0.994	-0.00247	-7.57	29.5	-3.35	29.7	353.54	31.8	-7.95	32.8	345.97
6.300E-02	0.991	-0.00391	-9.52	29.4	-4.21	29.7	351.85	31.1	-9.91	32.7	342.33
8.000E-02	0.986	-0.00630	-12.07	29.2	-5.33	29.7	349.66	30.0	-12.4	32.4	337.58
0.100	0.978	-0.00984	-15.05	28.9	-6.63	29.6	347.07	28.4	-15.1	32.1	332.02
0.130	0.963	-0.01658	-19.47	28.3	-8.54	29.5	343.20	25.4	-18.7	31.6	323.72
0.160	0.944	-0.02503	-23.82	27.5	-10.4	29.4	339.32	22.0	-21.6	30.8	315.50
0.200	0.914	-0.03886	-29.46	26.3	-12.8	29.3	334.16	16.9	-24.4	29.7	304.69
0.250	0.871	-0.06010	-36.22	24.5	-15.5	29.0	327.71	10.3	-26.1	28.0	291.49
0.320	0.801	-0.09646	-44.97	21.4	-18.8	28.5	318.69	1.65	-25.3	25.3	273.73
0.400	0.715	-0.14551	-53.68	17.3	-21.8	27.9	308.41	-5.82	-21.3	22.1	254.73
0.500	0.613	-0.21263	-62.34	11.6	-24.2	26.8	295.61	-10.9	-14.6	18.3	233.27
0.630	0.506	-0.29587	-70.20	3.98	-24.9	25.3	279.07	-12.4	-6.84	14.2	208.87
0.800	0.414	-0.38306	-78.10	-4.87	-22.3	22.8	257.67	-10.5	7.883E-02	10.5	179.57
1.00	0.330	-0.48103	-87.87	-11.8	-15.6	19.5	232.96	-5.87	4.10	7.16	145.09
1.30	0.229	-0.64019	-97.73	-13.8	-4.28	14.4	197.29	-0.608	3.61	3.66	99.56
1.60	0.158	-0.80017	-105.15	-9.28	2.59	9.63	164.41	0.866	1.46	1.69	59.26
2.00	0.136	-0.86605	-99.40	-3.00	3.67	4.74	129.31	0.621	0.357	0.716	29.91
2.50	0.225	-0.64726	184.69	-0.870	1.58	1.81	118.80	0.249	-0.376	0.451	303.49
3.20	0.256	-0.59138	16.91	-0.510	1.21	1.31	112.93	-0.239	0.286	0.372	129.84
4.00	0.224	-0.64880	213.68	0.177	0.822	0.840	77.84	7.680E-02	-0.195	0.209	291.52
5.00	0.224	-0.64908	-11.89	0.365	0.243	0.438	33.69	0.101	4.053E-02	0.109	21.80
6.30	0.113	-0.94595	37.01	0.517	3.852E-02	0.518	4.26	4.893E-02	4.294E-02	6.510E-02	41.27
			-54.80	-4.425E-02	-0.383	0.385	263.41	-7.461E-02	-3.737E-02	8.345E-02	206.61
										8.106E-02	180.00

Table IV

	INPUT DATA SHEET 1 OF 3 FREQUENCY RESPONSE FROM PULSE TEST DATA	BY _____ DATE _____

LIST 1: PROBLEM-SET IDENTIFICATION

‡ 1, I,S,A, S,2,6, F,R,E,Q,U,E,N,C,Y, R,E,S,P,O,N,S,E, F,R,O,M, P,U,L,S,E, T,E,S,T, D,A,T,A
 10   15   20   25   30   35   40   45   50   55   60   65   70

LIST 2: CASE IDENTIFICATION (Optional)

‡ 2, S,A,M,P,L,E, P,R,O,B,L,E,M, :, N,O,N, C,L,O,S,I,N,G, O,U,T,P,U,T, P,U,L,S,E
 10   15   20   25   30   35   40   45   50   55   60   65   70

LIST 3: DYNAMIC RESPONSE CONTROL DATA (Optional)

Description	Entry
List Number	‡ ___ 3 ___
Input Pulse Number (integer 1 through 9)	‡ ___ 1 ___
Output Pulse Number (integer 1 through 9)	2
Response Normalization Code (Enter 0 for normalization, otherwise enter 1)	0
Transform Print Code (Enter 0 to print transforms of both pulses, Enter 1 to print Input Pulse Transform only, Enter 2 to print Output Pulse Transform only, Enter 3 to suppress printing of both transforms)	0

LIST 4: FREQUENCY RANGE (Optional * Use List 5 if more appropriate.)

Description	Entry
List Number	‡ ___ 4 ___
Frequency Units (Enter 0 for radians/time, Enter 1 for cycles/time) (0 normal)	‡ ___ 1 ___
Number of Decades (5 normal, 5 max)	3
Number of Frequencies per Decade (10 normal, 10 max)	10
Initial Frequency (0.001 normal)	.01

LIST 5: FREQUENCY POINTS (Alternate to List 4) *

Description	Entry
List Number	‡ ___ 5 ___
Frequency Units (Enter 0 for radians/time, Enter 1 for cycles/time)	‡ ___
Number of Frequencies (50 max)	

		Frequencies		
‡				

* Note: The normal values in List 4 are used if both List 4 and List 5 are omitted.

‡ Start new card. Right justify numeric entries in 10 column fields.

Table IV (*continued*)

	INPUT DATA SHEET 2 OF 3 FREQUENCY RESPONSE FROM PULSE TEST DATA	
		BY _____ DATE _____

LIST 6: PULSE DATA (Repeat this list for each pulse using a unique pulse number, k)

Description	Entry
List Number and Pulse Identification Pulse Number, k (1 ≤ k ≤ 9)	‡ 6,RUN ,5,1,2,I ‡ 10 15 18
Pulse Height Prior to Excitation at Time 0⁻	0
Base Line Drift from Time 0 to T	0
Multiplier to Convert Pulse Height Units. Enter 1 if conversion not required.	/
Pulse Height Data Summary Code (0 normal. Enter 1 if printout desired.)	/
Pulse Closure Code. Enter 0 for closed pulses. Enter 2 for open pulses.	0
Number of Sections in Curve, M (1 ≤ M ≤ 10)	2

Section Number, m	1*	2*	3*	4*
Time Interval Between Points	‡ .125	.25		
Number of Last Point in Section	14	19		

Number of Pulse Height Data Points, N (2 ≤ N ≤ 1000)	‡ 19

Point No.	Pulse Height Data Starting at Time 0⁺ and Ending at Time T⁻				
1-5	‡ 0	.2	.9	2.3	4.5
6-10	6.7	8.3	9.7	10.7	11.4
11-15	11.8	10.2	7.9	5.4	2.9
16-20	1.5	.7	.2	.0	
21-25					
26-30					
31-35					
36-40					
41-45					
46-50					
51-55					
56-60					
61-65					
66-70					
71-75					
76-80					
81-85					
86-90					
91-95					
96-100					
101-105					
106-110					
111-115					
116-120					
121-125					

LIST 99: END OF CASE CONTROL	‡ 99

* Punch entries below column-wise.

‡ Start new card. Right justify numeric entries in 10 column fields.

Table IV   (*concluded*)

<table>
<tr><td></td><td colspan="2">INPUT DATA SHEET 3 OF 3
FREQUENCY RESPONSE FROM
PULSE TEST DATA</td><td>BY _____ DATE _____</td></tr>
</table>

LIST 6: PULSE DATA (Repeat this list for each pulse using a unique pulse number, k)

Description	Entry
List Number and Pulse Identification	‡ $\underset{10}{6RUM}\ \underset{2}{5,1,2,0}$
Pulse Number, k ($1 \le k \le 9$)	‡
Pulse Height Prior to Excitation at Time 0^-	0
Base Line Drift from Time 0 to T	1.3
Multiplier to Convert Pulse Height Units. Enter 1 if conversion not required.	1
Pulse Height Data Summary Code (0 normal. Enter 1 if printout desired.)	1
Pulse Closure Code. Enter 0 for closed pulses. Enter 2 for open pulses.	2
Number of Sections in Curve, M ($1 \le M \le 10$)	2

Section Number, m	1 *	2 *	3 *	4 *
Time Interval Between Points	‡ .125	.25		
Number of Last Point in Section	14	22		
Number of Pulse Height Data Points, N ($2 \le N \le 1000$)			‡ 22	

Point No.	Pulse Height Data Starting at Time 0^+ and Ending at Time T^-				
1-5	‡ 0	.1	.5	1.2	2.1
6-10	3.2	4.3	5.7	6.8	7.8
11-15	8.4	8.8	8.3	7.7	6.1
16-20	4.8	3.4	2.7	2.0	1.8
21-25	1.4	1.3			
26-30					
31-35					
36-40					
41-45					
46-50					
51-55					
56-60					
61-65					
66-70					
71-75					
76-80					
81-85					
86-90					
91-95					
96-100					
101-105					
106-110					
111-115					
116-120					
121-125					

LIST 99: END OF CASE CONTROL	‡ 99

* Punch entries below column-wise.

‡ Start new card. Right justify numeric entries in 10 column fields.

Table V
Input Data—Non Closing Pulse Example

```
1ISA - FREQUENCY RESPONSE FROM PULSE TEST DATA

   ISA S26 FREQUENCY RESPONSE FROM PULSE TEST DATA
   SAMPLE PROBLEM - NON CLOSING OUTPUT PULSE


   LIST 3 DATA        1.        2.        0.        0.


   LIST 4 DATA        1.        3.       10.    0.01000


   LIST 6 DATA    **********    PULSE NO. = 1  RUN S12I

      HEIGHT AT 0-    0.0000E+00   MULTIPLIER     1.000      CLOSURE CODE   0.
      BASE LINE DRIFT 0.0000E+00   SUMMARY CODE   1.         NO. SECTIONS   2.

      SECTION NUMBER        1           2
      TIME INTERVAL       0.125       0.250
      LAST POINT NO.       14.         19.

      PT. VALUE      PT. VALUE     PT. VALUE     PT. VALUE      PT. VALUE
       1  .000E+00    2  .200       3  .900       4  2.30        5  4.50
       6  6.70        7  8.30       8  9.70       9  10.7       10  11.4
      11  11.8       12  10.2      13  7.90      14  5.40       15  2.90
      16  1.50       17  .700      18  .200      19  .000E+00


   LIST 6 DATA    **********    PULSE NO. = 2  RUN S120

      HEIGHT AT 0-    0.0000E+00   MULTIPLIER     1.000      CLOSURE CODE   2.
      BASE LINE DRIFT 1.300        SUMMARY CODE   1.         NO. SECTIONS   2.

      SECTION NUMBER        1           2
      TIME INTERVAL       0.125       0.250
      LAST POINT NO.       14.         22.

      PT. VALUE      PT. VALUE     PT. VALUE     PT. VALUE      PT. VALUE
       1  .000E+00    2  .100E+00   3  .500       4  1.20        5  2.10
       6  3.20        7  4.30       8  5.70       9  6.80       10  7.80
      11  8.40       12  8.80      13  8.30      14  7.70       15  6.10
      16  4.80       17  3.60      18  2.70      19  2.00       20  1.80
      21  1.40       22  1.30
```

ISA S26 FREQUENCY RESPONSE FROM PULSE TEST DATA
SAMPLE PROBLEM - NON CLOSING OUTPUT PULSE

PULSE NO. = 1 RUN S121   INPUT
   TOTAL TIME SPAN = 2.875
   PULSE AREA   = 12.91

PULSE NO. = 2 RUN S120   OUTPUT
   TOTAL TIME SPAN = 3.625
   PULSE AREA   = 12.00
   LINEAR CORRECTION WAS MADE FOR BASE LINE DRIFT
   FINAL VALUE ASSUMED TO PERSIST TO INFINITY

STEADY STATE GAIN = 0.9293

FREQUENCY RESPONSE (OUTPUT/INPUT)				FOURIER TRANSFORM OF THE INPUT CURVE				FOURIER TRANSFORM OF THE OUTPUT CURVE			
FREQUENCY CYC./TIME	AMPLITUDE RATIO	LOG AMP. RAT	PH. DIF DEGREES	REAL PART	IMAG. PART	AMPLITUDE	PH.ANG DEG.	REAL PART	IMAG. PART	AMPLITUDE	PH.ANG DEG.
1.000E-02	0.929	-0.03198	-1.30	12.9	-0.953	12.9	355.76	11.9	-1.16	12.0	354.47
1.300E-02	0.929	-0.03208	-1.69	12.8	-1.24	12.9	354.49	11.9	-1.50	12.0	352.81
1.600E-02	0.929	-0.03221	-2.07	12.8	-1.52	12.9	353.22	11.8	-1.84	12.0	351.15
2.000E-02	0.928	-0.03243	-2.59	12.8	-1.90	12.9	351.53	11.7	-2.30	12.0	348.94
2.500E-02	0.927	-0.03276	-3.24	12.7	-2.37	12.9	349.41	11.6	-2.85	11.9	346.17
3.200E-02	0.926	-0.03336	-4.14	12.5	-3.01	12.9	346.45	11.3	-3.62	11.9	342.31
4.000E-02	0.924	-0.03422	-5.18	12.3	-3.74	12.8	343.06	11.0	-4.46	11.9	337.89
5.000E-02	0.921	-0.03557	-6.46	11.9	-4.62	12.8	338.83	10.4	-5.46	11.8	332.37
6.300E-02	0.917	-0.03778	-8.13	11.4	-5.70	12.7	333.34	9.57	-6.65	11.6	325.21
8.000E-02	0.909	-0.04142	-10.29	10.4	-7.00	12.6	326.16	8.21	-7.96	11.6	315.88
0.100	0.898	-0.04684	-12.80	9.18	-8.34	12.4	317.74	6.37	-9.12	11.1	304.94
0.130	0.877	-0.05724	-16.48	6.94	-9.86	12.1	305.15	3.38	-10.10	10.6	288.67
0.160	0.850	-0.07039	-20.02	4.48	-10.7	11.6	292.62	0.448	-9.88	9.89	272.60
0.200	0.809	-0.09219	-24.47	1.16	-10.9	11.0	276.05	-2.80	-8.42	8.88	251.59
0.250	0.748	-0.12607	-29.38	-2.49	-9.71	10.0	255.61	-5.19	-5.41	7.50	226.23
0.320	0.654	-0.18410	-34.54	-5.76	-6.31	8.54	227.65	-5.45	-1.27	5.59	193.11
0.400	0.554	-0.25662	-37.00	-6.52	-1.97	6.82	196.84	-3.54	1.30	3.77	159.84
0.500	0.472	-0.32622	-35.56	-4.57	1.64	4.85	160.30	-1.30	1.88	2.29	124.74
0.630	0.421	-0.37538	-36.07	-1.23	2.61	2.88	115.26	-0.228	1.19	1.22	79.19
0.800	0.311	-0.50712	-35.66	0.735	0.974	1.22	52.96	0.362	-0.113	0.380	17.31
1.00	0.505	-0.29705	-14.09	2.272E-02	-0.266	0.267	274.88	-2.154E-02	-0.133	0.135	260.79
1.30	0.195	-0.70992	-36.74	-6.740E-02	0.477	0.481	98.05	4.508E-02	8.236E-02	9.390E-02	61.31
1.60	0.111	-0.95536	-23.21	0.300	-7.499E-02	0.310	345.98	2.731E-02	-2.076E-02	3.430E-02	322.76

INPUT AMPLITUDE IS LESS THAN ONE PER CENT OF PULSE AREA. RESULTS AT HIGHER FREQUENCIES ARE QUESTIONABLE.

FREQUENCY CYC./TIME	AMPLITUDE RATIO	LOG AMP. RAT	PH. DIF DEGREES	REAL PART	IMAG. PART	AMPLITUDE	PH.ANG DEG.	REAL PART	IMAG. PART	AMPLITUDE	PH.ANG DEG.
2.00	0.259	-0.58615	-123.53	-6.079E-02	4.053E-02	7.306E-02	146.31	1.747E-02	7.337E-03	1.895E-02	22.78
2.50	0.304	-0.51694	-38.22	-2.588E-02	-4.272E-02	4.995E-02	238.79	-1.422E-02	-5.338E-03	1.519E-02	200.57
3.20	9.276E-02	-1.03265	163.94	1.645E-02	2.613E-02	3.087E-02	57.81	-2.126E-03	-1.907E-03	2.864E-03	221.75
4.00	0.253	-0.59710	-180.00	4.559E-02	-6.333E-08	4.559E-02	360.00	-1.153E-02	-1.378E-07	1.153E-02	180.00
5.00	0.501	-0.30018	-217.75	-2.530E-03	-7.197E-03	7.629E-03	250.63	3.210E-03	2.075E-03	3.822E-03	32.88
6.30	0.169	-0.77188	108.48	9.270E-03	1.377E-02	1.660E-02	56.06	-2.706E-03	7.484E-04	2.807E-03	164.54
8.00	1.00	0.00057	0.00	3.203E-14	1.864E-20	3.203E-14	0.00	3.207E-14	5.978E-20	3.207E-14	0.00
10.0	0.259	-0.58613	-123.53	-2.432E-03	1.621E-03	2.922E-03	146.31	6.988E-04	2.935E-04	7.579E-04	22.78

Table VII
Program Listing

```
//LVMSR444    JOB 3LV5020, EXECUTE PULSE ANALYSIS PROGRAM    R
RJE*PRIORITY R
//           CLASS=D,PRTY=6,MSGLEVEL=(1,1)
//DOPGM      EXEC XXXSCGE1
//C.SYSIN    DD *
C    TITLE        = FREQUENCY RESPONSE FROM PULSE TEST DATA (ISA)
C    AUTHOR       = A C PAULS
C    LOCATION     = CED, MONSANTO CO, ST LOUIS, MO
C    DATE WRITTEN = 7/7/67, 1/13/69, 12/1/69
C    COMPUTER     = USASI FORTRAN
C    KEYPUNCH     = IBM 029 (EBCDIC)
C    FILES        = FORTRAN 5 (INPUT), 6 (OUTPUT)
C    SUBPROGRAMS  = FTRAN
C
C ABSTRACT
C    THIS PROGRAM COMPUTES FOURIER TRANSFORMS FOR CLOSED OR OPEN PULSES
C    AND TRANSFER FUNCTIONS FOR PAIRS OF CURVES.
C
C DEFINITION SECTION
C    A      FOURIER TRANSFORM (FT) AMPLITUDE
C    AI     FT AMPLITUDE OF INPUT PULSE
C    AIN    ALPHANUMERIC WORD
C    AR     AMPLITUDE RATIO (OUTPUT/INPUT)
C    ARL    ALOG10(AR)
C    AZ     AREA UNDER PULSE (FT AMPLITUDE AT ZERO FREQUENCY)
C    AZI    AREA UNDER INPUT PULSE
C    BLANK  ALPAMERIC WORD
C    BN     NUMBER OF LAST POINT IN J-TH SECTION OF K-TH PULSE
C    CF     CHART FACTOR
C    CFI    CHART FACTOR FOR INPUT PULSE
C    DB     INPUT DATA BASE ARRAY
C    F      FREQUENCY ARRAY (CYCLES/TIME IF KFU=1, ELSE RAD/TIME)
C    FI     INITIAL FREQUENCY
C    FM     FREQUENCY ARRAY FOR ONE DECADE
C    FR     FREQUENCY (RAD/TIME)
C    FU     ALPHAMERIC WORD FOR FREQUENCY UNITS
C    GAIN   STEADY STATE GAIN (OUTPUT PULSE AREA / INPUT PULSE AREA)
C    H      TIME BETWEEN POINTS IN J-TH SECTION OF K-TH PULSE
C    IA     NUMBER OF THE FIRST POINT IN A PULSE SECTION
C    IB     NUMBER OF THE LAST  POINT IN A PULSE SECTION
C    K      PULSE NUMBER
C    KEE    PULSE CLOSURE CODE. 0=CLOSED, 2=OPEN
C    KEEI   PULSE CLOSURE CODE FOR INPUT PULSE
C    KFD    FREQUENCY DATA TYPE CODE
C    KFL    FREQUENCY LIMIT PRINT SWITCH
C    KFU    FREQUENCY UNITS CODE
C    KRN    RESPONSE NORMALIZATION CODE
C    KTP    TRANSFORM DATA PRINT CODE
C    LN     LIST NUMBER
C    NB     NUMBER OF LAST POINT IN J-TH SECTION OF A PULSE
C    NCI    PULSE NUMBER OF INPUT CURVE
C    NCO    PULSE NUMBER OF OUTPUT CURVE
C    ND     NUMBER OF DECADES
C    NF     NUMBER OF FREQUENCIES
C    NFD    NUMBER OF FREQUENCIES PER DECADE
C    NFP1   NF + 1
C    NP     NUMBER OF DATA POINTS
C    NS     NUMBER OF PULSE CURVE SECTIONS
C    OUT    ALPHAMERIC WORD
C    P      FT PHASE ANGLE
C    PA     ALPHAMERIC ARRAY FOR PULSE ID
```

Table VII (*continued*)

```
C      PD     PULSE DATA ARRAY
C      PH     PULSE HEIGHT DATA ARRAY
C      PI     FT PHASE ANGLE FOR INPUT PULSE
C      PL     PHASE LAG (OUTPUT - INPUT)
C      R      FT REAL PART
C      RI     FT REAL PART FOR INPUT PULSE
C      T      TIME
C      TEMP   TEMPORARY STORAGE FOR PROBLEM AND CASE TITLES
C      TT     TOTAL TIME
C      TTI    TOTAL TIME FOR INPUT PULSE
C      TTL    PROBLEM AND CASE TITLES
C      Y      PULSE HEIGHT ARRAY
C      YF     BASE LINE DRIFT FROM TIME 0 TO TT
C      YFI    BASE LINE DRIFT FROM TIME 0 TO TT FOR INPUT PULSE
C      YZ     PULSE HEIGHT AT TIME 0-
C      Z      FT IMAGINARY PART
C      ZI     FT IMAGINARY PART FOR INPUT PULSE
C
C      NONCE VARIABLES - AK, AN, ANFD, I, ID, J
C
C DECLARATIVE SECTION
       DIMENSION      DB(10000), TTL(30), TEMP(15)
       DIMENSION      BN(10,9), F(52), FM(10), H(10,9), PA(2,9),
      1               PD(10,9), PH(1000,9), R(52), RI(52),
      2               Y(1000), Z(52), ZI(52), NB(10)
       EQUIVALENCE    (DB(11),PD),    (DB(211),H),    (DB(311),BN),
      1               (DB(1001),PH)
       DATA   BLANK   /4H    /
       DATA   AIN/4H  IN/, OUT/4H OUT/, CYC/4HCYC./, RAD/4HRAD./
C
C FILE SECTION
  170 FORMAT(/ 3X, 15HHEIGHT AT 0-   , G11.4, 2X,
      1               12HMULTIPLIER  , G11.4, 2X,
      2               12HCLOSURE CODE, F4.0,
      3        / 3X, 15HBASE LINE DRIFT, G11.4, 2X,
      4               12HSUMMARY CODE, F4.0, 9X,
      5               12HNO. SECTIONS, F4.0)
  172 FORMAT(/ 3X, 18HSECTION NUMBER    , 10(I6,4X))
  174 FORMAT(  3X, 18HTIME INTERVAL     , 10G10.3)
  176 FORMAT(  3X, 18HLAST POINT NO.    , 10(F6.0, 4X))
  177 FORMAT(/ 3X, 5(15HPT. VALUE      ))
  178 FORMAT(  1X, I5, 1X, G9.3, I5, 1X, G9.3, I5, 1X, G9.3,
      1             I5, 1X, G9.3, I5, 1X, G9.3)
  181 FORMAT(//1X, 11HLIST 6 DATA, 4X, 10(1H*), 4X, 12HPULSE NO. = ,
      1             I1, 2X, 2A4)
  182 FORMAT(/ 1X, 12HPULSE NO. = , I1, 2X, 2A4, 2X, A4, 3HPUT)
  183 FORMAT(  10X, 20HTOTAL TIME SPAN  = , G13.4
      1        /10X, 20HPULSE AREA       = , G13.4)
  184 FORMAT(  10X, 46HLINEAR CORRECTION WAS MADE FOR BASE LINE DRIFT)
  185 FORMAT(  10X,42HFINAL VALUE ASSUMED TO PERSIST TO INFINITY)
  186 FORMAT(//20X, 50HFREQUENCY      REAL PART      IMAG. PART      AMPLI
      1        , 19HTUDE     PHASE ANGLE/
      2        20X, A4,  5H/TIME, 47X, 15HDEGREES (0-360))
  187 FORMAT( 18X, 4(1PG11.4,4X), 0PF8.2)
  190 FORMAT(//10X, 20HSTEADY STATE GAIN = , G13.4)
  191 FORMAT(/ 2X, 40H    FREQUENCY RESPONSE (OUTPUT/INPUT)   ,
      1             40H FOURIER TRANSFORM OF THE   INPUT CURVE ,
      2             38H FOURIER TRANSFORM OF THE OUTPUT CURVE/
      3        2X, 40HFREQUENCY  AMPLITUDE    LOG     PH. DIF ,
      4             40HREAL PART IMAG. PART   AMPLITUDE PH.ANG ,
      5             38HREAL PART IMAG. PART   AMPLITUDE PH.ANG/
      6        2X, A4, 36H/TIME    RATIO   AMP. RAT  DEGREES ,
      7             33X, 4HDEG., 36X, 4HDEG.)
  192 FORMAT(/ 2X, 40H    FREQUENCY RESPONSE (OUTPUT/INPUT)   ,
      1             25H FOURIER TRANSFORM OF THE, A4, 11HPUT CURVE  /
      2        2X, 40HFREQUENCY  AMPLITUDE    LOG     PH. DIF ,
```

Table VII   (*continued*)

```
      3                40HREAL PART IMAG. PART  AMPLITUDE PH.ANG   /
      4          2X, A4, 36H/TIME      RATIO   AMP. RAT  DEGREES  ,
      5               33X, 4HDEG.)
193 FORMAT(/ 2X, 40H   FREQUENCY RESPONSE (OUTPUT/INPUT)     /
    1           2X, 40HFREQUENCY  AMPLITUDE    LOG     PH. DIF  /
    2           2X, A4, 36H/TIME      RATIO   AMP. RAT  DEGREES  )
194 FORMAT(  1X, 50HINPUT AMPLITUDE IS LESS THAN ONE PER CENT OF PULSE
    1         , 50H AREA. RESULTS AT HIGHER FREQUENCIES ARE QUESTIONA
    2         , 4HBLE.)
195 FORMAT(2(1X,1PG10.3),0PF9.5,F9.2,2(3(1X,1PG10.3),0PF7.2))
900 FORMAT(  8X, I2, 15A4 )
901 FORMAT( 8F10.0 )
902 FORMAT( 6X, I4, 5F10.0, 9X, I1 )
903 FORMAT(  46H1ISA - FREQUENCY RESPONSE FROM PULSE TEST DATA
    1         //1X, 15A4/1X, 15A4)
904 FORMAT( //  12H LIST 3 DATA,     4F10.0   )
905 FORMAT( //  12H LIST 4 DATA, 3F10.0, F10.5  )
906 FORMAT(/)
907 FORMAT( 38H1JOB TERMINATED - BAD DATA BEFORE CARD // 8X, I2, 15A4)
C
C INPUT SECTION
C                               DETERMINE LIST TYPE
    5 READ   (5,900,END=6) LN,  (TEMP(I), I = 1,15)
      IF   (LN .GT. 0 .AND. LN .LE. 6)  GO TO (11,12,13,14,15,16), LN
      IF   (LN .EQ. 99) GO TO 30
      WRITE (6,907)  LN,  (TEMP(I), I = 1,15)
    6 CALL   EXIT
C
C                           READ PROBLEM-TITLE, INITIALIZE DATA BASE
   11 DO 21 I       = 1,15
         TTL(I) = TEMP(I)
   21    TTL(I+15) = BLANK
      DO 22 I     = 1,400
   22    DB(I) = 0.0
C                           SET 'DEFAULT' OPTIONS
         DB(7)  = 5.0
         DB(8)  = 10.0
         DB(9)  = 0.001
         KFD    = 4
      GO TO 5
C                               READ CASE-TITLE
   12 DO 23 I       = 1,15
   23    TTL(I+15) = TEMP(I)
      GO TO 5
C                           LIST 3 - FREQUENCY RESPONSE DATA
   13 READ   (5,901) (DB(I), I = 1,4)
      GO TO 5
C                           LIST 4 - FREQUENCY RANGE DATA
   14 READ   (5,901) (DB(I), I = 6,9)
      KFD    = 4
      GO TO 5
C                           LIST 5 - FREQUENCY POINTS
   15 READ   (5,901) DB(6), DB(10)
      NF     = DB(10)
      READ   (5,901) (DB(I+100), I=1,NF)
      KFD    = 5
      GO TO 5
C                           LIST 6 - PULSE DATA
   16 READ   (5,902) K, (PD(I,K), I = 2,6), NS
      PD(1,K) = K
      PD(7,K) = NS
      READ   (5,901)  (H(I,K), BN(I,K), I = 1,NS)
      READ   (5,902)  NP
      PD(8,K) = NP
      READ   (5,901)  (PH(I,K), I = 1,NP)
      PA(1,K) = TEMP(1)
```

Table VII   (*continued*)

```
                PA(2,K) = TEMP(2)
        GO TO 5
C
C INPUT DATA SUMMARY SECTION
C                               PRINT CONTROL AND FREQUENCY RANGE DATA
    30 WRITE (6,903) TTL
        IF     (DB(1) .NE. 0.0) WRITE (6,904) (DB(I), I = 1,4)
        IF     (KFD .EQ. 4) WRITE (6,905) (DB(I), I = 6,9)
C                               PRINT PULSE DATA
        DO 31 K        = 1,9
                AK     = K
        IF     (PD(1,K) .NE. AK) GO TO 31
        WRITE (6,181)  K, PA(1,K), PA(2,K)
        WRITE (6,170)  (PD(I,K), I = 2,6,2), (PD(I,K), I=3,7,2)
                NS     = PD(7,K)
        WRITE (6,172)  (I, I = 1,NS)
        WRITE (6,174)  (H(I,K), I = 1,NS)
        WRITE (6,176)  (BN(I,K), I = 1,NS)
C                               PRINT PULSE HEIGHT ON INITIAL ENTRY ONLY
        IF     (PD(5,K) .EQ. 0.0) GO TO 31
                PD(5,K) = 0.0
                NP     = PD(8,K)
        WRITE (6,177)
        WRITE (6,178)  (I, PH(I,K), I = 1,NP)
    31 CONTINUE
        WRITE (6,906)
C
C INITIALIZATION SECTION
C                               SET CONTROL AND FREQUENCY RANGE VARIABLES
                NCI    = DB(1)
                NCO    = DB(2)
                KRN    = DB(3)
                KTP    = DB(4) + 1.01
                KFU    = DB(6)
                ND     = DB(7)
                NFD    = DB(8)
                FI     = DB(9)
                NF     = DB(10)
                FU     = RAD
        IF     (KFU .EQ. 1) FU = CYC
C                               SET FREQUENCY ARRAY FROM RANGE DATA
        IF     (NF .GT. 0 .AND. KFD .EQ. 5) GO TO 35
                NF     = NFD*ND + 1
                ANFD   = NFD
                AN     = 0.0
        DO 32 I        = 1,NFD
                FM(I)  = 0.1*AINT(10.0**(1.0+AN/ANFD) + 0.6)
    32          AN     = AN + 1.0
        DO 33 J        = 1,NF
                ID     = (J - 1)/NFD
                I      = J - ID*NFD
    33          F(J+1) = FI*FM(I)*10.0**ID
        GO TO 38
C                               SET FREQUENCY ARRAY FROM INPUT POINTS
    35 DO 36 J         = 1,NF
    36          F(J+1) = DB(J+100)
C                               SET ZERO FREQUENCY FOR AREA CALCULATION
    38          F(1)   = 0.0
                NFP1   = NF + 1
C
C COMPUTATION SECTION
C                               COMPUTE TRANSFORMS ONLY
        IF     (NCI .GT. 0) GO TO 45
    41          K      = 0
    42 IF      (K .GE. 9) GO TO 5
                K      = K + 1
```

Table VII   (*continued*)

```
      IF    (IFIX(PD(1,K)) .NE. K) GO TO 42
      GO TO 50
C                                       COMPUTE TRANSFORMS AND FREQUENCY RESPONSE
   45       K     = NCI
C
C                                       SET PULSE PARAMETERS FOR K-TH CURVE
   50       CF    = PD(4,K)
            YZ    = PD(2,K)*CF
            YF    = PD(3,K)*CF
            KEE   = PD(6,K)
            NS    = PD(7,K)
            NP    = PD(8,K)
            TT    = 0.0
            IA    = 1
      DO 51 I     = 1,NS
            NB(I) = BN(I,K)
            TT    = TT + H(I,K)*FLOAT(NB(I) - IA)
   51       IA    = NB(I)
C                                       ADJUST PULSE FOR TIME 0- AND CHART FACTOR
      DO 53 I     = 1,NP
   53       Y(I)  = PH(I,K)*CF - YZ
C                                       ADJUST PULSE FOR LINEAR BASE LINE DRIFT
      IF    (YF .EQ. 0.0) GO TO 60
            T     = 0.0
            J     = 1
      DO 56 I     = 2,NP
      IF    (I .GT. NB(J))  J = J + 1
            T     = T + H(J,K)
   56       Y(I)  = Y(I) - YF*T/TT
C                                       COMPUTE FOURIER TRANSFORM
   60 DO 63 J     = 1,NFP1
            FR    = F(J)
      IF    (KFU .EQ. 1) FR = FR*6.283185
   63 CALL  FTRAN   (NS, NB, H(1,K), Y, KEE, FR, R(J), Z(J))
C                                       AREA UNDER PULSE
            AZ    = SIGN(SQRT(R(1)**2 + Z(1)**2),R(1))
C
C                                       TRANSFORMS ONLY - PRINT PULSE ID
      IF    (NCI .GT. 0) GO TO 80
      WRITE (6,903)  TTL
      WRITE (6,906)
      WRITE (6,182)  K, PA(1,K), PA(2,K)
      WRITE (6,183)  TT, AZ
      IF    (YF .NE. 0.0) WRITE (6,184)
      IF    (KEE .EQ. 2)  WRITE (6,185)
C                                       PRINT TABLE HEADING AND TRANSFORM DATA
      WRITE (6,186)  FU
      DO 77 J     = 2,NFP1
            A     = SQRT(R(J)**2 + Z(J)**2)
            P     = 57.29578*ATAN2(Z(J),R(J))
      IF    (P .LT. 0.0) P = P + 360.0
   77 WRITE (6,187)  F(J), R(J), Z(J), A, P
      GO TO 42
C
C                                       RESPONSE NEEDED - SAVE INPUT
   80 IF    (K .EQ. NCO) GO TO 85
            CFI   = CF
            YFI   = YF
            KEEI  = KEE
            TTI   = TT
            AZI   = AZ
      DO 81 J     = 1,NFP1
            RI(J) = R(J)
   81       ZI(J) = Z(J)
            K     = NCO
      GO TO 50
```

Table VII   (*continued*)

```
C                               STEADY STATE GAIN
   85          GAIN    = AZ/AZI
C
C                               PRINT RESPONSE TABLE
              KFL     = 0
       DO 120 J = 2,NFP1
       IF     (J .NE. 2 .AND. J .NE. 38) GO TO 100
       WRITE (6,903)  TTL
C                               PRINT INPUT PULSE ID
       IF     (J .NE. 2) GO TO 95
       WRITE (6,182)  NCI, PA(1,NCI), PA(2,NCI), AIN
       WRITE (6,183)  TTI, AZI
       IF     (YFI .NE. 0.0) WRITE (6,184)
       IF     (KEEI .EQ. 2)  WRITE (6,185)
C                               PRINT OUTPUT PULSE ID
       WRITE (6,182)  K, PA(1,K), PA(2,K), OUT
       WRITE (6,183)  TT, AZ
       IF     (YF .NE. 0.0) WRITE (6,184)
       IF     (KEE .EQ. 2)  WRITE (6,185)
       WRITE (6,190)  GAIN
C                               PRINT RESPONSE TABLE HEADINGS
   95 GO TO (96,97,98,99),KTP
   96 WRITE (6,191)  FU
      GO TO 100
   97 WRITE (6,192)  AIN, FU
      GO TO 100
   98 WRITE (6,192)  OUT, FU
      GO TO 100
   99 WRITE (6,193)  FU
C
C                               AMPLITUDE, AMP. RATIO (NORMALIZE IF KRN=0)
  100          AI      = SQRT(RI(J)**2 + ZI(J)**2)
              A       = SQRT(R(J)**2 + Z(J)**2)
              AR      = A/AI
       IF     (KRN .EQ. 0 .AND. MAX0(KEEI,KEE) .LT. 2) AR = ABS(AR/GAIN)
              ARL     = ALOG10(AR)
C                               PHASE ANGLES, PHASE DIFFERENCE
              PI      = 57.29578*ATAN2(ZI(J),RI(J))
              P       = 57.29578*ATAN2(Z(J),R(J))
       IF     (PI .LT. 0.0) PI = PI + 360.0
       IF     (P .LT. 0.0) P = P + 360.0
              PL      = P - PI
C                               PRINT INPUT AMPLITUDE WARNING
       IF     (KFL .NE. 0 .OR. AI .GE. 0.01*ABS(AZI)) GO TO 110
       WRITE (6,194)
              KFL     = 1
C                               PRINT J-TH FREQUENCY RESPONSE
  110 GO TO (111,112,113,114), KTP
  111 WRITE (6,195)  F(J), AR, ARL, PL, RI(J), ZI(J), AI, PI,
     1               R(J), Z(J), A, P
      GO TO 120
  112 WRITE (6,195)  F(J), AR, ARL, PL, RI(J), ZI(J), AI, PI
      GO TO 120
  113 WRITE (6,195)  F(J), AR, ARL, PL, R(J), Z(J), A, P
      GO TO 120
  114 WRITE (6,195)  F(J), AR, ARL, PL
C
  120 CONTINUE
      GO TO 5
      END
      SUBROUTINE     FTRAN  (NS, NB, H, Y, KEE, F, R, Z)
C
C     TITLE       = FOURIER TRANSFORM REAL AND IMAGINARY PARTS
C     AUTHOR      = A C PAULS
C     LOCATION    = CED, MONSANTO CO, ST LOUIS, MO
C     DATE WRITTEN = 7/7/67
```

Table VII   (*continued*)

```
C       COMPUTER      = USASI FORTRAN
C
C DEFINITION SECTION
C     F         FREQUENCY (RAD/TIME)
C     H         TIME BETWEEN POINTS IN J-TH SECTION OF PULSE
C     IA        NUMBER OF FIRST POINT IN A SECTION
C     IB        NUMBER OF LAST  POINT IN A SECTION
C     KEE       PULSE CLOSURE CODE.  0=CLOSED, 2=OPEN
C     NB        NUMBER OF LAST POINT IN J-TH SECTION OF PULSE
C     NS        NUMBER OF PULSE CURVE SECTIONS
C     R         FOURIER TRANSFORM REAL PART
C     TA        TIME AT FIRST POINT IN A SECTION
C     TB        TIME AT LAST  POINT IN A SECTION
C     Y         PULSE HEIGHT ARRAY
C     YF        PULSE HEIGHT AT INFINITE TIME
C     Z         FOURIER TRANSFORM IMAGINARY PART
C
C     NONCE VARIABLES - CA, CB, D, DH, FTA, FTB, FTI, IAP1, IBM1,
C           Q, R1, SA, SB, SYI, U, Z1
C
C
C DECLARATIVE SECTION
      DIMENSION      NB(1), H(1), Y(1)
C
C COMPUTATION SECTION
                R     = 0.0
                Z     = 0.0
                TA    = 0.0
                IA    = 1
C
C                              POSITIVE FREQUENCY
      IF    (F .EQ. 0.0) GO TO 60
      DO 50 N     = 1,NS
                IB    = NB(N)
C                              TIME AT END OF N-TH SECTION
                TB    = TA + H(N)*FLOAT(IB-IA)
C                              D AND Q PARAMETERS
                U     = F*H(N)
                D     = (SIN(U/2.0)/(U/2.0))**2
                DH    = D/2.0
      IF    (U .GT. 0.1) GO TO 32
                Q     = U*(1.0 - U**2/20.0)/6.0
      GO TO 35
   32           Q     = (1.0 - SIN(U)/U)/U
C
C                              CONTRIBUTION OF END POINTS
   35     FTA     = F*TA
          FTB     = F*TB
          SA      = SIN(FTA)
          CA      = COS(FTA)
          SB      = SIN(FTB)
          CB      = COS(FTB)
          R       = R + H(N)*(Y(IA)*(DH*CA - Q*SA) + Y(IB)*(DH*CB + Q*
     1             SB))
          Z       = Z - H(N)*(Y(IA)*(DH*SA + Q*CA) + Y(IB)*(DH*SB - Q*
     1             CB))
C
C                              CONTRIBUTION OF INTERIOR POINTS
          IAP1    = IA + 1
          IBM1    = IB - 1
      IF    (IAP1 .GT. IBM1) GO TO 45
          R1      = 0.0
          Z1      = 0.0
      DO 41 I     = IAP1,IBM1
          FTI     = F*(TA + H(N)*FLOAT(I-IA))
          R1      = R1 + Y(I)*COS(FTI)
```

Table VII (*concluded*)

```
    41        Z1      = Z1 + Y(I)*SIN(FTI)
              R       = R + H(N)*D*R1
              Z       = Z - H(N)*D*Z1
C                                    INITIALIZE FOR NEXT SECTION
    45        TA      = TB
              IA      = IB
    50 CONTINUE
C                                    ADJUST FOR NON-CLOSURE
       IF     (KEE .NE. 2) GO TO 55
              YF      = Y(IB)
              R       = R - YF*SIN(F*TB)/F
              Z       = Z - YF*COS(F*TB)/F
    55 RETURN
C
C                                    ZERO FREQUENCY
    60 DO 70 N        = 1,NS
              IB      = NB(N)
              SYI     = 0.0
              IAP1    = IA + 1
              IBM1    = IB - 1
       IF     (IAP1 .GT. IBM1) GO TO 65
       DO 61 I        = IAP1,IBM1
    61        SYI     = SYI + Y(I)
    65        R       = R + H(N)*(0.5*(Y(IA) + Y(IB)) + SYI)
    70        IA      = IB
C                                    ADJUST FOR NON-CLOSURE
       IF     (KEE .NE. 2 .OR. Y(IB) .EQ. 0.0) GO TO 80
              Z       = -R
              R       = SIGN(1.0E-37,R)
    80 RETURN
       END
/* PLACE THE  FORTRAN SOURCE DECK IN FRONT OF THIS CARD
//X.FT05F001   DD *
         1 ISA S26 FREQUENCY RESPONSE FROM PULSE TEST DATA
         2 SAMPLE PROBLEM TO TEST COMPUTER PROGRAM PROPERLY EXCIT
         3
         1         2         0         0
         4
         1         3        10       .01
         6RUN S12I
         1         0         0         1         1         0
       .05        18
        18
         0         4        11        43        56        69
        70        63        48        35        22        14
         1         0
         6RUN S120
         2         0         0         1         1         0
       .05        15       0.1        31
        31
         0         0         2         4         7        12
        30        37        40        44        42        41
        27        21        16        13        10         8
       3.5       2.5       1.7       1.1        .6        .2
        99
         2 SAMPLE PROBLEM TO TEST COMPUTER PROGRAM PROPERLY EXCIT
         3
         1         3         0         0
         6RUN S130
         3         0         0         1         1         0
       .05        45
        45
         0         0         2         4         7        12
        30        37        40        44        42        41
        32        27        24        21        18        16
        12        10         9         8         7         6
```

Index